VOLPE'S *UNDERSTANDING*
EVOLUTION

Seventh Edition

VOLPE'S *UNDERSTANDING*
EVOLUTION

Peter A. Rosenbaum
State University of New York at Oswego

VOLPE'S *UNDERSTANDING EVOLUTION,* SEVENTH EDITION

1 2 3 4 5 6 7 8 9 0 DOC/DOC 1 0 9 8 7 6 5 4 3 2 1 0

ISBN 978-0-07-338323-1
MHID 0-07-338323-6

Vice President & Editor-in-Chief: *Martin Lange*
Vice President, EDP: *Kimberly Meriwether David*
Publisher: *Janice Roerig-Blong*
Sponsoring Editor: *Margaret Kemp*
Marketing Manager: *Heather Wagner*
Development Editor: *Wendy Langerud*
Project Manager: *Joyce Watters*
Design Coordinator: *Margarite Reynolds*
Cover Designer: *Nicole Dean*
Photo Research: *Lori Hancock*
USE Cover Image: © *Tone Stone Images*
Production Supervisor: *Laura Fuller*
Compositor: *Laserwords Private Limited*
Typeface: *10/12 Times Roman*
Printer: *R. R. Donnelley*

Library of Congress Cataloging-in-Publication Data

Rosenbaum, Peter Andrew, 1952-
    Volpe's *Understanding evolution* / Peter A. Rosenbaum. — 7th ed.
        p. cm.
    Includes index.
    ISBN 978-0-07-338323-1 (alk. paper)
        1. Evolution (Biology)—Textbooks. I. Volpe, E. Peter (Erminio Peter).
    Understanding evolution. II. Title. III. Title: Understanding evolution.
    QH366.2.V64 2010
    576.8—dc22

                                                                                        2009052287

www.mhhe.com

To Samantha and Sophia, my evolutionary legacies.

BRIEF CONTENTS

Preface   x

Prologue   xiii

1 Variation in Populations   *1*

2 Darwinian Scheme of Evolution   *12*

3 Heritable Variation   *23*

4 Mutation   *43*

5 Genetic Equilibrium   *61*

6 Concept of Selection   *67*

7 Selection in Action   *82*

8 Balanced Polymorphism   *94*

9 Genetic Drift and Gene Flow   *110*

10 Races and Species   *122*

11 Instantaneous Speciation   *135*

12 Adaptive Radiation   *142*

13 Major Adaptive Radiations   *152*

14 Origin of Life   *166*

15 Molecular Evolution   *182*

16 History of Life   *197*

17 Emergence of the Human Species   *214*

18 Natural Selection, Social Behavior, and Cultural Evolution   *237*

Epilogue   *248*

Glossary   *256*

Index   *271*

CONTENTS

Preface   x

Prologue: Evidence for Evolution   xiii

CHAPTER **1**

VARIATION IN POPULATIONS   *1*

Environmental Modifications   *3*
Lamarckism   *5*
Heritable Variation   *7*
Hypothetico-Deductive Reasoning   *9*

CHAPTER **2**

DARWINIAN SCHEME OF EVOLUTION   *12*

Mechanism of Evolution   *15*
Darwin's Natural Selection   *17*
Differential Reproduction   *21*
Evolution Defined   *21*

CHAPTER **3**

HERITABLE VARIATION   *23*

Mendel's Principle of Segregation   *24*
Application of Mendelian Principles   *27*
Chromosomal Basis of Heredity   *28*
Mitosis and Meiosis   *31*
Mendel and the Chromosomal Theory of Heredity   *35*
Mendel's Principle of Independent Assortment   *36*
Three or More Gene Pairs   *36*
Significance of Genetic Recombination   *38*
Linkage and Crossing Over   *40*

CHAPTER **4**

MUTATION   *43*

Causes of Mutation   *43*
Chemical Nature of the Gene   *45*
Replication of DNA   *49*
Molecular Mechanism of Spontaneous Mutation   *50*
Language of Life   *50*
Transcription of DNA   *53*
Frequency of Mutations in Human Disorders   *55*
Evolutionary Consequences of Mutations   *57*
Chromosomal Aberrations and Pregnancy Loss   *58*

CHAPTER **5**

GENETIC EQUILIBRIUM   *61*

Mendelian Inheritance   *61*
Random Mating   *62*
Gene Frequencies   *64*
Hardy-Weinberg Equilibrium   *64*
Estimating the Frequency of Heterozygotes   *65*
Implications   *66*

CHAPTER **6**

CONCEPT OF SELECTION   *67*

Selection Against Recessive Defects   *67*
Significance of the Heterozygote   *70*
Interplay of Mutation and Selection   *70*
Partial Selection   *71*
Selection Against Dominant Defects   *73*
Concealed Variability in Natural Populations   *75*

Genetic Load in Human Populations   *75*
General Effects of Selection   *76*
Natural Selection and Pregnancy Loss   *77*
Sexual Selection   *78*

CHAPTER   **7**

SELECTION IN ACTION   *82*

Industrial Melanism   *82*
Cancer Cell Heterogeneity   *84*
Selection for Resistance   *85*
Multiple Antibiotic Resistance   *87*
Recombinant DNA   *89*
Prudent Use of Antibiotics   *91*
Evolutionary Implications   *92*

CHAPTER   **8**

BALANCED POLYMORPHISM   *94*

Sickle-Cell Anemia   *94*
Sickle-Cell Hemoglobin   *95*
Theory of Balanced Polymorphism   *97*
Superiority of the Heterozygotes   *100*
Relaxed Selection   *101*
Tay-Sachs Disease   *102*
Cystic Fibrosis   *103*
Selection Against the Heterozygote   *104*
The Control of Rh Disease   *108*
Implications of Balanced Polymorphism   *109*

CHAPTER   **9**

GENETIC DRIFT AND GENE FLOW   *110*

Role of Genetic Drift   *110*
Theory of Genetic Drift   *112*
Founder Effect   *113*
Religious Isolates   *114*
Amish of Pennsylvania   *116*
Consanguinity and Genetic Drift   *117*
Gene Flow   *118*
European Genes in African Americans   *119*

CHAPTER   **10**

RACES AND SPECIES   *122*

Variation Between Populations   *122*
Subspecies   *123*

The Species Concept   *125*
Formation of Species   *126*
Nomenclature   *128*
Reproductive Isolating Mechanisms   *129*
Origin of Isolating Mechanisms   *131*
Humans: A Single Variable Species   *132*
Microevolution and Macroevolution   *132*
Punctuated Equilibrium   *133*

CHAPTER   **11**

INSTANTANEOUS SPECIATION   *135*

Polyploidy in Nature   *135*
Wheat   *138*
Origin of Wheat Species   *139*
Mechanism of Speciation of Polyploidy   *139*
Experimental Verification   *141*

CHAPTER   **12**

ADAPTIVE RADIATION   *142*

Galápagos Islands   *142*
Darwin's Finches   *144*
Factors in Diversification   *145*
Competitive Exclusion   *148*
Coexistence   *148*
Classification   *150*

CHAPTER   **13**

MAJOR ADAPTIVE RADIATIONS   *152*

Transition to Land   *152*
Conquest of Land   *155*
Adaptive Radiation of Reptiles   *159*
Extinction and Replacement   *163*

CHAPTER   **14**

ORIGIN OF LIFE   *166*

What Is Life?   *167*
Primitive Earth   *167*
Experimental Synthesis of Organic Compounds   *168*
Life's Beginnings   *169*
Autotrophic Existence   *172*
Prokaryotic and Eukaryotic Cells   *172*
Organelles and Evolution   *174*

Classification of Organisms   *176*
Deep-sea Vents   *179*
Extraterrestrial Origins   *180*

CHAPTER **15**

MOLECULAR EVOLUTION   *182*

Neutral Theory of Molecular Evolution   *182*
Cytochrome *c*   *183*
Molecular Clock   *184*
Gene Duplication   *185*
Evolutionary History of Hemoglobin   *187*
Pseudogenes   *188*
Homeotic Genes   *189*
Ocular Malformations and Evolution   *192*
Multilegged Frogs Revisited   *193*
Mitochondrial Eve   *194*
Y-Chromosome Adam   *195*

CHAPTER **16**

HISTORY OF LIFE   *197*

Geologic Ages   *198*
Fossil Record of Plants   *200*
Fossil Record of Animals   *201*
Convergence   *203*
The Plains-Dwelling Mammals   *207*
Evolutionary Stability   *207*
Mass Extinction   *208*
Holocene Extinction   *209*

CHAPTER **17**

EMERGENCE OF THE HUMAN SPECIES   *214*

Primate Radiation   *214*
Adaptive Radiation of Humans   *217*

Forerunners of the Great Apes   *217*
Early Hominoids   *218*
Australopithecines: The First Hominids   *221*
Humans Emerge: *Homo Habilis*   *225*
Gathering-Hunting Way of Life   *225*
Homo Erectus: The Explorer   *226*
Flores Man   *228*
Evolution of Human Society   *228*
Status of the Neanderthals   *230*
Modern Humans: The Cro-Magnons   *231*
Origin of Modern Humans   *232*
Genomes of the Chimpanzees and Humans   *233*

CHAPTER **18**

NATURAL SELECTION, SOCIAL BEHAVIOR, AND CULTURAL EVOLUTION   *237*

Group Selection   *237*
Kin Selection   *238*
Kin Selection in Humans   *240*
The Selfish Gene   *242*
Reciprocal Altruism   *242*
Parental Reproductive Strategies   *243*
Parent-Offspring Conflict   *245*
Memes   *245*
Cultural Evolution   *246*

EPILOGUE: THE CRUCIBLE OF EVOLUTION   *248*

Changing Worldviews   *250*
Scientific Literacy   *250*
Darwin in the Courts   *251*
Closing Commentary   *254*

Glossary   *256*
Index   *271*

PREFACE

This new edition of Volpe's *Understanding Evolution,* like earlier editions, is addressed to college students with no previous experience with the subject. It is meant to stand in good stead for use as a primary text for an introductory-level evolution course or as a supplement to a basic biology or anthropology course. In revising this time-honored volume, every effort has been made to provide a clear, simple, concise account of the scope, significance, and principles of evolution for the liberally educated citizen without burdening the reader with exhaustive detail. No apology is offered for drawing many of the examples used in this text from our own species. This edition has been rewritten and expanded to transmit a wealth of new ideas in evolution fostered by new discoveries in many fields of science. To accommodate the changing body of knowledge, all chapters have been extensively revised, new chapters have been added, and many new illustrations have been added. In addition, this edition includes a concise glossary to aid the reader with the challenges of scientific terminology.

The publication of Charles Darwin's *On the Origin of Species by Means of Natural Selection,* or the *Preservation of Favoured Races in the Struggle for Life,* Darwin's "big book," is one of the great landmarks in the history of science and laid the foundation for the theory of evolution via natural selection. This groundbreaking tome sold out on the first day of publication and was revised and reissued six times during Darwin's lifetime. Unlike other scientific ideas, evolution provides a framework for much of human knowledge spanning all of the natural sciences, the social sciences, and beyond and brought about one of the most remarkable and far-reaching revolutions in human thought.

The Origin of Species shocked Victorian readers and forever changed science. It profoundly altered and forever changed our conception of ourselves and our world, of the origins and history of life on our planet, and in doing so became and remains controversial to this day. Before Darwin's *Origin of Species,* we presumed that humans occupied an exalted position in the world. In its aftermath, we were no longer at the center stage of the Earth and were instead made part of the fabric of nature. Volpe's *Understanding Evolution* is an effort to demystify evolution and provide the reader with a clear and concise grasp of this pivotal and groundbreaking idea. Although Darwin's monumental *Origin of Species* was initially greeted with a storm of criticism, his findings are now universally accepted.

Darwin's thesis has been fortified by the ever-expanding knowledge of the gene. One of the finest triumphs of modern science has been the elucidation of the chemical makeup of the gene. The fundamental chemical component of the gene is the remarkable molecule *deoxyribonucleic acid* (DNA). This molecule contains a coded blueprint for all life in its molecular structure and is the universal alphabet for the book of life. The integration of genetics and evolution occurred in the early part of the twentieth-century and is often called "the synthesis" because of the unity of knowledge the melding of these two great bodies of knowledge brought to each other. Due to the importance of genetics to the study of evolution, Volpe's *Understanding Evolution* rehearses those facets of genetics that are fundamental to the study and understanding of evolution. Today, most of the important contributions to evolution are coming from genetics, developmental genetics, and the comparison of proteins *(proteomics)* and whole sets of chromosomes *(genomics)* among and between organisms. Modern DNA technology has so sharpened the precision of molecular analysis that it has found application in virtually all fields of scientific inquiry, from medicine to paleontology to the evolutionary history of the human species. Hence, the reader is asked to integrate the principles of genetics with basic Darwinian tenets in order to understand how important "the synthesis" was and is to our modern conceptualization of evolution.

The year 2009 marked the 200th birthday of Charles Darwin (February 12, 1809) and the 150th anniversary of the first publication of *The Origin of Species* (November 24, 1859). These dates remind us that the theory of evolution has a long and complex history filled with extraordinary discoveries and public controversies. Excellent companion books to this volume are Darwin's original books *The Origin of Species* (1859) and *The Voyage of the Beagle* (1839). Both were written for the laymen, the everyday person, living in Victorian England in Darwin's day, and both were bestsellers. *The Origin of Species* is one of the 300 most important books ever published in the history of the Western world—a collection of works spanning literature, philosophy, and science and including such giants as Homer, Dante, Hippocrates, Chaucer, Dickens, Melville, Aristotle, Shakespeare, Plato, Euripides, Copernicus, Euclid, Einstein, Galileo, Newton, Galen, Locke, Cervantes, Shaw, Orwell, and Dobzhansky.

Evolution is a central concept in science. In the 150-plus years since the publication of *The Origin of Species,* Darwin's thesis has been fortified from many branches of science. No other scientific idea has had a more unifying impact on the sciences, enabling scientists from a myriad of disciplines to better understand and integrate their science. In 1973 the renowned Columbia University evolutionist Theodosius Dobzhansky wrote a now-famous essay entitled "Nothing in Biology Makes Sense except in the Light of Evolution." This essay could as easily have been entitled "Nothing in the *Natural Sciences* Makes Sense except in the Light of Evolution." More than any other scientific concept, evolution serves as a lattice for the natural sciences, providing both explanation and venues for prediction and hypothesis testing.

The word evolution is derived from the Latin *evolutio,* meaning "an unraveling" or "an unfolding." The term reminds us that throughout the history of the universe, change has been a constant motif, especially of the living world. In Darwin's day, evolution was conceptualized by the word *transmutation* and it was only later that the term *evolution* gained acceptance. Throughout life's history, organisms have constantly and endlessly changed. Change is the leitmotif, the recurring theme and idea, of all living things.

Coupled with the ever-increasing and rapid advances in science and technology is the need for scientific literacy of the citizenry. Scientific literacy is an essential part of a liberal arts education. With recent polls revealing that nearly half of U.S. citizens do not *accept* the tenets of evolutionary biology, the rationale for Volpe's *Understanding Evolution* have never been more important. Many otherwise well-educated people cannot describe the principles of evolution or articulate the vast body of evidence built up over the last 150 plus years since the publication of Charles Darwin's *Origin of Species*. In this book effort has been made to leave the reader a better educated member of society capable of engaging in informed discussions in all manner of venues.

NEW TO THIS EDITION

- A Prologue outlining the evidence for evolution.
- An Epilogue describing the controversy of teaching evolution in the school, scientific literacy and the legal history of teaching evolution in U.S. law.

- A Glossary to aid in learning the vocabulary associated with evolutionary biology.
- New and updated graphics throughout the chapters to clarify the concepts presented in the text.
- Supporting text website that provides instructor tools, including slides of all figures and tables as PowerPoint lectures and test questions for each chapter. Chapter outlines are available to students.

Chapter Revisions Include

- A new section on the scientific method (Chapter 1).
- A new section of Darwin's reluctance to publish the "Origin of Species" and a major revision of Darwin's explanatory model for natural selection (Chapter 2).
- Several new figures and a new section of Gregor Mendel (Chapter 3).
- New discussion of mutation supported by two new figures (Chapter 4).
- A revised description of the Hardy-Weinberg equilibrium (Chapter 5).
- A new section on sexual selection and several new figures (Chapter 6).
- A revised section on the evolution of bacterial resistance to antibiotics (Chapter 7).
- A new section on polyploidy in nature with several new figures and tables (Chapter 11).
- Major revisions and updating of material on adaptive radiation, including new sections and many new and revised figures (Chapter 13).
- Two new sections on the origin of life along with several new figures and tables (Chapter 14).
- New section of mass extinctions, revised text and many new figures (Chapter 16).
- Extensive revision and updates on human evolution including several new figures and a new discussion on Flores man and the genomic and chromosomal evolution of chimps and humans (Chapter 17).
- Integration of three previous chapters, a revised section on cultural evolution and a new section on memes (Chapter 18).

ACKNOWLEDGEMENTS

In the mid 1960s when science education was undergoing dramatic restructuring and innovation, Volpe's *Understanding Evolution* was originally conceived as *The Process of Evolution* but ultimately went to press in 1967 under the title *Understanding Evolution*. Erminio "Peter" Volpe's graceful use of language and mastery as a teacher shows through in each paragraph and page. Much of the art used in the first edition was drawn by Peter's wife, Carolyn Thorpe Volpe, some of which remains in the pages that follow. My involvement with this text began roughly a decade ago in the preparation of the sixth edition. Volpe's *Understanding Evolution* has enabled generations of students to better appreciate the basic tenets of evolutionary biology, and thus this new edition rests on broad shoulders with high expectations. This edition is the culmination of a promise I made to Peter Volpe, my professor, mentor, colleague, and friend, to keep *Understanding Evolution* in press for as long as I could. I owe an enormous debt of gratitude to the inspiration, mentorship, and trust that Peter Volpe placed in me to continue his extraordinary contribution to science education.

I am indebted to the reviewers whose comments have helped improve this and previous editions.

Ronald Edwards, *DePaul University*
Joseph L. Fail Jr., *Johnson C. Smith University*
Teresa L. Felton, *Southeastern University*
Michael L. Foster, *Eastern Kentucky University*
Gerard M. Nadal, *Dominican College*

I would also like to thank the staff at McGraw-Hill, including Janice Roerig-Blong, Marge Kemp, Wendy Langerud, Heather Wagner, Lori Hancock, and Joyce Watters.

I acknowledge and appreciate the enthusiastic support of my colleagues at the State University of New York–Oswego who have been most supportive of my efforts toward this revision. Lastly, I owe a debt of gratitude to my spouse, Robin, who has bestowed love, emotional support, and understanding throughout this revision process.

Peter A. Rosenbaum
Oswego, New York
January 2010

There is no shortage of "evidence for evolution," and in the pages that follow the reader will be exposed to nearly all of the many major groupings of information that form the basis of this topic. The following is a brief guide to help the reader better appreciate the many examples of evidence for evolution provided herein.

ARTIFICIAL SELECTION

Chapter 2 describes several examples of artificial selection. Darwin used insight from his observations of artificial selection to develop his concept of natural selection. He reasoned that if humans could easily select different traits, so could nature.

NATURAL SELECTION

Natural selection is the mechanism of evolution. Several examples of natural selection are provided in this volume. In chapter 6 the concept of selection is reviewed, and in chapter 7 several illustrations of natural selection, such as the tale of the peppered moth, are provided. Chapter 8 includes the story of balanced polymorphism in humans, another example of natural selection.

COMMON DESCENT

Sprinkled throughout the chapters is the concept of *common descent. Descent with modification* is the most fundamental line of evidence in support of evolution. Exemplars of common descent come from genetics, comparative studies of anatomy, embryology, and molecular genetics. The occurrence of vestigial organs and evolutionary opportunism further strengthen these arguments.

GENES AND GENOMICS

In many respects, the single most powerful evidence for common descent (and therefore evolution) comes from genetics. The genetic code (chapter 4) is the same in all living organisms. There is no reasonable explanation for this universality except for inheritance from a common ancestor. As DNA sequencing has become commonplace, scientists have utilized these tools to compare organisms in the same way one would look at anatomy or any other observable feature. This is an especially exciting new field of study and has revolutionized the way in which we organize life (see chapters 15 and 16). As researchers come to compare genomes, it has become apparent that many basic kinds of patterns of development (for example) are highly conserved across broad ranges of organisms. As seen in chapter 15, it is extraordinary that nearly the same gene sequences code for eye development or body part development in organisms as overtly divergent as fruit flies and humans. Similarly, in chapter 17, comparison of chimp and human chromosomes demonstrate our common descent. This is hard evidence that humans are simply another branch on the tree of life.

Likewise, as seen in the new three-domain system of organizing life (chapter 14), data from the comparison of genomes has helped us better understand our relationship with the rest of life on the planet. Furthermore, these comparative studies have aided in our quest to better understand the origins of life (chapter 14).

PROTEINS AND PROTEOMICS

Common descent is also observed in organic molecules such as proteins. *Homology* of protein structure (chapter 15) provides an important link in the reasoning of common descent. As with genes, why else would proteins be homologous and show homology over time (e.g., hemoglobin and myoglobin in chapter 15) if not for their common derivation from a common ancestor? The detailed comparison of proteins and linking their structure to their function *(proteomics)* is another new venue for documenting evolution.

CELLULAR HOMOLOGIES

The fact that all life is based on cells is another supporting line of evidence for common descent. The identical organelles and other cellular structures in plant and animal cells bespeak of common descent (fig. P.1). Homologies found between prokaryotes and eukaryotes (Chapter 14) could only be true if these cells were linked through common ancestry.

COMPARATIVE ANATOMY

The entire field of comparative anatomy is based on homology due to common descent. Why else would organisms have similar structure, if they were not related by common ancestry? The

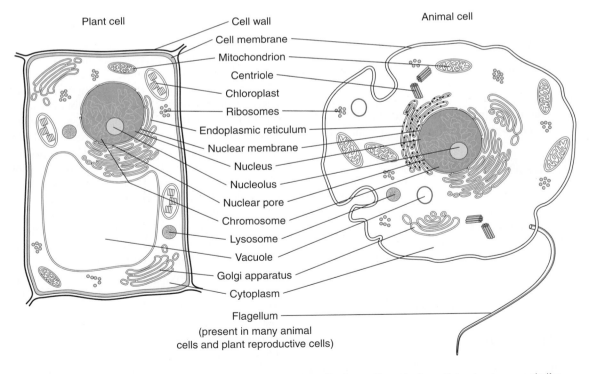

Plant cell — Cell wall
Cell membrane
Mitochondrion
Centriole
Chloroplast
Ribosomes
Endoplasmic reticulum
Nuclear membrane
Nucleus
Nucleolus
Nuclear pore
Chromosome
Lysosome
Vacuole
Golgi apparatus
Cytoplasm
Flagellum
(present in many animal cells and plant reproductive cells)

Animal cell

Figure P.1 Cellular homology between plant and animal cells. Organelles and other cellular structures are similar due to common descent.

exquisite example of the vertebrate limb is in chapter 13. The comparison of the skeletal elements shows that, bone for bone, the composition of these vastly dissimilar-looking limbs are derived from the same building blocks.

COMPARATIVE EMBRYOLOGY

In chapter 15, developmental homology is illustrated with the account of homeotic genes such as the *Hox* and the *Pax-6* genes. Homology is also seen in the common extra-embryonic membrane found in the vertebrate egg (chapter 13). Another example of a developmental homology is the anatomical similarities between young vertebrate embryos (fig. P.2). These images do not require a trained eye to appreciate. As with other homologies, what explanation can be offered to explain these obvious correspondences than other descent from a common ancestor?

VESTIGIAL ORGANS

Many modern organisms retain rudiments or vestiges of structures that became obsolete over the

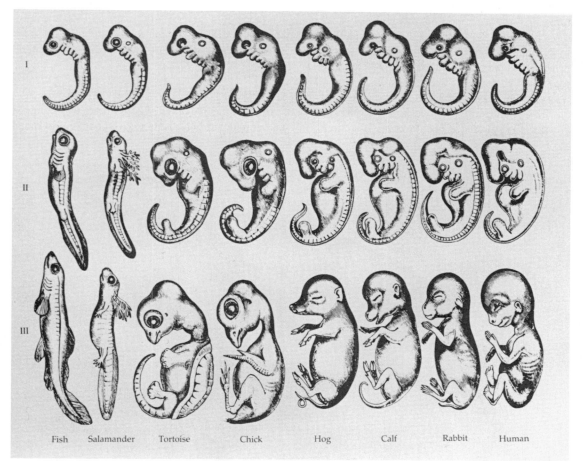

| Fish | Salamander | Tortoise | Chick | Hog | Calf | Rabbit | Human |

Figure P.2 Embryonic homology. Comparisons of early vererbate embryos reveal striking similarity. Each early embryo has the same number of gill arches (pouches below the head) and similar vertebral columns. During development, modifications associated with adult forms become apparent.

course of evolutionary history. Where these structures are found, they provide further evidence of previously evolved structures that lost their utility. The existence of *vestigial organs,* anatomical structures with no apparent function but that resemble structures found in a presumed ancestor, is another line of evidence in support of evolution. As organisms adapt to new environments, they carry with them the structures that evolved for some purpose that is no longer necessary. These structures are evolutionary baggage that reflects the journey from the past. Natural selection allots less energy to the maintenance of extraneous structures than to structures that enhance reproductive success. Furthermore, structures that have lost their original function may in fact be harmful to their owners and hence be selected against. With the passage of time, structures without function tend to diminish until only traces of their former size and function remain.

Excellent examples of vestiges from a past life are the pelvic girdles and rudimentary hind limbs in whales, harkening back to their ancestry as terrestrial mammals (fig. P.3). Again the rhetorical question must be asked: Why else would a whale have an embedded pelvic bone in the place its pelvis might be, had it not evolved from a terrestrial mammal that once needed a pelvis to support its legs?

Examples of vestigial structures abound. Cave-dwelling creatures tend to have reduced or absent eyes. Some snakes have rudimentary limbs during development, and some, like boas and pythons retain limb buds as adults. Vestigial structures found in humans and the great apes include the greatly reduced tail bones, a few remnants of tail muscles, the third molar (wisdom teeth), and the vermiform appendix. All of these remnants and rudiments speak to common ancestry and descent with modification.

EVOLUTIONARY OPPORTUNISM

Another line of evidence in support of evolution are anatomical structures that have been appropriated

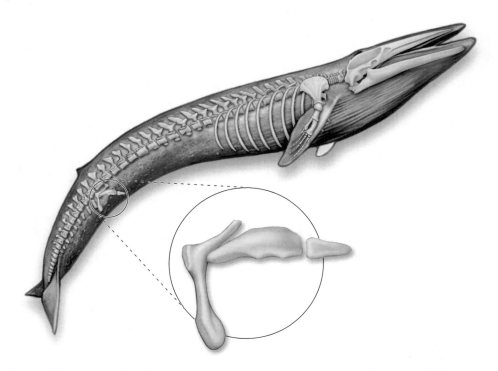

Figure P.3 Vestigial structures. The retention of pelvic bones in the modern whale reveals their common ancestry with the terrestrial mammals from which they evolved.

for novel purposes over the course of time. Imperfectly designed body parts exemplify the fact that evolution is based on contingency and that nature makes due with whatever starting materials are available. Perhaps the most famous such anatomical anomaly in evolutionary lore is the commandeering of a wrist bone fashioned to become the panda's "thumb."

Giant pandas are exclusive to the high-elevation bamboo forests of western China. Unlike the other members of the Order Carnivora (the other bears, cats, dogs, weasels, etc.), giant pandas survive by exploiting the abundant bamboo in the forests they inhabit as their sole source of nutrition. Dexterous thumbs are a trademark of primates, best exemplified in humankind. By contrast, carnivores have forelimbs adapted for mobility and cutting, and their fingers lack dexterity for fine-motor manipulations.

When examined by comparative anatomists, the mystery of the panda's thumb is revealed (fig. P.4). Anatomically, this "thumb" is not a true finger but an enlarged wrist bone, the radial sesamoid bone. Opportunistically, over evolutionary time and under selective pressure from its environment, this wrist bone and associated muscles have

been enlarged to adapt to the task of stripping bamboo leaves from their stalks. As such, the panda's "thumb" is a hallmark of how sometimes the best evidence for evolution is found in the imperfections left behind by evolution and not in marvels of apparent perfection of design.

FOSSILS

Fossils are another line of evidence in support of evolution. Their existence provided important insights to Darwin, who brought back several fossils from South America (chapter 2). The sequence of fossils in the strata of rocks is impossible to explain except in the light of evolution. As seen in table 16.1, the events that are chronicled in the geological strata are in the sequence predicted by evolutionary theory. As expected, no human remains have ever been found among the dinosaur fossils.

TRANSITIONAL FOSSILS

Transitional fossils are fossils that fill the "missing links" in the fossil record. Over time, the number of discoveries of transitional fossils has

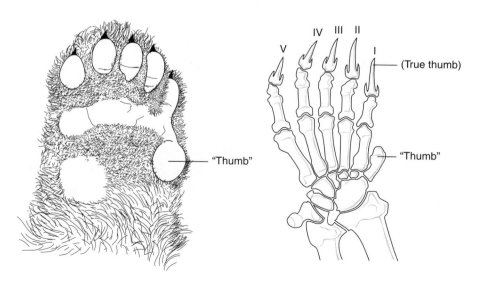

Figure P.4 **The panda's thumb.** Giant pandas have enlarged wrist bones, the radical sesamoid bone, that appears to be a sixth-digit thumb.

increased, and in many instances it is difficult to find true missing links. Much of chapter 13 is devoted to discussions of transitional fossils such as *Archaeopteryx,* the transitional form between birds and dinosaurs, and *Tiktaalik,* an intermediary between aquatic and terrestrial tetrapods.

Ancestral human transitional forms are described in chapter 17.

A recently described series of transitional fossil closes many of the gaps between hoofed mammals and modern whale (fig. P.5). Three transitional cetacean (the mammalian Order of whales,

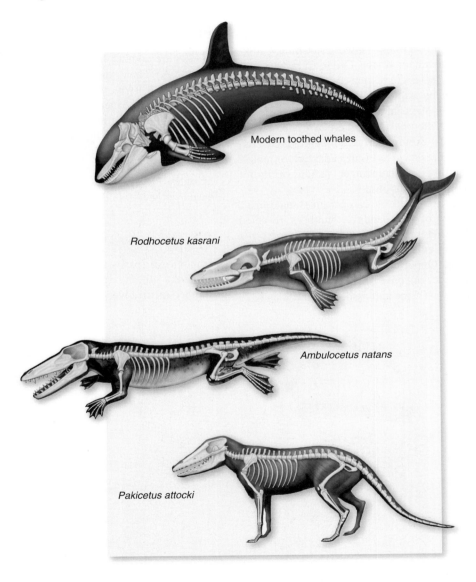

Modern toothed whales

Rodhocetus kasrani

Ambulocetus natans

Pakicetus attocki

Figure P.5 Transitional cetaceans. Fossil whale ancestors that fill in many gaps in the evolutionary transition from hoofed mammal to modern whales.

dolphins, and porpoises) fossils have collectively changed the way in which scientists view whale evolution.

The fossil cetacean *Pakicetus* was first discovered in 1983 (a complete skeleton was unearthed in 2001). This wolf-size whale ancestor lived on land but showed many features found in modern whales, such as a cetacean inner ear. Crocodile-like *Ambulocetus* (meaning "walking whale") fossils were discovered in the mid-1990s. *Ambulocetus* has structural features that suggest that it lived on land while also being well adapted for swimming. *Rodhocetus* fossils were discovered in 2001. They have even more whale-like features than either *Ambulocetus* or *Paticetus,* such as greatly reduced hind limbs that allowed them to swim with a motion similar to that of modern whales.

BIOGEOGRAPHY

Biogeography, the distribution of life over geographical areas, forms the basis of another line of evidence for evolution. Species that are closely related, such as the chimps, gorillas, and humans, are all known to have come from Africa. As species colonize new habitats, they take on the distributions we see today. The adaptive radiations outlined in chapters 12 and 13 exemplify the connection between geography and evolution.

On the whole, species that are distributed around the globe are generally most related to the other species in that vicinity unless there are good reasons for this not to be the case, such as possessing the capacity for great mobility (such as sea animals and birds) or being organisms recently redistributed by humans. Another reason for the observed distributional relationships has to do with the historical movement of the continents know as *continental drift.*

An illustration of this concept is the modern distribution of marsupials, which predominate in Australia, where they evolved before the evolution of placental mammals. Some marsupials are found in South and North America due to the fact that millions of years ago, South America, Antarctica, and Australia were once part of a large massive continent called *Gondwanaland* (fig. P.6). As the continents separated, marsupials were stranded in Australia and South America, where they then underwent their own adaptive radiation. Although largely outcompeted by placental mammals in the Americas, some highly adapted marsupials survive in South America and in North America, such as the opossum, a supreme generalist that can successfully exploit a wide variety of habitats.

CLOSING

In summary, as you read through the pages ahead, use this guide and your own mind to take note of the various lines of evidence supporting the fact of evolution. As discussed in chapter 1, a scientific theory is a large body of tested ideas and not just a notion someone thought up one morning.

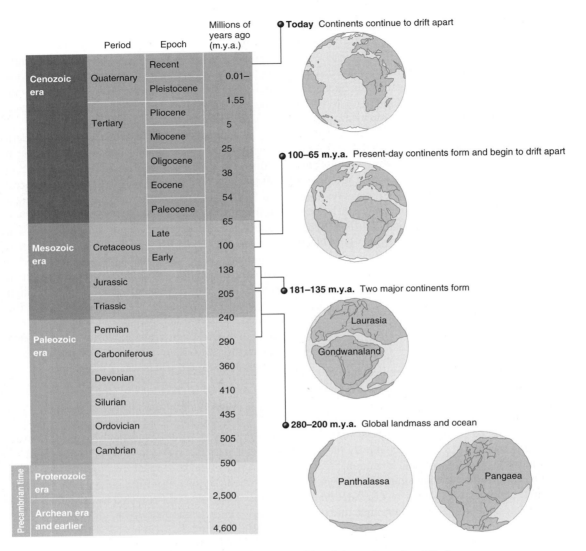

Figure P.6 Changing continents. Over geologic time, the position of the continents has shifted.

VARIATION IN POPULATIONS

In the fall of 1958, the folks of a quiet, rural community in the southern United States were startled and dismayed to find large numbers of multilegged frogs in an 85-acre artificial lake on a cotton farm. Widespread newspaper publicity of this strange event attracted the attention of university scientists, curiosity seekers, and gourmets. The lake supported a large population of bullfrogs, estimated at several thousand. Although reports tended to be exaggerated, there were undoubtedly in excess of 350 multilegged deviants. As illustrated in figure 1.1, the extra legs were oddly positioned, but they were unmistakably copies of the two normal hind limbs. Incredibly, the extra limbs were functional, but their movements were perceptibly not in harmony with the pair of normal legs. The bizarre multilegged frogs were clumsy and graceless.

All the multilegged frogs appeared to be of the same age, approximately two years old, and of the same generation. These atypical frogs were found only during the one season and were not detected again in subsequent years. The multilegged frogs disappeared almost as dramatically as they had appeared.

History repeated itself in August 1995, almost four decades later, with the appearance, in the wetlands of Minnesota and neighboring states, of many frogs with misshapen hind limbs (fig. 1.2). In this instance, these multilegged frogs gained instant notoriety, because information and pictures were heralded on the Internet. The Minnesota assemblage of variant frogs consisted of several different species. The most severely affected was the highly aquatic mink frog (taxonomically speaking, *Lithobates septentrionalis* (formerly *Rana septentrionalis*)), known for its distinct musky odor. Well over 200 malformed mink frogs were captured at one site in one year. Many different malformations have been encountered—multiple legs, missing limbs (whole or in part), distorted limbs, missing eyes, and deformed jaws. Disconcertingly, the disfigured frogs in the Minnesota environs have reappeared in successive years—as many as four years in several populations. This differs noticeably from the aforementioned multilegged bullfrogs (*Lithobates catesbeiana*) found in the Mississippi farm locality for only a relatively brief period.

Strange and exceptional events of this kind are of absorbing interest, and challenge us for an explanation. How do such oddities arise, and

Figure 1.1 Two multilegged frogs, each viewed from the back (dorsal) and front (ventral) surface. These bizarre bullfrogs were discovered in October 1958 in a lake near Tunica, Mississippi. Several hundred frogs with extra hind limbs were found at this locality. How does such an abnormality arise? Two reasonable interpretations are set forth in chapter 1.

(Photographs by E. Peter Volpe, then at Tulane University.)

(a)   (b)

Figure 1.2 Dorsal (a) and ventral (b) views of a multilegged mink frog (*Rana septentrionalis*) collected in the wetlands of Minnesota by David Hoppe of the University of Minnesota. Several extra hind limbs protrude from the sides.
Source: Photo by David Hoppe, The University of Minnesota (Morris campus).

what factors are responsible for their ultimate disappearance in a natural population? In ancient times, bodily deformities evoked reverential awe and inspired some fanciful tales. Early humans constructed a number of myths to explain odd events totally beyond their control or comprehension. One legend said that deformities arose when masses of skeletons were reanimated and the bones of different animals mixed together. Another old idea is that grotesquely shaped frogs are throwbacks to some remote prehistoric ancestor. These accounts are, of course, novelistic and illusory, but they do reveal the uniquely imaginative capacity of the human mind. Nevertheless, we should seek a completely different cause-and-effect sequence, relying more on our faculty for logical analysis.

Worldwide reports of massive amphibian extinctions and myriad instances and kinds of gross abnormalities point to a constellation of causes and synergies rather than any single causal factor. Many believe that recently observed extinction events and deformity syndromes reflect the "canary in the mineshaft," foretelling deep-seated environmental problems facing our planet. They point to the exquisitely sensitive nature of amphibians with their wet skin that is in constant contact with its surrounding aquatic ecosystem. For our purposes in this book, we wish to limit our discussion to the question of whether such anomalous variations might be inherited, and leave the broader discussion of the environment to books and articles specifically devoted to that subject. This does not mean that these matters are unimportant—quite the contrary. Rather, our goal here is to explore the nature and sources of variation, as Darwin did in his first chapters of his most famous book, The *Origin of Species,* which is also the subject of chapter 2.

Environmental Modifications

Inspection alone cannot reveal the underlying cause of the multilegged anomaly. We may thoroughly dissect the limbs and describe in detail the anatomy of each component, but no amount of dissection can tell us how the malformation arose. The deformity either was preordained by heredity or originated from injury to the embryo at a vulnerable stage in development. We shall direct our attention first to the latter possibility and its implications.

Some external factor in the environment may have adversely affected the pattern of development of the hind limb region. The cotton fields around the lake in Mississippi were periodically sprayed with pesticides to combat infestations of insects. Purulent airborne pollutants may have settled onto the landscape in Minnesota. It is also possible that excess ultraviolet radiation leaked through a thinning protective ozone layer of the atmosphere and caused the limb deformities. It is likely that the widespread deformities and amphibian declines are the result of multiple causes, most likely induced by humans. In essence, it is not inconceivable that the pesticides, pollutants, and even ultraviolet rays were potent *teratogens*—that is, chemical and physical agents capable of causing marked distortions of normal body parts.

That chemical substances can have detrimental effects on the developing organism is well documented. For example, the geneticist Walter Landauer demonstrated in the 1950s that a wide variety of chemicals, such as boric acid, pilocarpine, and insulin (normally a beneficial hormone), can produce abnormalities of the legs and beaks when injected into chick embryos. This type of finding cannot be dismissed complacently as an instance of a laboratory demonstration without parallel in real life. Indeed, in 1961 medical researchers discovered with amazement amounting to incredulity that a purportedly harmless sleeping pill containing the drug thalidomide, when taken by a pregnant woman, particularly during the second month of pregnancy, could lead to a grotesque deformity in the newborn baby, a rare condition in humans called *phocomelia*—literally, "seal limbs." The arms are absent or reduced to tiny, flipperlike stumps (fig. 1.3).

More than 6,000 thalidomide babies were born in West Germany and at least 1,000 in other countries. The United States was largely spared the thalidomide disaster by the astute scientific sense of Frances O. Kelsey of the U.S. Food and Drug Administration (FDA): she blocked general distribution of the drug. Despite her efforts,

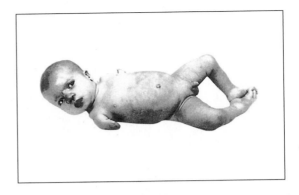

Figure 1.3 Armless deformity in the newborn infant, resulting from the action of thalidomide, a sedative taken by his mother in her second month of pregnancy.
Source: W. Lenz and K. Knapp, "Thalidomide Embryopathy," *Archives of Environmental Health* 5 (1962):100–105. Courtesy of the publisher.

some American women did obtain the drug from European sources. The thalidomide tragedy changed the attitude of the medical profession toward drug-induced malformations. It also awakened the mass media; congenital defects rapidly became a subject of compelling interest to the public.

Thalidomide was banned worldwide in 1962. Strange as it may seem, the world's most infamous drug has resurfaced in recent years. Commercial laboratories have been testing the drug for possible beneficial effects. In September 1997 the FDA approved the use of thalidomide for treatment of a severe complication of leprosy, for which no good alternative therapies are presently available. Ongoing research suggests that thalidomide may be useful in the treatment of certain types of cancerous growths (brain tumors) and the wasting syndrome associated with AIDS, but there is gnawing apprehension about the distribution and marketing of the drug.

In different organisms—fish, frogs, mice, and humans—anomalies can be caused by such diverse agents as extremes of temperature, alcohol consumption, x rays, viruses, drugs, dietary deficiencies, and lack of oxygen. There is no longer the slightest doubt that environmental factors may be causal agents of specific defects. A variation that arises as a direct response to some external change in the environment, and not by any change in the genetic makeup of the individual, is called an environmental modification.

If the multilegged anomaly in the bullfrog was environmentally induced, we may surmise that (as depicted in fig. 1.4) the harmful chemical or other causative factor acted during the sensitive early embryonic stage. Moreover, if some external agent had brought about the abnormality in the bullfrog, this agent must have been effective only once, because the multilegged condition did not occur repeatedly over the years in the Mississippi region. It may be that the harmful environmental factor did not recur, in which case we would not expect a reappearance of the malformation.

Our suppositions could be put to a test by controlled breeding experiments, the importance of which cannot be overstated. Only through breeding tests can the basis of the variation be firmly established. If the anomaly constituted an environmental modification, then a cross of two multilegged bullfrogs would yield all normal progeny, as illustrated in figure 1.4. In the absence of any disturbing environmental factors, the offspring would develop normal limbs. This breeding experiment was not actually performed; none of the malformed frogs survived to sexual maturity. Nonetheless, we have brought into focus an important biological principle: environmentally induced (nongenetic) traits cannot be passed on to another generation.

The elements that are transmitted to the next generation are two tiny cells, the egg and the sperm. These two specialized cells are often referred to as *germ cells*, because they are the beginnings, or germs, of new individuals. The germ cells are the only connecting thread between successive generations. Accordingly, the mechanism of hereditary transmission must operate across this slender connecting bridge. The hereditary qualities of the offspring are established when the sperm unites with the egg. The basic hereditary determiners, the genes, occur in pairs in the fertilized egg. Each inherited characteristic is governed by at least one pair of *genes,* with one member of each pair coming from the male parent and the other from the female parent. Stated another way, the sperm and egg each have half the number of genes of each parent. Fertilization restores the original number and the paired condition (fig. 1.4).

The underlying assumption in the cross depicted in figure 1.4 is that the genes that influence the development of the hind limbs are normal. Of course, the multilegged parents themselves possess normal genes. It might seem strange that an abnormal character can result from a perfectly sound set of genes. However, normal genes cannot be expected to act normally under all environmental circumstances. A gene may be likened to a photographic negative. A perfect negative (normal gene) may produce an excellent or poor positive print (normal or abnormal trait), depending on such (environmental) factors as the quality or concentration of the chemical solutions used in preparing the print. The environment thus affects the expression of the negative (gene), but the negative (gene) itself remains unaffected throughout the making of the print (trait).

We shall learn more about the nature of genes and the mechanism of inheritance in chapter 2. For the moment, the important consideration is that a given gene prescribes a potentiality for a trait and not the trait itself. Genes do not act in a vacuum. Genes always act within the conditioning framework of the environment. The materialization of a trait represents the interplay of genetic determinants and environmental factors.

LAMARCKISM

Few people would expect bodily deformities caused by harmful environmental factors to be heritable. And yet many people once believed that favorable or beneficial bodily changes acquired or developed during one's lifetime are transmitted to the offspring—for example, that athletes who exercised and developed large muscles would pass down their powerful muscular development to their children. This is the famous theory of the inheritance of acquired characteristics, or Lamarckism (after Jean-Baptiste de Lamarck, a French naturalist of the late 1700s and early 1800s). Lamarckism has no foundation of factual evidence. We know, for example, that a woman who has altered her body by having laser treatments to remove body hair does not pass this alteration on to her daughters. Circumcision is still a requisite in the newborn male, despite the fact that the rite that has been practiced for well over 4,000 years. It is sufficient to state that the results of countless laboratory experiments testing the possibility of the inheritance of acquired, or environmentally induced, bodily traits have been negative.

The notion of the inheritance of acquired characteristics became the cornerstone of Lamarck's comprehensive, although incorrect, explanation of evolution. Lamarck's theory is exemplified

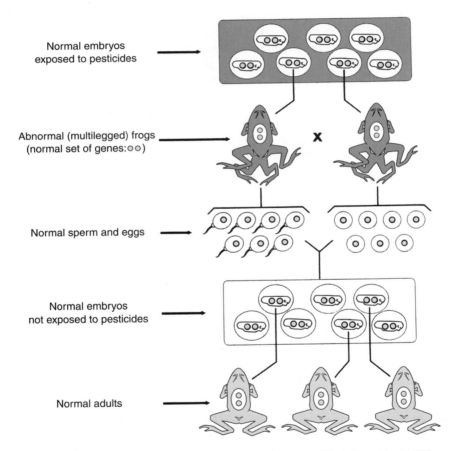

Normal embryos exposed to pesticides

Abnormal (multilegged) frogs (normal set of genes: ○ ○)

Normal sperm and eggs

Normal embryos not exposed to pesticides

Normal adults

Figure 1.4 Test for the noninheritance of an environmentally induced trait. If the malformation of the hind limbs arose from without by the action of an adverse environmental agent (pesticide) and not from within by genetic change, then the abnormality would *not* be transmitted from parent to offspring. The multilegged parents, although visibly abnormal, are both genetically normal. Normal hind limbs are expected if, during development, the offspring are not exposed to the same injurious environmental agents to which the parents were exposed.

by a fabled quotation from his book *Philosophie Zoologique* (1809):

> The giraffe lives in places where the ground is almost invariably parched and without grass. Obliged to browse on trees it is continually forced to stretch upwards. This habit sustained over long periods of time by every individual of the race has resulted in the forelimbs becoming longer than the hind ones, and the neck so elongated that a giraffe can lift his head to a height of six metres [about 20 feet] without taking his forelimbs off the ground.

Lamarck advocated that the organs of an animal became modified in appropriate fashion in direct response to a changing environment. The various organs became greatly improved through use or reduced to vestiges through disuse. Such bodily modifications, he argued, could in some manner be transferred and impressed on the germ cells to affect future generations. Thus, Lamarck was arguing that the whale lost its hind limbs as a consequence of the inherited effects of disuse, and the wading bird developed its long legs through generations of sustained stretching to keep the body above the water level. Inheritance, in his view, was simply the direct transmission of those superficial bodily changes that arose within the lifetime of the individual owing to use or disuse. Lamarck's views are unacceptable because we know of no mechanism that allows changes in the body to register themselves on the germ cells.

Although Lamarck was incorrect about the mechanism of evolution, he deserves great credit for formulating the first truly well-thought-out theory of organic evolution. In Lamarck's day, like for Darwin, ideas that diverged from the biblical account of creation and predestination were heretical and counter to the prevailing *zeitgeist* or worldview. Darwin properly credits Lamarck, noting that:

> Lamarck was the first man whose conclusions on the subject excited much attention. This justly celebrated naturalist first published his views in 1801 . . . he first did the eminent service of arousing attention to the probability of all changes in the organic, as well as in the inorganic world,

being the result of law, and not of miraculous interposition.

The foregoing considerations bear importantly on our bullfrog population. We have not been able to perform the critical breeding test to prove or disprove that the multilegged trait is a nonheritable modification. However, let us for the moment accept the thesis that the multilegged condition was environmentally induced and examine the implications of this view on the population as a whole. Outwardly, a striking change in the bullfrog population seems to have occurred. But in actuality the composition of the population has remained essentially unaltered, because the basic hereditary materials, the genes, were not affected. In other words, there was no change of evolutionary significance in the population. Evolution can occur only where there is heritable variation.

HERITABLE VARIATION

We shall now examine the possibility that the multilegged frogs were genetically abnormal. One or more of their genes may have been impaired. Unlike teratogens, which damage an already conceived offspring and affect only a single generation, deleterious genes are in the germ plasm before conception and may lie dormant for several generations.

We may assume that the multilegged trait is due to an alteration in a single gene, which is transmitted in simple Mendelian fashion. We infer the existence of two alternative forms of the gene—namely, a normal version and a variant version. Geneticists use the Greek word *allele* to denote the normal and variant forms of a gene. In other words, the normal allele of a gene conveys appropriate instructions, whereas the variant allele of the gene conveys improper information.

We shall postulate that the variant allele for the multilegged condition is *recessive* to the normal allele. That is to say, the expression of the variant allele is completely masked or suppressed by the normal (or *dominant*) allele when the two are present together in the same individual (fig. 1.5). Such an

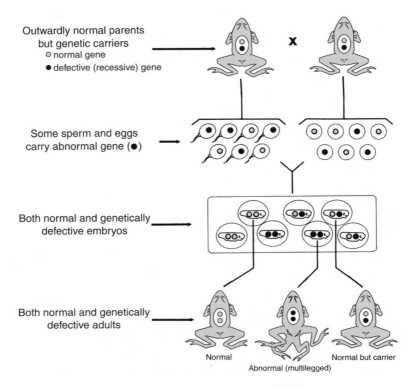

Outwardly normal parents but genetic carriers
○ normal gene
● defective (recessive) gene

Some sperm and eggs carry abnormal gene (●)

Both normal and genetically defective embryos

Both normal and genetically defective adults

Normal

Abnormal (multilegged)

Normal but carrier

Figure 1.5 Mating of two "carrier" parents resulting in the emergence of multilegged offspring. The multilegged trait is assumed to be controlled by a defective recessive gene. Both parents are normal in appearance, but each carries the defective gene, masked by the normal gene. The recessive gene manifests its detrimental effect when the offspring inherits one abnormal gene from each of its parents.

individual, normal in appearance but harboring the detrimental recessive allele, is said to be a *carrier.* The recessive allele may be transmitted without any outward manifestation for several generations, continually being sheltered by its dominant partner. However, as seen in figure 1.5, the deleterious recessive allele ultimately becomes exposed when two carrier parents happen to mate. Those progeny that are endowed with two recessive alleles, one from each parent, are malformed. On the average, one-fourth of the offspring will be multilegged.

Genetic defects transmitted by recessive genes are not at all unusual. Pertinent to the present discussion is an inherited syndrome of abnormalities in humans, known as the Ellis–van Creveld syndrome. Afflicted individuals are disproportionately dwarfed (short-limbed), have malformed hearts, and possess six fingers on each hand (fig. 1.6). The recessive gene that is responsible for this complex of defects is exceedingly rare. Yet, as shown by the late geneticist Victor A. McKusick formerly of Johns Hopkins University, the Ellis–van Creveld anomaly occurs with an exceptionally high incidence among the Amish people in Lancaster County, Pennsylvania. The variant recessive allele apparently was present in one member of the original Old Order Amish immigrants from Europe two centuries ago. For a few generations,

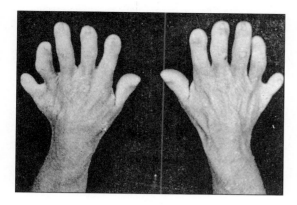

Figure 1.6 Six-digited hands, one of the manifestations of the Ellis–van Creveld syndrome in humans, a rare recessive deformity. Affected offspring generally come from two normal, but carrier, parents, each of whom harbors the abnormal recessive gene.
Courtesy of Victor A. McKusick, Johns Hopkins University.

the harmful allele was passed down unobserved, masked by its normal partner allele. Since 1860, the Ellis–van Creveld deformity has appeared in at least 50 offspring. Ordinarily, it is uncommon for both members of a married couple to harbor the defective recessive allele. However, in the Amish community, marriages have been largely confined within members of the sect with a resulting high degree of *consanguinity*. Marriages of close relatives have tended to promote the meeting of two normal, but carrier, parents.

We know very little about the past history or breeding structure of the Mississippi bullfrog population in which the multilegged trait appeared with an exceptionally high frequency. We may suspect that all 350 or more multilegged frogs were derived from a single mating of two carrier parents. In contrast to humans, a single mated pair of frogs can produce well over 10,000 offspring. The similarity in age of the multilegged bullfrogs found in nature adds weight to the supposition that these frogs are members of one generation, and probably of one mating.

Much of the preceding discussion on the multilegged bullfrog is admittedly speculative. However, one aspect is certain: *previously concealed harmful genes are brought to light through the mechanism of heredity.* A trait absent for many generations can suddenly appear without warning. Once a variant character expresses itself, its fate will be determined by the ability of the individual displaying the trait to survive and reproduce in its given environment.

It is difficult to imagine that the grotesquely shaped frogs could compete successfully with their normal kin. However, we shall never know whether or not the multilegged frogs were capable of contending with the severities of climatic or seasonal changes, or of successfully escaping their predators, or even of actively defending themselves. What we do know is that the multilegged bullfrogs did not survive to reproductive age. Despite diligent searches by many interested investigators in the Mississippi locale, no sexually mature abnormal frogs have been uncovered in the natural population. It seems that this unfavorable variant has been eliminated. In the next chapter, we will explore this situation in light of Darwin's tenet of natural selection.

HYPOTHETICO-DEDUCTIVE REASONING

We have acted as objective and open-minded spectators in observing and interpreting the events surrounding the origin and demise of the multilegged frogs. We formulated a few *hypotheses* or, in the vernacular, educated guesses or hunches. One of the provisional conjectures for the Mississippi assemblage of bullfrogs was that a chance mating of two carrier parents resulted in genetically defective offspring. This hypothesis permitted the *deduction* that the unfavorable trait would not likely resurface in a subsequent breeding season. In essence, we rendered a hypothesis consistent with the observations, and deduced predictions or consequences from the hypothesis.

We will accept, reject, or modify the hypothesis in accordance with the degree of fulfillment of the predictions. If the hypothesis is to become a scientific explanation, we are obliged to test

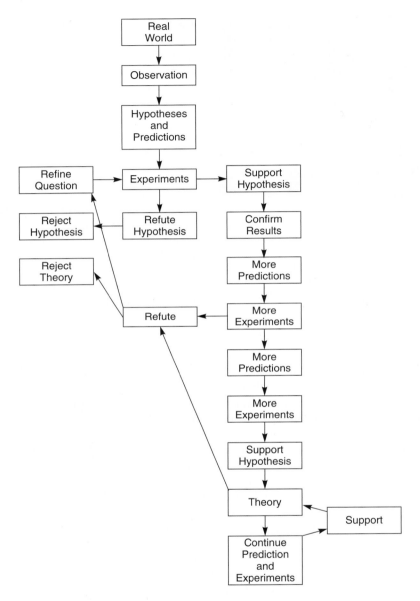

Figure 1.7 Hypothetico-deductive reasoning, also called the scientific method. This conceptual diagram illustrates how hypotheses and theories are generated in science. Scientists start with observations of the natural world, from which they generate a hypothesis (or several hypotheses). The hypothesis is then experimentally verified or rejected. At any point the hypothesis may be redefined, based on new data or falsification of the original hypothesis. A hypothesis that is supported by experimentation is then repeatedly subjected to prediction and further experimentation. Only with the extensive support of numerous experiments are hypotheses considered to constitute a theory. Theories are continually used to make predictions that are repeated tested. When experiments fail to support predictions, the hypothesis or theory may again be redefined and retested. If, after experimentation, hypotheses or theories are no longer supported, they are abandoned. The theory of evolution via natural selection has been repeatedly used to make predictions and to guide experimentation and has withstood 150 years of scrutiny.

the deductions by additional observations or by performing experiments. This mode of inquiry is the *hypothetico-deductive* style of reasoning that is the cornerstone of investigative sciences, particularly the natural sciences (fig. 1.7). This method of establishing explanations for observed phenomena has the goal of developing the habit of scientific thinking, characterized by objectivity, open-mindedness, skepticism, and the willingness to suspend judgment if there is insufficient evidence.

Our prediction was fulfilled for the 1958 population of malformed bullfrogs in the Mississippi area: no such frogs were found again in subsequent breeding seasons. The supposition that a heritable transmission of a faulty gene has been in play in the more recent Minnesota populations of mink frogs is less inviting, particularly in light of a fruitful experiment undertaken by David Hoppe of the University of Minnesota. He examined the outcome of dividing into two batches a cluster of freshly deposited frog eggs from a natural pond where malformed frogs had occurred. One batch was left at its site of collection and subject to the vicissitudes of nature. The second batch was reared in the laboratory in treated water known to be suitable for the growth of tadpoles. Accordingly, the two batches of eggs had the same genetic heritage but were submitted to different environmental experiences. The informative result was that some of the eggs left in nature developed into froglets having extra legs and other malformations, whereas all the laboratory-reared eggs developed into normal froglets. This suggests strongly that some enduring noxious environmental agent, in or around the pond, has persisted in inducing abnormal development of the frogs through the years.

We are not yet justified in accepting any hypothesis as the most probable explanation. If we were able to formulate critical tests that continually substantiate our deductions, our confidence would grow that a particular thesis is the most probable explanation. This explanatory statement would then be regarded as "true" and would become part of the arsenal of scientific knowledge.

Acceptance of a scientific statement is the attainment of a high degree of confidence in its accuracy. But even the most definitive statement is subject to revision. A statement acceptable at one moment of time is still open to possible exception and modification. The natural sciences disallow absolute certainty. Stated another way, there is *no eternal or absolute truth in the natural sciences.* This view offers shallow comfort to those persons who are uncomfortable with, or unaccustomed to, uncertainty.

When a hypothesis is recognized as having far-reaching implications and is continually verified by observation and experimentation, the hypothesis may then attain the status of a *theory.* A theory is, accordingly, a hypothesis sufficiently tested for a scientist to have complete confidence in its probability. A theory is not merely an untested opinion, as implied by the oft-heard recantation: "After all, evolution is *only* a theory!" The theory that life has evolved and is ever changing is founded on as much evidence as the broadly accepted generalization that the Earth is round.

Occasionally, but hesitatingly, a theory advances to a law. In the natural sciences, laws have questionable utility. A law tends to connote that a phenomenon is absolutely certain. The famous Mendelian laws of inheritance remain fundamentally sound, but they are not, in fact, inviolate. Several dramatic exceptions have recently been uncovered, which Gregor Mendel could not have dared envision. Modern molecular biologists have attached several significant codicils to Mendel's bequest.

2

DARWINIAN SCHEME OF EVOLUTION

More than a century and a half ago, in 1859, Charles Robert Darwin (fig. 2.1) gave the biological world the master key that unlocked all previous perplexities about evolution. His thesis of natural selection can be compared only with such revolutionary ideas as Newton's law of gravitation and Einstein's theory of relativity. The concept of natural selection was set forth clearly and convincingly by Darwin in his monumental treatise *The Origin of Species.* This epoch-making book was the fruition of more than twenty years of meticulously sifting through the voluminous data he had accumulated.

In 1831, Charles Darwin, then 22 years old and fresh from Cambridge University, accepted the post of naturalist, without pay, aboard H.M.S. *Beagle,* a ship commissioned by the British Admiralty for a surveying voyage around the world. Although Darwin was an indifferent student at Cambridge, he did show an interest in the natural sciences. He was an ardent collector of beetles, enjoyed bird watching and hunting, and was an amateur geologist.

It took the *Beagle* nearly five years—from 1831 to 1836—to circle the globe (fig. 2.2). When Darwin first embarked on the voyage, he did not dispute the dogma that every species of organism had come into being at the same time and had remained permanently unaltered. He shared the views of his contemporaries that all organisms had been created about 4000 B.C.—more precisely, at 9:00 A.M. on Sunday, October 23, in 4004 B.C., according to the bold pronouncement of Archbishop James Ussher in the seventeenth century. Darwin had, in fact, studied for the clergy at Cambridge University. But he was to make observations on the *Beagle's* voyage that he could not reconcile with accepted beliefs.

Darwin's quarters on the *Beagle* were cramped, and he took only a few books on board. One of them was the newly published first volume of Charles Lyell's *Principles of Geology,* a parting gift from his Cambridge mentor, John Henslow, professor of botany. Lyell challenged the prevailing belief that the Earth had been created by a divine plan merely 6,000 years ago. On the contrary, according to Lyell, the Earth's age could be measured in hundreds of millions of years. Lyell argued that the Earth's mountains, valleys, rivers, and coastlines were shaped not by Noah's Flood

Figure 2.1 **Charles Darwin** at the age of 31 (1840), four years after his famous voyage around the world as an unpaid naturalist aboard H.M.S. *Beagle*.
Department of Library Services American Museum of Natural History.

but by the ordinary action of the rains, the winds, earthquakes, volcanoes, and other natural forces. Darwin was impressed by Lyell's emphasis on the great antiquity of the Earth's rocks and gradually came to perceive that the characteristics of organisms, as well as the face of the Earth, could change over a vast span of time.

The Voyage of the Beagle (1839) chronicles Darwin's five-year voyage on H.M.S. *Beagle*, which set sail from Plymouth on the drizzling morning of December 27, 1831, the day after Christmas, with a crew of 75, under the leadership of Captain Robert Fitzroy. Fitzroy employed the young Mr. Darwin as his companion, and not, as many think, as the ship's naturalist (Robert McCormick was the original ship's naturalist and surgeon, but soon he jumped ship in Brazil). In those days, aristocrats of Captain Fitzroy's status could not dine or converse with lowly crewmembers and often took along passengers of their own social class as companions, so as to not go mad on long, lonesome voyages (the first captain of the *Beagle* committed suicide). After McCormick's hasty departure from the *Beagle*,

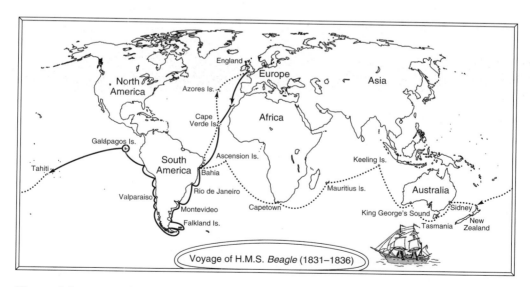

Voyage of H.M.S. *Beagle* (1831–1836)

Figure 2.2 **Five-year world voyage of the H.M.S.** *Beagle*. Darwin's observations on this voyage convinced him of the reality of evolution. Particularly impressive to him were the fossil skeletons of mammals unearthed in the Argentine pampas (see fig. 2.3) and the variety of tortoises and birds in the small group of volcanic islands, the Galápagos Islands (see fig. 12.4).

Charles did indeed assume the duties of the ship's naturalist. The young Charles Darwin was plagued with severe seasickness and took every opportunity to leave the *Beagle* to explore and collect whenever it made port. His youthful vitality and keen personality were extraordinary assets to Darwin, who easily made friends with aristocrats and native people alike. The thousands of specimens he shipped back to England helped establish Darwin as a man of science at the young age of 27.

The living and extinct organisms that Darwin observed in the flat plains of the Argentine pampas and in the Galápagos Islands sowed the seeds of Darwin's views on evolution. From old river beds in the Argentine pampas, he dug up bony remains of extinct mammals. One fossil finding was the massive *Toxodon,* whose appearance was likened to a hornless rhinoceros or a hippopotamus (fig. 2.3). Another fossil remain that attracted Darwin's attention was the skeleton of *Macrauchenia,* which he erroneously thought was clearly related to the camel because of the structure of the bones

of its long neck. Other remarkable creatures were the huge *Pyrotherium,* resembling an elephant, and the light and graceful single-toed *Thoatherium,* rivaling the horse. The presence of *Thoatherium* testified that a horse had been among the ancient inhabitants of the continent. It was the Spanish settlers who reintroduced the modern horse, *Equus,* to the continent of South America in the sixteenth century. Darwin marveled that a native horse should have lived and disappeared in South America. This was one of the first indications that species gradually became modified with time and that not all species survived through the ages.

When Darwin collected the remains of giant armadillos and sloths on an Argentine pampa, he reflected on the fact that, although they clearly belonged to extinct forms, they were constructed on the same basic plan as the small living armadillos and sloths of the same region. This experience started him thinking of the fossil sequence of a given animal species through the ages and the causes of extinction. He wrote: "This wonderful

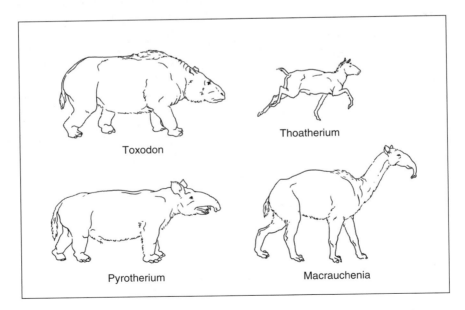

Figure 2.3 Curious hoofed mammals (ungulates) that flourished on the continent of South America some 60 to 70 million years ago and have long since vanished from the scene. Bones of these great mammals were found by Darwin on the flat treeless plains of Argentina.

relationship in the same continent between the dead and the living will, I do not doubt, hereafter throw more light on the appearance of organic beings on our Earth, and their disappearance from it, than any other class of facts."

What Darwin had realized was that living species have ancestors. This fact is commonplace now, but it was a revelation then. On traveling from the north to the south of South America, Darwin observed that one species was replaced by similar, but slightly different, species. In the southern part of Argentina, Darwin caught a rare species of ostrich that was smaller and differently colored from the more northerly, common American ostrich, *Rhea americanus*. This rare species of bird was later named after him, *Rhea darwini*. Lore holds that Darwin and his shipmates had consumed much of the only specimen of this rare bird before Darwin realized what it was. Darwin preserved the uneaten parts and sent them back to England, where Sir Robert Owen would later identify the bird as a new species. It was scarcely imaginable to Darwin that several minor versions of a species would be created separately, one for each locality. It appeared to Darwin that species change not only in time but also with geographical distance. He later wrote: "It was evident that such facts could only be explained on the supposition that species gradually became modified; and the subject haunted me."

In the Galápagos Islands, Darwin's scientific curiosity was sharply prodded by the many distinctive forms of life. The Galápagos consist of an isolated cluster of rugged volcanic islands in the eastern Pacific, on the equator about 600 miles west of Ecuador. One of the most unusual animals is the giant land-dwelling tortoise, which may weigh as much as 550 pounds, grow to 6 feet in length, and attain an age of 200 to 250 years (fig. 2.4). The Spanish name for tortoise, *galápago*, gives the islands their name. Darwin noticed that the tortoises were clearly different from island to island, although the islands were only a few miles apart. Darwin reasoned that, in isolation, each population had evolved its own distinctive features. Yet, all the island tortoises showed basic resemblances

Figure 2.4 **The giant tortoise,** *Geochelone elephantopus,* found only on the Galápagos Islands.
Source: © Tess & David Young/Tom Stack & Assoc.

not only to each other but also to relatively large tortoises on the adjacent mainland of South America. All this revealed to Darwin that the island tortoises shared a common ancestor with the mainland forms. The same was true of a group of small black birds, known today as *Darwin's finches.* Darwin observed that the finches were different on the various islands, yet they were obviously closely related to each other. Darwin reasoned that the finches were derived from an ancestral stock that had emigrated from the mainland to the volcanic islands and had undergone profound changes under the different conditions of the individual islands. Apparently, a single ancestral group could give rise to several different varieties or species.

MECHANISM OF EVOLUTION

When Darwin left England in 1831, he had accepted the established theory that different species had arisen all at once and remained unchanged and constant. His scientific colleagues did not deny the presence of fossil remains of species no longer extant. Such extinct species were thought to represent organisms that had been periodically annihilated by sudden and violent catastrophes. Each cataclysmic destruction was said to be accompanied by the creation of new forms of life, each distinct from the fossilized species. The theory of successive episodes of catastrophic destruction and creation, formulated by the French anatomist Georges Cuvier, was accepted by most eighteenth-century scholars.

Upon his return to England in October 1836, Darwin was convinced of the truth of the idea of *descent with modification*—that all organisms, including humans, are modified descendants of previously existing forms of life. Darwin's observations during his voyage assured him that animals and plants are not the products of special creation, as Victorian society believed. We can discern two stages in the development of Darwin's thought: the first was the realization that the world of organisms is not fixed and unchangeable; the second was an explanation of the process of evolutionary change.

At home, Darwin assembled data relevant to the mechanism of evolution. He became intrigued by the extensive domestication of animals and plants brought about by conscious human efforts. Throughout the ages, humans have been a powerful agent in modifying wild species of animals and plants to suit our needs and whims. Through careful breeding programs, we determine which characteristics or qualities will be incorporated or discarded in our domesticated stocks. By conscious *selection,* we perfected the toylike Shetland pony, the Great Dane dog, the sleek Arabian race horse, and vast numbers of cultivated crops and ornamental plants.

Many clearly different domestic varieties have evolved from a single species through our efforts. As an example, we may consider the many varieties of domestic fowl (fig. 2.5), all of which are derived from a single wild species of red jungle fowl *(Gallus gallus)* present at one time in northern India. The red jungle fowl dates back to 2000 B.C. and has been so modified as to be nonexistent today. The variety of fowls perpetuated by fowl fanciers range from the flamboyant ceremonial cocks (the Japanese *Onaga-dori*) to the leghorns, bred especially to deposit spotless eggs. In the plant world, horticulturists have evolved the astonishing range of begonias shown in figure 2.6. The wild begonia, from which most of the modern varieties have been bred, was found in 1865 and still exists today in the Andes.

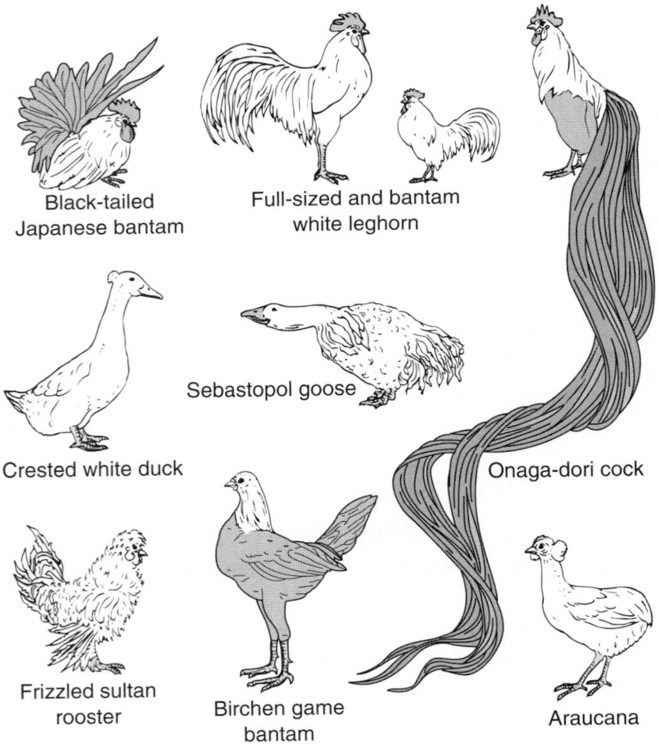

In the late 1830s, Darwin attended the meetings of animal breeders and intently read their publications. Animal breeders were conversant with the variability in their pet animals and utilized the technique of *artificial selection.* That is, the breeders selected and perpetuated those variant types that interested them or seemed useful to them. The breeders, however, had only vague notions as to the origin, or inheritance, of the variable traits.

Darwin acknowledged the unlimited variability in organisms but was

Black-tailed
Japanese bantam

Full-sized and bantam
white leghorn

Sebastopol goose

Crested white duck

Onaga-dori cock

Frizzled sultan
rooster

Birchen game
bantam

Araucana

Figure 2.5 Evolution under domestication. A variety of domestic fowls have evolved from a single wild species (now extinct) as a result of the continual practice by humans of artificial selection.

never able to explain satisfactorily how a variant trait was inherited. He and other naturalists were unaware of Gregor Mendel's contemporaneous discovery. Rather, Darwin followed Lamarck in assuming that biological variation is chiefly conditioned by direct influences of the environment on the organism. He believed that the changes induced by the environment became inherited; that bodily changes in the parents leave an impression or mark on their germ cells, or gametes. Darwin thought that all organs of the body sent contributions to the germ cells in the form of particles, or *gemmules*. These gemmules, or minute delegates of bodily parts, were supposedly discharged into the blood stream and became ultimately concentrated in the gametes. In the next generation, each gemmule reproduced its particular bodily component. The theory was called *pangenesis,* or "origin from all," since all parental bodily cells were supposed to take part in the formation of the new individual.

Having explained the origin of variation (although incorrectly), Darwin wondered how artificial selection (a term familiar then only to animal breeders) could be carried on in nature. There was no breeder in nature to pick and choose. In 1837, Darwin wrote: "How selection could be applied to organisms living in a state of nature remained a mystery to me." Slightly more than a year later, Darwin found the solution. In his autobiography, Darwin explained: "In October 1838, that is fifteen months after I had begun my systematic enquiry, I happened to read for amusement *Malthus on Population . . .* at once it struck me that under these circumstances favourable variations would tend to be preserved and unfavourable ones destroyed. The result of this would be the formation of a new species." The circumstances mentioned by Darwin were those associated with the assertions of an English clergyman, Thomas Robert Malthus, that population is necessarily limited by the means of subsistence.

Figure 2.6 Artificial selection. In the hands of skilled horticulturists, several varieties of the begonia have been established.

DARWIN'S NATURAL SELECTION

Malthus's writings provided the germ for Darwin's theory of *natural selection.* In his famous *An Essay on the Principle of Population,* Malthus expressed the view that the reproductive capacity of humankind far exceeds the food supply available to nourish an expanding human population. Humans compete among themselves for the necessities of life. This unrelenting competition engenders vice, misery, war, and famine. It thus occurred to Darwin that competition exists among all living things. Darwin then envisioned that the "struggle for existence" might be the means by which the well-adapted individuals survive and the ill-adjusted are eliminated. Darwin was the first to realize that perpetual selection existed in nature in the form of *natural selection.* In natural selection, as contrasted to artificial selection, the animal breeder or horticulturist is replaced by the conditions of the environment that prevent the survival and reproduction of certain individuals. The process of natural selection occurs without a conscious plan or purpose. Natural selection was an entirely new concept, and Darwin was its principal proponent.

It was not until 1844 that Darwin developed his idea of natural selection in an essay, but not for publication. He showed the manuscript to the geologist Charles Lyell, who encouraged him to prepare a book. Darwin still took no steps toward publishing his views. It appears that Darwin might not have prepared his famous volume had not a fellow naturalist, Alfred Russel Wallace (in the Dutch East Indies), independently conceived of the idea of natural selection. Wallace had spent many years exploring and collecting in South America and the West Indies. Wallace was also inspired by reading Malthus's essay, and the idea of natural selection came to him in a flash of insight during a sudden fit of malarial fever. In June of 1858, Wallace sent Darwin a brief essay on his views. The essay was entitled *On the Tendencies of Varieties to Depart Indefinitely from the Original Type.* With the receipt of this essay, Darwin was then induced to make a statement of his own with that of Wallace.

Wallace's essay and a portion of Darwin's manuscript, containing remarkably similar views, were read simultaneously before the Linnaean Society in London on July 1, 1858. The joint reading of the papers stirred little interest. Darwin then labored for eight months to compress his voluminous notes into a single book, which he modestly called "only an Abstract." Wallace shares with Darwin the honor of establishing the mechanism by which evolution is brought about, but it was the monumental *The Origin of Species,* with its impressive weight of evidence and argument, that left its mark on humankind. The full title of Darwin's treatise was *On the Origin of Species by Means of Natural Selection, or the Preservation of Favoured Races in the Struggle for Life.* The first edition, some 1,250 copies, was sold out on the very day it appeared, November 24, 1859. The book was immediately both acidly attacked and effusively praised. Today, *The Origin*

of Species remains the one book to be read by all serious students of nature.

Darwin's reasons for not publishing *The Origin of Species* for over 20 years after returning from the voyage of the *Beagle* were complex. He described his anxiety to one friend by saying, "It is like confessing a murder." Some of the specific reasons for his delay were these:

- Darwin believed that his theory was still lacking sufficient evidence, and he wanted to gather more data.
- He also feared that he would be ridiculed by the public and the established church, as others with unconventional ideas were, and was determined to avoid that fate.
- Darwin suffered from ill health, grief and depression brought on by the unexpected death of his 10-year-old daughter, Annie.
- Darwin's wife, Emma, a devout Christian, feared that her husband's ideas would prevent them from being eternally united in heaven.

The idea of evolution—that organisms change—did not originate with Darwin. There were many before him—notably Lamarck, Georges Buffon, and Charles Darwin's grandfather, Erasmus Darwin—who recognized or intimated that animals and plants have not remained unchanged through time but are continuously changing. Indeed, early Greek philosophers, who wondered about everything, speculated on the gradual progression of life from simple to complex. It was reserved for Darwin to remove the doctrine of evolution from the domain of speculation. Darwin's outstanding achievement was his discovery of the principle of natural selection. In showing *how* evolution occurs, Darwin endeavored to convince skeptics that evolution *does* occur.

As a whole, the principle of natural selection stems from five important observations made by Darwin and three deductions that logically follow from them. The *first observation* was that all living things tend to increase in number at a prolific rate. This observation was not unique to Darwin but was most clearly articulated by him, though it was surely influenced by his early readings of William Paley and Thomas Robert Malthus. A single oyster may produce as many as 100 million eggs at one spawning; one tropical orchid may form well over a million seeds; a single salmon can deposit 28 million eggs in one season. It is equally apparent (*the second observation*) that no group of organisms swarm uncontrollably over the surface of the Earth. In fact, the actual size of a given population of any particular organism remains relatively constant over long periods of time. Darwin's *third observation* was that there are only a limited amount of resources available for organisms to use as they strive to live and reproduce. If we accept these readily confirmable observations, the conclusion necessarily follows that not all individuals that are produced in any given generation can survive. There is inescapably in nature an intense "struggle for existence" (*deduction 1*).

Darwin's *fourth observation* was that individuals in a population are not alike but instead differ from one another in various features. That all living things vary is indisputable. Likewise, although the exact source of the variation between individuals was unknown to Darwin, he and the rest of the people of his day had the vague notion that variation was in large part due to inherited factors ("*observation*" 5). Darwin concluded (*deduction* 2) that individuals endowed with the most favorable variations would have the best chance of passing their favorable characteristics on to their progeny. This differential survival or "survival of the fittest" was termed natural selection (*deduction* 2). It was the British philosopher Herbert Spencer who proposed the expression *survival of the fittest,* and Darwin accepted the notion of survival of the fittest as being equivalent to natural selection. Over many generations, natural selection leads to changes in populations, which eventually leads to evolution (*deduction* 3). A conceptual model of Darwin's tenet of natural selection is presented in figure 2.7.

Darwin's book appeared at a time when England embraced the Victorian economic doctrine of *laissez-faire.* This French phrase expresses the notion of letting people do what they choose.

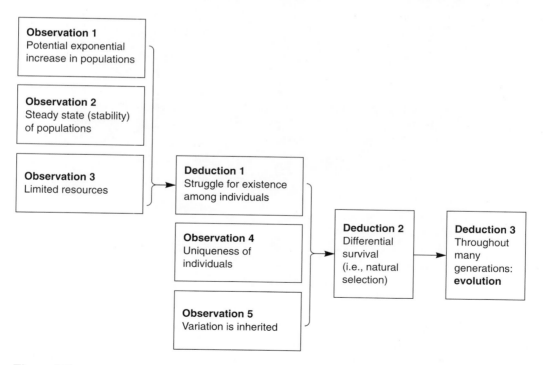

Figure 2.7 Darwin's explanatory model of evolution via natural selection.

As used in economics, the doctrine deprecates any form of governmental interference intended to regulate labor, manufacturing, commerce, and financial institutions. The Victorian era was one of fierce competition among manufacturers, merchants, and bankers. Herbert Spencer advocated that Darwin had discovered not merely the laws of biological evolution but also those governing human societies. Spencer gave science's blessing to the selection process that operated in the ruthless English society where only the fittest could survive the intolerable working and living conditions. Victorian England was also staunchly biblical. Science was linked in many devout minds with atheistic thinking, and Darwin's book was denounced from fundamentalist pulpits. The Victorian press assailed Darwin for his arrogance in removing humans from an exalted position in the world of life (fig. 2.8).

Figure 2.8 Cartoon in Punch's Almanack (1882). The cartoon portrays the transition of worms into apes and gnomes, culminating in a gentleman with sartorial splendor paying facetious homage to Charles Darwin.

Source: Punch's Almanack

DIFFERENTIAL REPRODUCTION

The survival of favorable variants is one facet of the Darwinian concept of natural selection. Equally important is the corollary that unfavorable variants do not survive and multiply. Consequently, natural selection necessarily embraces two aspects, as inseparable as the two faces of the same coin: the negative (elimination of the unfit) and the positive (perpetuation of the fit). In its negative role, natural selection serves as a conservative or stabilizing force, pruning out the aberrant forms from a population.

The superior, or fit, individuals are popularly extolled as those who emerge victoriously in brutal combat. Fitness has often been naively confused with physical, or even athletic, prowess. This glorification is traceable to such seductive catch phrases as the "struggle for existence" and the "survival of the fittest." But what does fitness actually signify?

The true gauge of fitness is not merely survival but also the organism's capacity to leave offspring. An individual must survive to reproduce, but not all individuals that survive do, or are able to, leave descendants. We have seen that the multilegged bullfrogs were not successful in propagating themselves. They failed to make a contribution to the next or succeeding generations. Therefore, they were unfit. Hence, individuals are biologically unfit if they leave no progeny. They are also unfit if they do produce progeny, none of whom survives to maturity. The less spectacular normal-legged frogs did reproduce, and to the extent that they are represented by descendants in succeeding generations, they are the fittest. *Darwinian fitness, therefore, is measured as reproductive effectiveness.* Natural selection can thus be thought of as *differential reproduction,* rather than differential survival.

EVOLUTION DEFINED

Any given generation is descended from only a small fraction of the previous generation. It should be evident that the genes transmitted by those individuals who most successfully reproduce will predominate in the next generation. Because of unequal reproductive capacities of individuals with different hereditary constitutions, the genetic characteristics of a population become altered each successive generation. This is a dynamic process that has occurred in the past, occurs today, and will continue to occur as long as heritable variation and differing reproductive abilities exist. Under these circumstances, the composition of a population can never remain constant. This, then, is evolution—*changes in the genetic composition of a population with the passage of each generation.*

The outcome of the evolutionary process is the adaptation of an organism to its environment. Many of the structural features of organisms are marvels of construction. It is, however, not at all remarkable that organisms possess particular characteristics that appear to be precisely and peculiarly suited to their needs. This is understandable because the individuals who leave the most descendants are most often those who are best equipped to cope with special environmental conditions. In other words, the more reproductively fit individuals tend to be those who are better adapted to the environment.

An important consideration is that evolution is a property of populations. A population is not a mere assemblage of individuals; it is a breeding community of individuals. Moreover, it is the *population* that evolves in time, not the *individual.* An individual cannot evolve, for he or she survives for only one generation. An individual, however, can ensure continuity of a population by leaving offspring. Life is perpetuated only by the propagation of new individuals who differ in some degree from their parents. Natural selection is a term serving to inform us that some individuals leave more offspring than others. Natural selection is not purposeful or guided by a specific aim; it does not seek to attain a specific end. Natural selection is solely the consequence of the differences between individuals with respect to their capacity to produce progeny under a particular set of environmental conditions.

Throughout the ages, appropriate adaptive structures have arisen as the result of gradual changes in the hereditary endowment of a population. Admittedly, past events are not amenable to direct observation or experimental verification. There are no living eyewitnesses of very distant events. So, the process of evolution in the past must be inferred. Nevertheless, we may be confident that the same evolutionary forces we witness in operation today have guided evolution in the past. The basis for this confidence is imparted in the chapters ahead.

HERITABLE VARIATION

Darwin recognized that the process of evolution is inseparably linked to the mechanism of inheritance. But he could not satisfactorily explain how a given trait is transmitted from parent to offspring, nor could he account adequately for the sudden appearance of new traits. Unfortunately, Darwin was unaware of the great discovery in heredity made by a contemporary, the unpretentious Austrian monk and mathematics teacher, Gregor Johann Mendel.

Mendel carried out carefully controlled breeding experiments on the common pea plant in his small garden at the monastery of Saint Thomas at Brünn in Moravia (now Brno, Czech Republic) (fig. 3.1). He worked with large numbers of plants so that statistical accuracy could be assured. In 1866, Mendel's results appeared in a publication of the Brünn Society of Natural History. In the small compass of 44 printed pages, one of the most important principles of nature ever discovered is clearly revealed. Yet his paper elicited little interest or comment. No one praised or disputed Mendel's findings. It is likely that nineteenth-century biologists were baffled by Mendel's mathematical, or statistical, approach to the study of hybridization.

Botany and mathematics were strange bedfellows for that period.

Mendel was an excellent experimental scientist but was also very lucky. His research was successful for a number of reasons, including the following:

- He used excellent experimental techniques.
- For years he kept detailed records of his crosses.
- The data he collected was numerical. Ultimately, the ratios of the crosses confirmed his theories.
- He was ahead of his time in his use of mathematical analysis.
- He picked a simple system with a few well-defined characters to study. Specifically, because his characters all showed dominant and recessive traits and behaved as though there was no *linkage,* his analysis was unambiguous.
- Fortunately, even though some of his characters were on the same chromosome (i.e., linked), they were sufficiently far enough apart from each other that they appeared to assort independently.

(a)   (b)

Figure 3.1 (a) **George Mendel, 1822–1884,** Augustinian monk and scientist, father of genetics. (b) Mendel conducted his experiments on pea plants in a small garden adjacent to his monastery.

Mendel's valuable contribution lay ignored or unappreciated until 1900, sixteen years after his death, when his obscure 1866 manuscript resurfaced. Mendel's findings were independently rediscovered and verified almost simultaneously by three botanists: Hugo de Vries in Holland, Carl Correns in Germany, and Erich Tschermark in Vienna. Since then, Mendel has received scientific acclaim. The rediscovery of Mendel's obscure manuscript at the turn of the twentieth century ushered in the science of genetics and ultimately led to our current understanding of heritable variation.

MENDEL'S PRINCIPLE OF SEGREGATION

Mendel's profound inference was that traits are passed on from parent to offspring through the gametes, in specific discrete units, or *factors*. The individual factors do not blend and do not contaminate one another. The hereditary factors of the parents can, therefore, be reassorted in varied combinations in different individuals at each generation. Today, Mendel's hereditary units are called *genes*. The almost infinite diversity that exists among individuals is attributable largely to the shuffling of tens of thousands of discrete genes that occurs in sexual reproduction.

Mendel reasoned that an individual has two copies of each gene, having received one copy of the gene from the gamete of the female parent and the other copy from the gamete of the male parent. Today, we refer to the contrasting forms of a given gene as *alleles*. Expressed in another way, all alternative sequences of a gene, including the normal and all variant sequences, are called alleles. In the most elementary case of inheritance, the trait is governed by a single gene that exists in two allelic forms—the normal (or typical) allele and the variant (or mutant) allele. Albinism in humans may be taken as an example of a trait under the simple control of a pair of alleles.

The word *albino* is derived from the Latin *albus,* meaning "white," and refers to the paucity or absence of pigment (melanin) in the skin, hair, and eyes. The skin is often very light ("milk white"), and the hair whitish yellow. The eyes appear pinkish because the red blood vessels give the otherwise colorless iris a rosy cast. Contrary to popular opinion, albinos are neither blind nor mentally retarded. They often have poor vision, are acutely sensitive to sunlight, and are prone to life-threatening skin cancer. In some cultures, as among the natives of the Archipelago Coral Islands, albinos are believed to have a mesmeric aura (fig. 3.2).

The absence of melanin is the consequence of a variant allele. Specifically, albinism arises only when the individual is endowed with two variant

Figure 3.2 Albino person showing the complete absence of pigment in the eyes, hair, and skin.
Natives of the Archipelago Coral Islands (100 miles east of the New Guinea mainland) are said to take care
of their "pure white" relatives who can hardly see in the glare of the tropical sun.
(Wide World Photos)

alleles, one from each parent. A cardinal fact of
inheritance is that each gamete (egg or sperm) contains *one, and only one,* member of a pair of alleles.
Thus, when the egg and sperm unite, the two alleles
are brought together in the new individual. During the production of the gametes, the members of
a pair of alleles separate, or *segregate,* from each
other. It bears emphasizing that a given gamete can
carry only one allele but not both. This fundamental
concept that only one member of any pair of alleles
in a parent is transmitted to each offspring is Mendel's first principle—the *principle of segregation.*

As seen in figure 3.3, there are nine possible
matings with respect to the pair of alleles determining the presence or absence of melanin. If both parents are perfectly normal—that is, each possesses

a pair of normal alleles—then all offspring will be
normally pigmented (cross 1). When both parents
are albino and accordingly each has a pair of variant
genes, all offspring will be afflicted with albinism
(cross 2). The persons involved in either of these
two crosses are said to be *homozygous,* since they
each possess a pair of similar, or like, alleles. The
normal individuals in cross 1 are homozygous for
the normal allele, and the albino individuals in
cross 2 are homozygous for the variant allele.

Looking at cross 3, we notice that one of the
parents (the female) carries an unlike pair of alleles:
one member of the pair is normal; its partner allele is
variant. This female parent, possessing unlike alleles,
is *heterozygous.* This parent has normal coloration,
which leads us to deduce that the expression of the

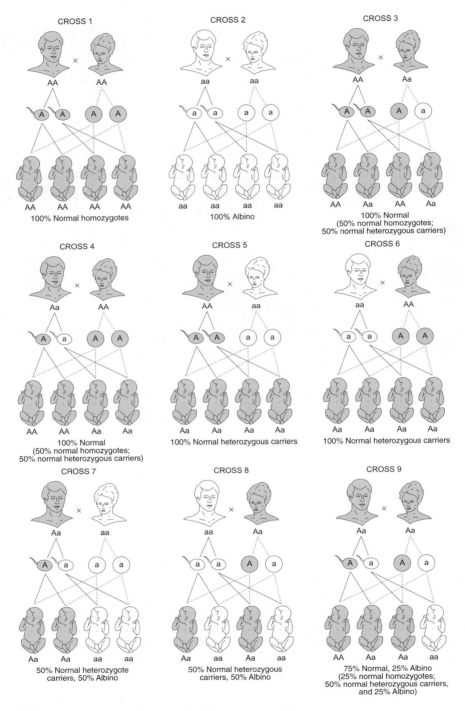

Figure 3.3 The segregation of a single pair of alleles. The nine possible matings with respect to the recessive trait for albinism show the operation of Mendel's principle of segregation.

variant allele is completely masked or suppressed by the normal allele. In genetical parlance, we say that the normal allele is *dominant* over its partner, the *recessive* allele. The dominant allele is customarily symbolized by a capital letter (in our case, *A*); the recessive allele, by a corresponding small letter *(a)*. The heterozygous parent is depicted as *Aa*, and is referred to as *heterozygote* or a *carrier.*

When the heterozygote *(Aa)* mates with a dominant homozygote *(AA),* all the offspring will be outwardly normal, but half the offspring will be carriers *(Aa)* like one of the parents (cross 3). The ratio of *AA* to *Aa* offspring may be expressed as 1/2 *AA* : 1/2 *Aa* or simply 1 *AA* : 1 *Aa.* Cross 4 differs from cross 3 only in that the male parent is the carrier *(Aa)* and the female parent is the dominant homozygote *(AA).* The outcome is the same: all children are normal in appearance; but half of them are carriers. Thus, with respect to albinism, the males and females are equally likely to have or to transmit the trait. (There are traits, such as hemophilia and red-green color blindness, in which transmission is influenced by the sex of the parents.)

All offspring will be carriers from the mating (crosses 5 and 6) of a dominant homozygote *(AA)* and an albino parent *(aa).* Examining crosses 5 and 6 (fig. 3.3), we again notice that males and females are equally affected. The mating of a heterozygous person *(Aa)* and an albino *(aa)* gives rise to two types of progeny, in equal numbers (crosses 7 and 8). Half the offspring are normal but carriers *(Aa);* the remaining half are albino *(aa).*

The outcome of the last of the crosses (no. 9) in figure 3.3 is comprehensible in terms of Mendel's law of segregation. Both parents are normal in appearance, but each carries the variant allele *(a)* masked by the normal allele *(A).* The male parent *(Aa)* produces two kinds of sperm cells; half the sperm carry the *A* allele; half carry the *a* allele. The same two kinds occur in equal proportions among the egg cells. Each kind of sperm has an equal chance of meeting each kind of egg. The random meeting of gametes leads to both normal and albino offspring. One-fourth of the progeny are normal and completely devoid of

the recessive allele *(AA);* one-half are normal but carriers like the parents *(Aa);* and the remaining one-fourth exhibit the albino anomaly *(aa).* The probabilities of occurrence are the same as those illustrated by the tossing of a coin. If two coins are repeatedly tossed at the same time, the result to be expected on the basis of pure chance is that the combination "head and head" will fall in one-fourth of the cases, the combination "head and tail" in one-half of them, and the combination "tail and tail" in the remaining one-fourth. The ratio of genetic constitutions among the offspring may be expressed as 1/4 *AA* : 1/2 *Aa* : 1/4 *aa*, or 1 *AA* : 2 *Aa* : 1 *aa.*

Geneticists continually have to make a distinction between the external, or observable, appearance of the organism and its internal hereditary constitution. They call the former the *phenotype* (visible type) and the latter the *genotype* (hereditary type). In our last cross (no. 9), there are three different genotypes: *AA, Aa,* and *aa.* The dominant homozygote *(AA)* cannot be distinguished by inspection from the heterozygote *(Aa).* Thus, *AA* and *Aa* genotypes have the same phenotype. Accordingly, on the basis of phenotype alone, the progeny ratio is 3/4 normal : 1/4 albino, or 3 normal : 1 albino.

If we disregard the sexes of the parents, the nine crosses in figure 3.3 become reducible to six possible types of matings. The six possible crosses and their offspring are set forth in table 3.1.

APPLICATION OF MENDELIAN PRINCIPLES

Our interpretation of the mode of inheritance of the multilegged trait in the bullfrog was based on Mendel's principles. In chapter 1, we postulated a cross between two normal, but heterozygous, frogs (see fig. 1.5). The expectation for multilegged offspring from two heterozygous parents is 25 percent, in keeping with Mendel's 3:1 ratio. This cross, using appropriate genetic symbols, is shown graphically by "checkerboard" analysis in figure 3.4. The convenient checkerboard method of analysis, called

TABLE 3.1	Simple Recessive Mendelian Inheritance, Involving a Single Pair of Alleles (Normal Pigmentation vs. Albinism)

Mating Types		Gametes				Offspring	
		First Parent		Second Parent			
Genoptypes	Phenotypes	50%	50%	50%	50%	Genotypes	Phenotypes
$AA \times AA$	Normal × Normal	A	A	A	A	100% AA	100% Normal
$AA \times Aa$	Normal × Normal	A	A	A	a	50% AA	100% Normal
						50% Aa	
$Aa \times Aa$	Normal × Normal	A	a	A	a	25% AA	75% Normal
						50% Aa	
						25% aa	25% Albino
$AA \times aa$	Normal × Albino	A	A	a	a	100% Aa	100% Normal
$Aa \times aa$	Normal × Albino	A	a	a	a	50% Aa	50% Normal
						50% aa	50% Albino
$aa \times aa$	Albino × Albino	a	a	a	a	100% aa	100% Albino

the Punnett square, was devised by the British geneticist R. C. Punnett. The eggs and sperm are listed separately on two different sides of the checkerboard, and each square represents an offspring that arises from the union of a given egg cell and a given sperm cell. The "reading" of the squares discloses the classical Mendelian phenotypic ratio of 3:1.

It should be understood that the 3:1 phenotypic ratio resulting from the mating of two heterozygous persons, or the 1:1 ratio from the mating of a heterozygote and a recessive person, are expectations based on probability, and are not invariable outcomes. The production of large numbers of offspring increases the probability of obtaining, for example, the 1:1 progeny ratio, just as many tosses of a coin improve the chances of approximating the 1 head: 1 tail ratio. If a coin is tossed only two times, a head on the first toss is not invariably followed by a tail on the second toss. In like manner, if only two offspring are produced from a mating of heterozygous *(Aa)* and recessive *(aa)* parents, it should not be thought that one normal offspring is always accompanied by one recessive offspring. With small numbers of progeny, as is characteristic of humans, any ratio might arise in a given family.

Stated another way, the 3:1 and 1:1 ratios reveal the risk or odds of a given child having the particular trait involved. If the first child of two heterozygous parents is an albino, the odds that the second child will be an albino remain 1 out of 4. These odds hold for each subsequent child, irrespective of the number of previously affected children. Each birth is an entirely independent event. If this still appears puzzling, consider once again the tossing of a coin. The first time a coin is tossed, the chance of obtaining either a head or a tail is 1 in 2. Whether the toss is repeated immediately or nine months later, the chance of the coin falling head or tail is still 1 in 2, or 50 percent.

CHROMOSOMAL BASIS OF HEREDITY

Each organism has a definite number of *chromosomes,* ranging in various species from 2 to more than 1,000. Humans, for example, have 46 microscopically visible, threadlike chromosomes in the nucleus of every body cell (fig. 3.5). Not

Normal (heterozygous) parents

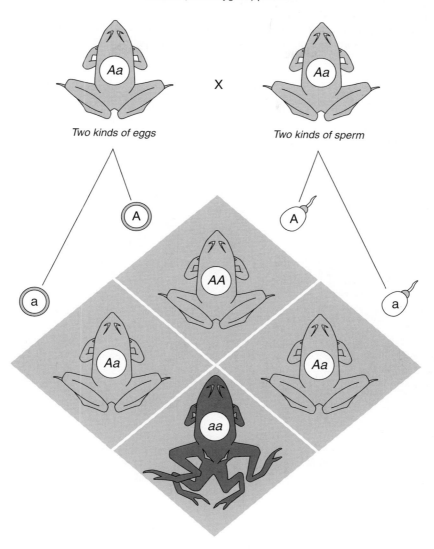

$^3/_4$ Normal : $^1/_4$ Multilegged

Figure 3.4 Cross of two heterozygous carrier frogs. One-fourth of the offspring
are normal and completely devoid of the recessive allele *(AA);* one-half are normal
but carriers, like the parents *(Aa);* and the remaining one-fourth exhibit the multilegged
anomaly *(aa).*

only are the numbers of chromosomes constant for a given species, but the shapes and sizes of the chromosomes in a given species differ sufficiently for us to recognize particular chromosomes

and to distinguish one kind from the other. An important consideration is that chromosomes in a *zygote* occur in pairs. The significance of the double, or paired, set of chromosomes cannot be

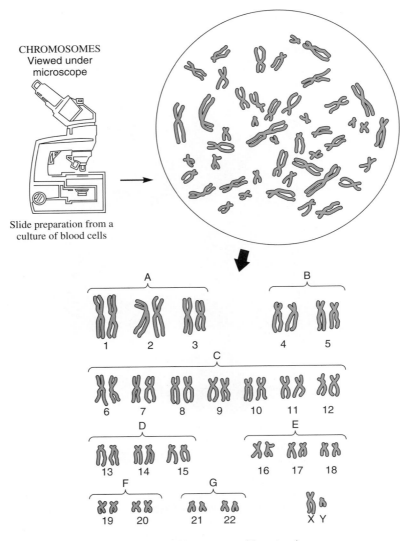

Figure 3.5 Normal complement of 46 chromosomes prepared from a culture of blood cells of a human male. Each metaphase chromosome consists of two strands (chromatids) joined at the centromere. When the chromosomes are arranged in a sequence (karyotype), the male set consists of 23 pairs, including the XY.

overstated: *each parent contributes one member of each pair of chromosomes to the offspring.*

The chromosomes can be systematically arranged in a sequence known as a *karyotype* (fig. 3.5). In preparing the karyotype, the chromosomes are paired and classified into groups according to their length. There are 22 matching pairs of human chromosomes that are called *autosomes*. The autosome pairs are numbered 1 to 22 in descending order of length, and further classified into seven groups, designated by capital letters, A through G. In addition, there are two sex chromosomes, which are unnumbered. As seen in figure 3.5, the male has one X chromosome and one unequal-sized Y chromosome, called *sex chromosomes.* The X is of medium size, and the Y is one of the smallest chromosomes of the complement. The human

female has two X chromosomes of equal size and no Y chromosome. Thus, the complement of 46 human chromosomes comprises 22 pairs of autosomes plus the sex chromosome pair, XX in normal females and XY in normal males. In current nomenclature, the female is described as "46,XX" and the male, "46,XY."

MITOSIS AND MEIOSIS

The formation of new cells by *mitosis* is a continual process in the human body, although certain cells divide more often than others, and some not at all. Mitosis is the means by which two new cells receive identical copies of the chromosome complement of the pre-existing cell (fig. 3.6). If gametes were to be produced by the usual mitotic

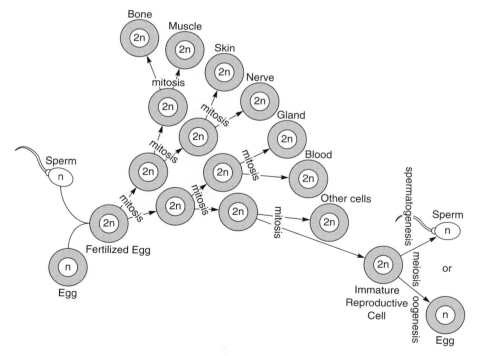

Figure 3.6 Mitosis ensures that each new cell of the developing organism has a complete set of the chromosomes (diploid, or 2*n***) found in the fertilized egg cell.** Through the process of meiosis, gametes (egg and sperm) come to possess only half the number of chromosomes (haploid, or 1*n*).

process, and, accordingly, if they had the same number of chromosomes as body cells, then the union of sperm and egg would necessarily double the number of chromosomes in each generation. Such progressive increase is forestalled, however, by a special kind of division that guarantees that the gametes contain *half* the number of chromosomes characteristic of body cells. This distinctive type of division is referred to as *meiosis*. The term "meiosis" is derived from the Greek word *meion,* meaning "to make smaller." Meiosis is thus a diminution process, referring specifically to a reduction in the number of chromosomes. In humans, the number of chromosomes is reduced from 46 to 23 during the formation of gametes. Figure 3.7a shows the relationship between mitosis and meiosis as part of the generalized life cycle. Figure 3.7b applies these concepts to the more familise human.

Somatic (or body) cells, possessing twice the number of chromosomes as gametes, are said to be *diploid* (Greek, *diploos,* "twofold" or "double"). In contrast, the gametic number of chromosomes is *haploid* (Greek, *haploos,* "single") or, as some investigators prefer, *monoploid* (Greek, *monos,* "alone"). This can be expressed in abbreviated form by saying that body cells contain $2n$ chromosomes and germ cells contain $1n$ chromosomes,

where n stands for a definite number. As earlier mentioned, a given chromosome has a mate that is identical in size and shape. The two identical members of a pair are called *homologous chromosomes,* or simply *homologues.* Duplicated chromosomes after replication are called *sister chromatids.* Figure 3.8 is a graphic representation of the concepts of homologous chromosomes and sister chromatids. With this orientation, we can explore the mechanism of meiosis.

Meiosis is not a single division but rather two successive ones, the first and second meiotic divisions. Figure 3.9 reveals the essential features in a simplified "stick" representation involving merely one pair of homologous chromosomes. The process begins with each member of the pair duplicating itself to form two sister strands (technically, called *sister chromatids*). Then, an event that is *unique* to meiosis occurs: the homologous chromosomes actually come to lie alongside one another. The mutual attraction, or pairing, of homologues is known as *synapsis.*

Because two homologous chromosomes pair, and because each chromosome of a homologous pair has duplicated itself, the result is a quadruple chromatid structure known as a *tetrad* (Greek, *tetras,* "a collection of four"). For the four

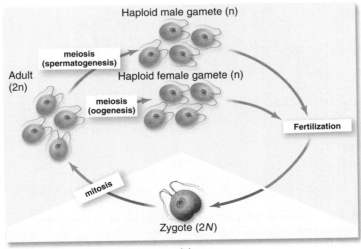

(a)

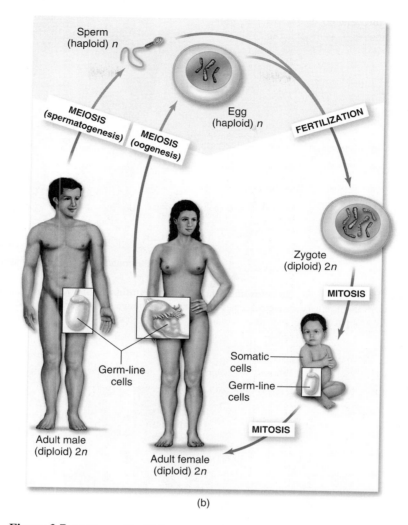

(b)

Figure 3.7 (a) **Generalized life cycle:** The relationship between mitosis (growth) and meiosis (gamete formation) and fertilization. (b) **Sexual cycle in humans:** In humans and many other animals, the fertilized diploid (2*n*) egg (zygote) undergoes mitotic divisions (growth), giving rise to the adult. Germ cells produce gonads (ovaries and testes), which produce haploid (*n*) eggs and sperms via meiosis.

chromatids to be distributed individually into four haploid cells, two divisions would be required. This is exactly what takes place. The two divisions accomplish (1) separation of the homologous chromosomes that had paired and (2) separation of the duplicated strands of each chromosome.

It would be instructive to amplify these remarks as we re-examine figure 3.9. During the first meiotic division, it may be seen that the tetrad is separated into two *dyads* (Greek, *dyas,* "a couple"). The first meiotic division is concerned with the separation of the original synaptic mates, or the original

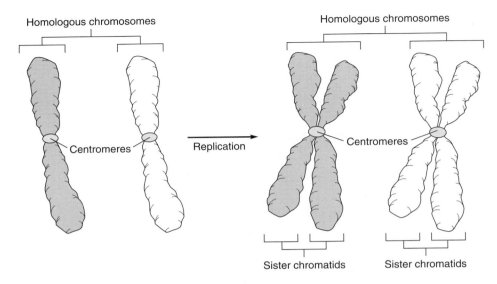

Figure 3.8 Homologous chromosomes and sister chromatids during mitosis: Homologous chromosomes are maternal and paternal copies of the same chromosomes (e.g., chromosome 4). Sister chromatids are replicas of single chromosomes held together by their centromere after DNA replication.

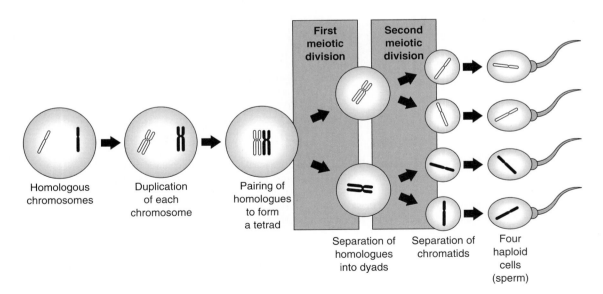

Figure 3.9 The process of meiosis depicted for only two chromosomes. Duplicated members of the pair of homologous chromosomes come to lie side by side in a four-stranded configuration (tetrad). Two successive divisions then result in formation of four haploid cells (sperm, in this case).

homologous members of the chromosomal pairs. The second meiotic division involves the separation of the chromatids, which had arisen earlier by duplication of the chromosomes. The overall outcome of meiosis is the formation of four haploid cells from one diploid cell. Each of the four haploid cells has half as many chromosomes as the original parent cell.

MENDEL AND THE CHROMOSOMAL THEORY OF HEREDITY

We shall now endeavor to relate Mendel's work to our present knowledge of chromosomes. Mendel, of course, was unaware that his hereditary factors were carried in the chromosomes. It was Walter Sutton in 1902 who first called attention to the parallelism in the behavior of genes and chromosomes. We have learned that chromosomes occur in pairs and that the members of a pair separate, or segregate, from each other during meiosis. Mendel's experiments reveal that the hereditary factors occur in pairs and that the members of each pair of factors separate from each other in the production of gametes. Evidently, then, the most reasonable inference is that the genes are located in chromosomes. Extensive studies have provided confirmation that the chromosomes do carry the genes. Indeed, chromosomes are the vehicles for transmitting the blueprints of traits to the offspring.

Figure 3.10 shows the separation of the pair of homologous chromosomes during meiosis of a heterozygote carrying the A and a alleles. The allele A occupies a particular site, or *locus,* in one of the chromosomes. The

alternative allele, *a,* occurs at the identical locus in the other homologous chromosome. The alternative forms of a given gene, occupying a given locus, are termed *alleles.* Thus, A is an allele of *a,* or we may also say that A is allelic to *a.* In summary, an *allele* is one of two or more alternative forms of a *gene,* which is a sequence of DNA nucleotides on a chromosome that encodes a protein, whereas the *locus* is the position or place on a chromosome where an allele is located.

The heterozygote *(Aa)* produces two types of gametes, A and *a,* in equal numbers. It should be

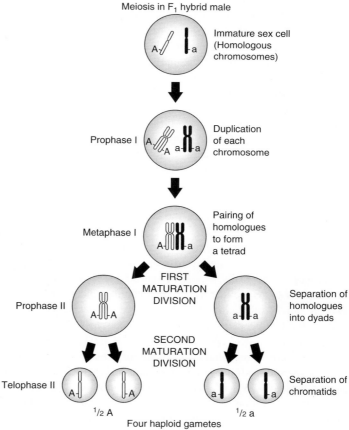

Meiosis in F_1 hybrid male

Immature sex cell (Homologous chromosomes)

Prophase I — Duplication of each chromosome

Metaphase I — Pairing of homologues to form a tetrad

FIRST MATURATION DIVISION

Prophase II — Separation of homologues into dyads

SECOND MATURATION DIVISION

Telophase II — Separation of chromatids

$^{1}/_{2}$ A $^{1}/_{2}$ a

Four haploid gametes

Figure 3.10 Chromosomal basis of monohybrid inheritance. When one chromosome of a pair carries a given gene (*A,* in this case) and its homologue carries the alternative form of the gene *(a),* then meiosis results in production of two distinct kinds of gametes in equal proportion, A and *a.*

evident from figure 3.10 that the gametic ratio of 1*A*:1*a* is a consequence of the separation of the two homologous chromosomes during meiosis, one chromosome containing the *A* allele and its homologue carrying the *a* allele. The behavior of chromosomes also explains another feature of Mendel's results—the independent assortment of traits.

MENDEL'S PRINCIPLE OF INDEPENDENT ASSORTMENT

The fruit fly *(Drosophila melanogaster),* approximately one-fourth the size of a house fly, has long been a favorite subject for genetical research. Numerous varieties have been found, affecting all parts of the body. For example, the wings may be reduced in length to small vestiges, the result of action of a recessive *vestigial* allele. Two heterozygous normal-winged flies, each harboring the recessive allele for vestigial wings, can produce two types of offspring: normal-winged and vestigial-winged, in a ratio of three-to-one. Now suppose that these same parents have a second concealed recessive allele, namely, *ebony.* This recessive allele, when homozygous, modifies the normal gray color of the body to black, or ebony.

Two normal-winged, gray-bodied parents, each heterozygous *(NnGg),* can give rise to four types of offspring phenotypically, as seen in figure 3.11. Two of the phenotypes are like the original parents, and two are new combinations (normal-winged, ebony and vestigial-winged, gray). The four phenotypes appear in the classical Mendelian ratio of 9:3:3:1. For such a ratio to be obtained, each parent *(NnGg)* must have produced four kinds of gametes in equal proportions: *NG, Ng, nG,* and *ng.* Stated another way, the segregation of the members of one pair of alleles must have occurred independently of the segregation of the members of the other allelic pair during gamete formation. Thus, 50 percent of the gametes received *N,* and of these same gametes, half obtained *G* and the other half *g.* Accordingly, 25 percent of the gametes were *NG* and 25 percent were *Ng.* Likewise, 50 percent of the gametes

carried *n,* of which half contained *G* and half *g* (or 25 percent *nG* and 25 percent *ng).* The four kinds of egg cells and the four kinds of sperm cells can unite in 16 possible ways, as shown graphically in the "checkerboard" in figure 3.11.

The above cross illustrates Mendel's *principle of independent assortment,* which states that one trait (one allelic pair) segregates independently of another trait (another allelic pair). The separation of the *Nn* pair of alleles and the simultaneous independent separation of the *Gg* pair of alleles occurs because the two kinds of genes are located in two different pairs of chromosomes. Figure 3.12 shows how the four types of gametes are produced with equal frequency in terms of the behavior of two pairs of chromosomes during meiosis. When the two pairs of alleles *N,n* and *G,g* are inserted at appropriate loci in the two pairs of chromosomes, it becomes apparent that four types of gametes, *NG, Ng, nG,* and *ng* are produced in equal numbers. The parallelism in the behavior of genes and chromosomes is striking.

THREE OR MORE GENE PAIRS

When individuals differing in three characteristics are crossed, the situation is naturally more complex, but the principle of independent assortment still is true. An individual heterozygous for three independently assorting pairs of alleles (for example, *Aa, Bb,* and *Cc)* produces eight different types of gametes in equal numbers, as illustrated in figure 3.13. Random union among these eight kinds of gametes yields 64 equally possible combinations. Stated another way, Punnett's checkerboard method of analysis would have 64 (8×8) squares. An analysis of the offspring reveals that, among the 64 possible combinations, there are only eight visibly different phenotypes. You may wish to verify that the phenotypic ratio is 27:9:9:9:3:3:3:1.

The use of the Punnett square for three or more independently assorting genes is cumbersome. Here we will introduce one of the cardinal rules of probability: *The chance that two or more independent events will occur together is the*

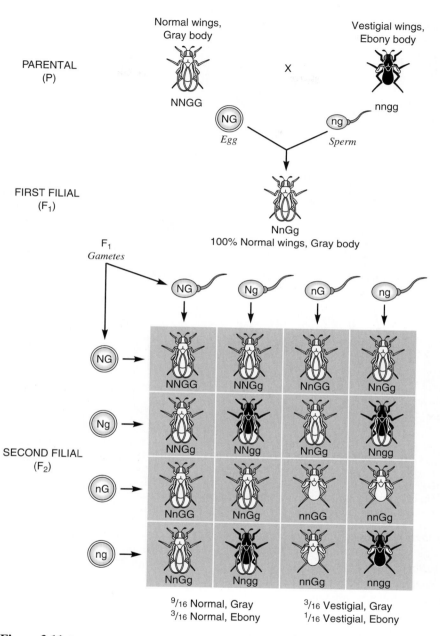

PARENTAL (P)

Normal wings, Gray body

NNGG

X

Vestigial wings, Ebony body

nngg

NG
Egg

ng
Sperm

FIRST FILIAL (F₁)

NnGg
100% Normal wings, Gray body

F₁ Gametes

SECOND FILIAL (F₂)

NG   Ng   nG   ng

NG

Ng

nG

ng

NNGG	NNGg	NnGG	NnGg
NNGg	NNgg	NnGg	Nngg
NnGG	NnGg	nnGG	nnGg
NnGg	Nngg	nnGg	nngg

9/16 Normal, Gray
3/16 Normal, Ebony

3/16 Vestigial, Gray
1/16 Vestigial, Ebony

Figure 3.11 Mendel's principle of independent assortment, as revealed in the inheritance of two pairs of characters in the fruit fly. A pure normal-winged fly with gray body mated with a vestigial-winged, ebony-bodied fly produces all normal-winged, gray-bodied flies in the first generation *(F₁)*. When these *F₁* flies are inbred, a second generation *(F₂)* is produced that displays a phenotypic ratio of 9:3:3:1.

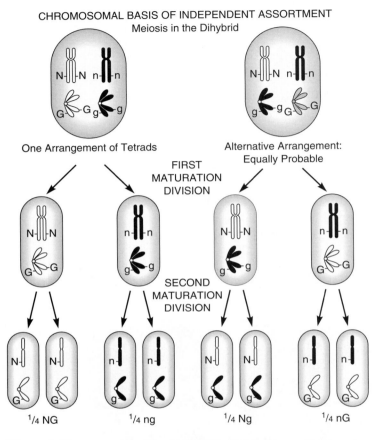

CHROMOSOMAL BASIS OF INDEPENDENT ASSORTMENT
Meiosis in the Dihybrid

One Arrangement of Tetrads

Alternative Arrangement: Equally Probable

FIRST MATURATION DIVISION

SECOND MATURATION DIVISION

¹/₄ NG   ¹/₄ ng   ¹/₄ Ng   ¹/₄ nG

Figure 3.12 Chromosomal basis of dihybrid inheritance. The two pairs of genes, *N, n* and *G, g* are located in two different pairs of chromosomes. When a dihybrid *(NnGg)* undergoes meiosis, there are two equally likely ways the tetrads may line up. The outcome is the formation of four kinds of gametes, in equal numbers *(NG, Ng, nG,* and *ng).*

product of their chances of occurring separately. As an example, what proportion of the offspring of the cross *AaBbCc × AaBbCc* would be expected to have the genotype *AaBBcc*? The individual computations are as follows, treating each character (or event) independently:

1. The chance that an individual will be *Aa* = 1/2.
2. The chance that an individual will be *BB* = 1/4.
3. The chance that an individual will be *cc* = 1/4.

Assuming independence among these three pairs of alternatives, the chance of the three combining together *(AaBBcc)* is the product of their separate probabilities, as follows:

$$1/2 \times 1/4 \times 1/4 = 1/32.$$

SIGNIFICANCE OF GENETIC RECOMBINATION

The Mendelian principles permit a genuine appreciation of the source of heritable variation in natural populations of organisms. An impressively large array of different kinds of individuals can arise from the simple processes of segregation and recombination of independently assorting genes. Figure 3.14 considers a few of the independently assorting traits in the fruit fly. It is seen that the number of visibly different classes (phenotypes) of offspring is doubled by each additional heterozygous pair of independently assorting alleles. Each additional heterozygous allelic pair multiplies the number of visibly different combinations by a factor of two. Thus, if the parents are each heterozygous for ten pairs of genes (when dominance is complete), the number of different phenotypes among the progeny becomes 2^{10}, or 1,024. For humans with 23 pairs of chromosomes, the number of different phenotypes among the progeny is 2^{23} or 4,194,304.

Other mathematical regularities are evident as we increase the number of heterozygous pairs (table 3.2). As the number of heterozygous traits increases, the chance of recovering one of the parental types becomes progressively less. Thus, when a single heterozygous pair *(Aa)* is involved, 1 in 4 will resemble one of the original parents in both

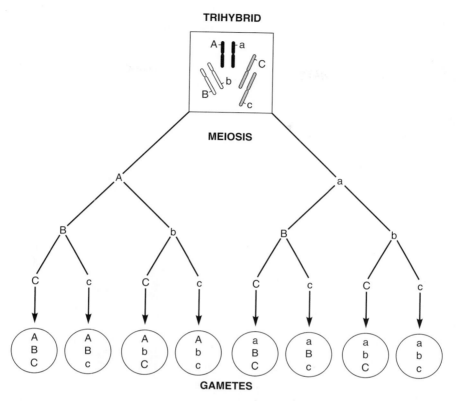

Figure 3.13 Gamete formation in an individual heterozygous for three pairs of alleles ("tri-hybrid"). The members of the three different pairs of alleles assort independently of each other when gametes are formed. The separation of the *Aa* pair, *Bb* pair, and the *Cc* pair, respectively, can be treated as independent events. The outcome is eight genetically different gametes.

TABLE 3.2	Characteristics of Crosses Involving Various Pairs of Independently Assorting Alleles			
Number of heterozygous pairs involved in the cross	Number of different types of gametes produced by the heterozygote	Number of visibly different progeny (that is, phenotypes)*	Number of genotypically different combinations (that is, genotypes)	Chance of recovering an individual that is either homozygous dominant or recessive for all traits
1	2	2	3	1 in 4
2	4	4	9	1 in 16
3	8	8	27	1 in 64
4	16	16	81	1 in 256
10	1,024	1,024	59,049	1 in 1,048,576
n	2^n	2^n	3^n	1 in 4^n

*Assumes complete dominance in each pair.

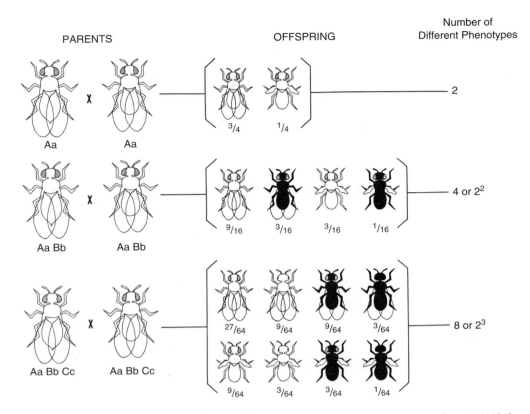

Figure 3.14 Inheritance of traits in the fruit fly, when the parents are each heterozygous for one *(Aa)* independently assorting trait and for two *(AaBb)* and three *(AaBbCc)* independently assorting traits. The allele for long wings *(A)* is dominant over the allele for vestigial wings *(a)*; the allele for gray body *(B)* is dominant over the allele for ebony *(b)*; and the allele for normal eyes *(C)* is dominant over the allele producing the eyeless condition *(c)*. The number of visibly different classes (phenotypes) of offspring increases progressively with each additional pair of traits. The generalized formula for determining the number of different phenotypes among the offspring (when dominance is complete) is 2^n, where *n* stands for the number of different heterozygous pairs of alleles.

appearance and genotype *(AA* or *aa)*. When four heterozygous traits are involved, only 1 in 256 will be genotypically like either of the original parents. Evidently, no single genetic constitution is ever likely to be exactly duplicated in a person (save in identical twins). Each individual is genetically unique.

LINKAGE AND CROSSING OVER

Mendel's principle of independent assortment is based on the assumption that each gene for a trait is carried on a different chromosome. However,

the number of chromosomes is appreciably less than the number of genes. Accordingly, many genes must be carried in a single chromosome. Genes located together in any one chromosome are said to be *linked,* and the linked genes tend to remain together during their hereditary transmission. It was a most fortuitous circumstance that Mendel did not encounter linked genes in his breeding experiments. If Mendel had happened to work on two linked genes, he certainly would have come upon some puzzling ratios among the offspring.

Some modern authors assert that Mendel likely encountered linkage but never reported those data. Interestingly, we now know that some of Mendel's pea plant characters are in fact linked but, fortuitously, are sufficiently far enough apart from each other that crossing makes them appear to assort independently. Perhaps had Mendel reported all of his data, he would have also discovered linkage!

The principle of linkage was firmly established in the early 1900s by T. H. Morgan and his students through studies on the fruit fly. The fruit fly *(Drosophila melanogaster)* has only four pairs of chromosomes. Several hundred different genes have been identified in the fruit fly, and all fall into four linkage groups corresponding to the four pairs of chromosomes. Humans have 23 linkage groups.

The effect of linkage is to decrease variability, but an intricate mechanism exists for rearranging the linked genes. As a result of the mechanism of *crossing over*—the exchange of segments of homologous chromosomes during meiosis—the original combination of genes in a particular chromosome can be broken up. In essence, crossing over permits different alleles of one gene to be tested in new combinations with alleles of other genes. In an illustrative example (fig. 3.15), if the two pairs of genes had been so closely linked as to preclude any crossing over between them, only two

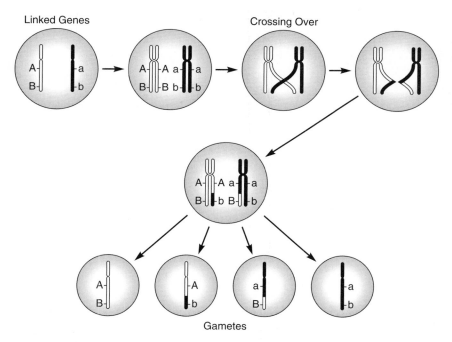

Figure 3.15 **Crossing over the chromosomal material during the tetrad stage of meiosis.** If the two pairs of genes were not linked, the hybrid individual *(AaBb)* would produce four genetically different gametes in equal proportions *(AB, Ab, aB,* and *ab).* A hybrid containing the two pairs of linked genes also produces four genetically different gametes as a consequence of crossing over, but the new combinations of linked genes, or recombination types *(Ab* and *aB),* arise less frequently than the original combination types *(AB* and *ab).* The magnitude of deviation from a 1:1:1:1 ratio depends on the frequency with which crossing over occurs between the linked genes.

classes of gametes would have been formed; one gamete containing *AB,* and the other, *ab.* However, as a consequence of crossing over, four genetically different types of gametes are produced. Thus, crossing over provides a mechanism for increasing variability—that is, rearranging genes in new combinations.

The reshuffling of genes by independent assortment, coupled with the recombination of genes through crossing over, may be compared, on a very modest scale, to the shuffling and dealing of playing cards. One pack of 52 cards can yield a large variety of hands. And, just as in poker a full house is far superior to three of a kind, so in organisms certain combinations of genes confer a greater reproductive advantage to their bearer than other combinations. Natural selection favors the more reproductive genotypes.

MUTATION

In the opening pages of *The Origin of Species,* Charles Darwin declared, "Any variation which is not inherited is unimportant for us." This and other incisive statements show that Darwin understood the importance of inherited variability. He was, however, eternally assailed by doubts as to how new varieties initially arise. Today we know that the ultimate source of genetic variation is *mutation.*

Each gene of an organism may assume a variety of forms. A normal gene may change to another form and produce an effect on a trait different from that of the original gene. An inheritable change in the structure of the gene is known as a mutation. Variant traits, such as albinism in humans and vestigial wings in the fruit fly, are traceable to the action of altered, or mutant, genes. *All differences in the genes of organisms have their origin in mutation.*

CAUSES OF MUTATION

New mutations arise from time to time, and the same mutations may occur repeatedly. Mutations are not uncommon and may occur each time the billions of building blocks of DNA, called *nucleotides,* are copied during cell division. Thus, as we age, mutations inevitably build up in our cells. It is often difficult to distinguish between new mutations and old ones that occurred previously and were carried concealed in ancestors. A recessive mutant gene may remain masked by its normal dominant allele for many generations and reveal itself for the first time only when two heterozygous carriers of the same mutant gene happen to mate.

Each gene runs the risk of changing to an alternative form. The causes of naturally occurring, or *spontaneous,* mutations are complex (table 4.1). In addition to spontaneous mutations, mutations may be induced by chemical and physical agents. The environment contains a background of inescapable radiation from natural sources (such as radon gas, radioactive elements, cosmic rays, and gamma rays) and human made sources (such as medical x rays and scans) as well as an increasingly large number of chemical agents that are common in our daily lives (table 4.2). Background radiation and other physical agents (table 4.2) account for only a small fraction of spontaneous mutations. We may

TABLE 4.1 | Some Common Causes of Gene Mutations

Common causes of mutations	Description
Spontaneous:	
Errors in DNA replication	A mistake by DNA polymerase may cause a point mutation.
Toxic metabolic products	The products of normal metabolic processes may be a reactive chemicals such as free radicals that can alter the structure of DNA.
Spontaneous changes in nucleotide structure	On rare occasions, the linkage between purines and deoxyribose can spontaneously break. Also, changes in base structure (isomerization) may cause mispairing during DNA replication.
Transposons	Transposons are small segments of DNA that can insert at various sites in the genome. If they insert into a gene, they may inactivate the gene.
Induced:	
Chemical agents	Chemical substances, such as benzo(a)-pyrene, a chemical found in cigarette smoke, may cause changes in the structure of DNA.
Physical agents	Physical agents such as UV (ultraviolet) light and x rays can damage DNA.

TABLE 4.2 | Commonly Encountered Mutagens

Mutagen	Source
Aflatoxin B	Fungi growing on peanuts and other foods
2-amino 5-nitrophenol	Hair dye components
2,4-diaminoanisole	"
2,5-diaminoanisole	"
2,4-diaminotoluene	"
p-phenylenediamine	"
Furylfuramide	Food additive
Nitrosamines	Pesticides, herbicides, cigarette smoke
Proflavine	Antiseptic in veterinary medicine
Sodium nitrite	Smoked meats
Tris (2,3-dibromopropyl phosphate)	Flame retardant in children's sleepwear

say that spontaneous variations are random and unpredictable.

In 1927 the late Nobel laureate Hermann J. Muller of Indiana University announced that genes are highly susceptible to the action of x rays. By irradiating fruit flies with x rays, he demonstrated that exposure to radiation can enormously speed up the process of mutation. The production of mutations is directly proportional to the total dosage of x rays. Stated another way, the greater the dosage, the greater the mutation rate. In this linear, or straight-line, relationship, the mutation rate increases proportionately to the amount of exposure to x rays.

It has long been held that the mutagenic effect is the same whether the dose is given in a short time or spread over a long period. In other words, low intensities of x rays over long periods of time produce as many mutations as the same dose administered in high intensities in a short period of time. Recent experiments on mice have cast some doubt on this view, for it has been shown (at least in mice) that the mutagenic effect of a single exposure to the germ cells is greater than the effect of the same exposure administered in several smaller doses separated by intervals of time. Nevertheless, there does not appear to be a critical, or *threshold,* dose of radiation below which there is no effect. In essence, no dose is so low (or so "safe") as to carry no risk of inducing a mutation. Modern workers stress that any amount of radiation, no matter how little, can cause a mutation. Natural and man-made sources of radiation are shown in table 4.3.

In discussing the sources (table 4.3) and hazards of x rays and other ionizing radiations (fig. 4.1), we must be careful to distinguish between *somatic damage* and *genetic damage* (fig. 4.2). Injury to the body cells of the exposed individuals themselves constitutes somatic damage. On the other hand, impairment of the genetic apparatus of the sex cells (germ-line cells) represents genetic damage or *germinal mutations*. Typically, the genetic alterations do not manifest themselves in the immediate persons but present a risk for their descendants in the next or succeeding generations. In this way, germinal mutations are more important to evolution because heritable mutation serves as the fuel of evolution.

TABLE 4.3	Sources of Radiation	

Natural Sources	(%)
Radon gas	54
Human body	11
Rocks and soil	8
Cosmic ray from space	8
	81
Man-Made Sources	
Medical x rays	11
Nuclear medicine	4
Consumer products	3
Other (nuclear fallout, occupational)	1
	19

CHEMICAL NATURE OF THE GENE

During the early decades of the twentieth century, the picture of genes that had emerged was that of discrete entities, strung together along the chromosome, like individual beads on a string.

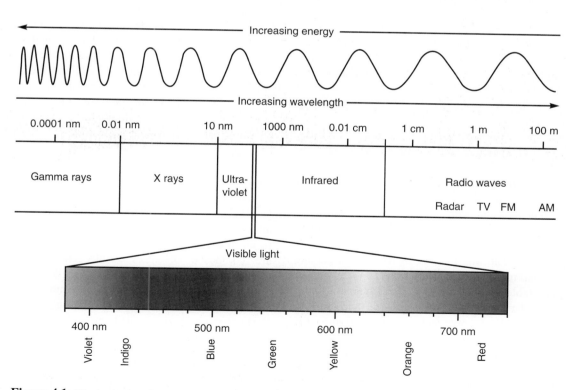

Figure 4.1 Electromagnetic spectrum. The higher-energy portions of the spectrum, such as x rays and gamma rays, are potent mutagens.

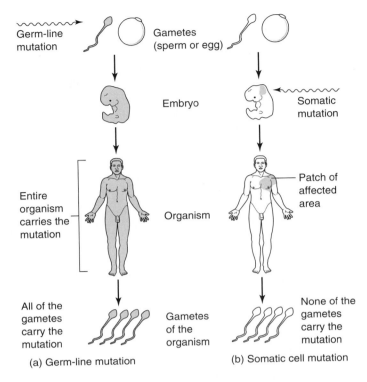

Germ-line mutation

Gametes (sperm or egg)

Embryo

Somatic mutation

Entire organism carries the mutation

Organism

Patch of affected area

All of the gametes carry the mutation

Gametes of the organism

None of the gametes carry the mutation

(a) Germ-line mutation

(b) Somatic cell mutation

Figure 4.2 Germ-line versus somatic mutation. Somatic mutations damage body cells, whereas germ-line mutations injure gonads (ovaries or testes) or gametes (eggs and sperm).

This portrayal carried with it the implication that genes were not pure abstractions, but actual particles of living matter. Yet, until the 1950s the chemical nature of the gene remained enigmatic. Biochemists wondered whether there existed an elaborate organic molecule with a simple string-like form.

One of the masterly feats of modern science has been the elucidation of the chemical nature of the gene. The transmission of traits from parents to offspring depends on the transfer of a specific giant molecule that carries a coded blueprint in its molecular structure. This complex molecule, the basic chemical component of the chromosome, is *deoxyribonucleic acid,* or DNA (figs. 4.3 and 4.4). The information carried in the DNA molecule can be divided into a number of separable units, which we now recognize as the genes. Stated

simply, chromosomes are primarily long strands of DNA, and genes are coded sequences of the DNA molecule. Genes specify the structure of individual proteins. The diverse genes in a single fertilized human egg can code an estimated 20,000 to 30,000 proteins. The once theoretical gene has finally been placed on a material (chemical) level.

The two scientists who worked together in the early 1950s to propose a configuration for the DNA molecule were Francis H.C. Crick, a biophysicist at Cambridge University, and James D. Watson, an American student of virology who was then studying chemistry at Cambridge on a postdoctoral fellowship. With the invaluable aid of x-ray diffraction pictures of DNA crystals prepared by Maurice H. F. Wilkins and Rosalind Franklin at King's College in London, Watson and Crick built an inspired metal model of the DNA's configuration. This achievement won Watson, Crick, and Wilkins the coveted Nobel Prize for physiology or medicine in 1962. The other deserving scientist, Rosalind Franklin, unfortunately died of cancer before the Nobel Prize (awarded only to living persons) was announced.

The remarkable feature of DNA is its simplicity of design. The double-stranded DNA molecule is shaped like a twisted ladder (fig. 4.3). The two parallel supports of the ladder are made up of alternating units of sugar (*deoxyribose*) and phosphate molecules, while the cross-links, or rungs, are composed of specific nitrogen-containing ring compounds, or *nitrogenous bases*. Each rung of the ladder consists of one pair of nitrogenous bases, held together by specific hydrogen bonds. Hydrogen bonds are weak bonds. However, the sum total of the hydrogen bonds between the two strands assures that the two strands of the double helix are firmly associated with each other under conditions commonly found in living cells.

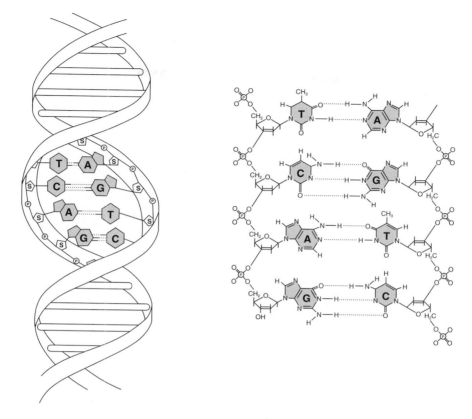

Figure 4.3 **Watson-Crick double-stranded helix configuration of deoxyribonucleic acid (DNA).** The backbone of each twisted strand consists of alternating sugar (S) and phosphate (P) residues. Enlarged view on right shows that the larger, two-ring purines (adenine or guanine) lay opposite the smaller, one-ring pyrimidines (cytosine or thymine). The nitrogenous bases are held together by hydrogen bonds, three between cytosine (C) and guanine (G), and two between adenine (A) and thymine (T).

There are two classes of nitrogenous bases, the larger, two-ring *purines* and the smaller, one-ring *pyrimidines* (fig. 4.4). In making up the model, Watson and Crick found that the nitrogenous bases could be accommodated best if the double-ring purines (adenine or guanine) lay opposite the single-ring pyrimidines (thymine or cytosine). The arrangement of the bases is not haphazard: adenine (A) in one chain is typically coupled to thymine (T) in the other chain, and guanine (G) is preferentially linked with cytosine (C). Along one chain, any sequence of the bases is possible, but if the sequence

along one chain is given, then the sequence along the other is automatically determined, because of the precise pairing rule (A=T and G≡C). The combination of one purine and one pyrimidine to make up each cross connection is conveniently called a *base pair.*

The DNA molecule has a structure that is sufficiently versatile to account for the great variety of different genes. The four bases (A, C, G, and T) may be thought of as a four-letter alphabet or code. The uniqueness of a given gene would then relate to the specific order or arrangement of the

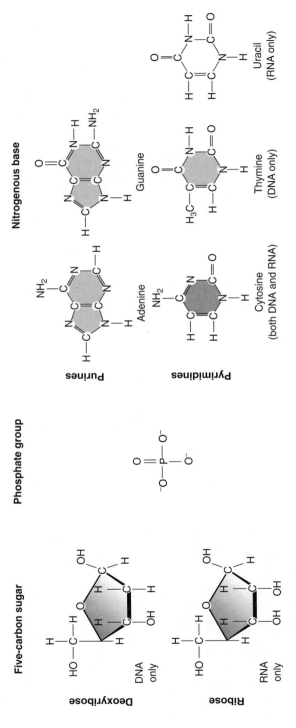

Figure 4.4 DNA and RNA are composed of three subunits. A five-carbon sugar (deoxyribose in DNA and ribose in RNA), a phosphate group, and a nitrogenous base. Purines pair with pyrimidines to form complementary base pairs, (A-T, G-C in DNA; A-U, C-G in RNA).

bases, just as words in our language differ according to the sequence of letters of the alphabet or as a telegraphic message becomes comprehensible by the varied combinations of dots and dashes. The number of ways in which the nitrogenous bases can be arranged in the DNA molecule is exceedingly large. The fertilized human egg contains 46 chromosomes with over 6 billion base pairs of DNA. As a generalization, a single gene is a linear sequence of approximately 1,500 base pairs.

REPLICATION OF DNA

One fundamental property that has long been ascribed to the gene is its ability to make an exact copy of itself. If the gene is really a linear sequence of base pairs, then the Watson-Crick model of DNA must be able to account for the precise reproduction of the sequences of the bases. The replication machinery is quite complicated, but conceptually the self-copying of the double helix can be visualized with little difficulty.

The DNA molecule consists of two chains, each of which is the complement of the other. Accordingly, each single chain serves as a mold, or *template,* to guide the formation of a new companion chain (fig. 4.5). The two parallel chains separate, breaking the hydrogen bonds that hold together the paired bases. Each chain then attracts new base units from among the supply of free units always present in the cell. As disassociation of the strands takes place, each separated lengthwise portion of the chain can begin to form a

portion of the new chain. Eventually, the entire original double helix has produced two exact replicas of itself. If the original two chains are

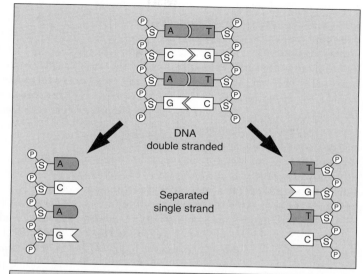
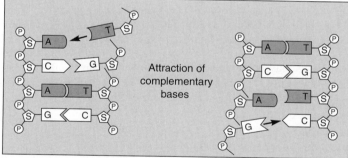
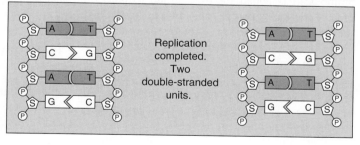

Figure 4.5 Replication of DNA. The parental strands separate, each serving as a guide or template for the synthesis of a new daughter strand. Each new DNA molecule contains one parental and one daughter strand.

designated *A* and *B,* the old *A* will direct the formation of a new *B* and the old *B* will guide the production of a new *A.* Where one *AB* molecule existed previously, two *AB* molecules, exactly like the original, exist afterward. This complex process is carried out by specific proteins called *enzymes.* Enzymes are proteins that catalyze chemical reactions.

Although the replication of a DNA strand is remarkably accurate, errors do occur in the positioning, or insertion, of the nitrogenous bases in a growing chain. The cell has several repair ("proofreading") mechanisms by which incorrectly inserted bases, or otherwise altered bases, are replaced by the appropriate bases. Occasionally, however, the cell's DNA repair program fails, and an accidental mistake endures. Such a sustained error qualifies as a base substitution or *point mutation.* Indeed, the vast majority of single base changes occur during DNA replication. (See Molecular Mechanism of Spontaneous Mutation.)

MOLECULAR MECHANISM OF SPONTANEOUS MUTATION

Mutations may be thought of as spontaneous mishaps or mistakes that occur as part of the copying of DNA from one generation to the next. In its simplest form, mutations alter DNA base pairs, which in turn alter amino acids, which thus alter proteins. As we will see later in this chapter, each amino acid is coded for by a group (triplet) of three nitrogenous bases. As shown in figure 4.6, the code is read linearly, starting at one end, in consecutive blocks of three bases. Accordingly, the reading of the code could be modified if an error occurred, such as a substitution, addition, or elimination (deletion) of a nitrogenous base. A substitution of one of the DNA bases constitutes a *gene (point) mutation.*

Point mutations have a more profound effect on the protein function if they alter the conformational folding property of the protein or cause the protein sequence to prematurely terminate

or interfere with the active sites for binding substances. For example, in the hemoglobin molecule, alterations in the amino acid sequence can impact on the shape of the red blood cell itself or on the ability of hemoglobin to bind oxygen.

As indicated in figure 4.6, a point mutation that results in the replacement of one amino acid for another is called a *missense mutation.* If the triplet coding of the amino acid changes to a termination codon (which signals the termination of the formation of the polypeptide chain), the change is called a *nonsense mutation.*

Additions or *deletions* of bases in the DNA molecule typically result in an altered sequence of amino acids in the protein. For example, if the base sequence ATGCTTCTC is normally read [ATG] [CTT] [CTC], then an insertion of the base C between [ATG] and [CTT] would lead to a shift in the "reading frame" as follows: [ATG] [CTT] [TCT] [C..]. Such modifications are called *frameshift mutations.*

Some point mutations are said to be "silent" because they do not alter the amino acid sequence, due to the duplication or redundancy in the genetic code. For example, the DNA triplet TCG codes for the amino acid serine. If base G changes to A, the triplet TCA will still code for serine. In fact, the amino acid serine can be specified by six DNA triplets (or six RNA codons). One of the surprises in recent years has been the discovery that the number of silent base substitutions during the evolutionary divergence of some protein molecules has been as high as 40 percent!

LANGUAGE OF LIFE

Cellular proteins are key components of living matter. Proteins are composed of 20 basic building blocks, the *amino acids,* arranged in long chains, called *polypeptides.* Each polypeptide chain may be several hundred amino acid units in length. The number of possible arrangements of the different amino acids in a given protein is unbelievably large. In a polypeptide chain made up of

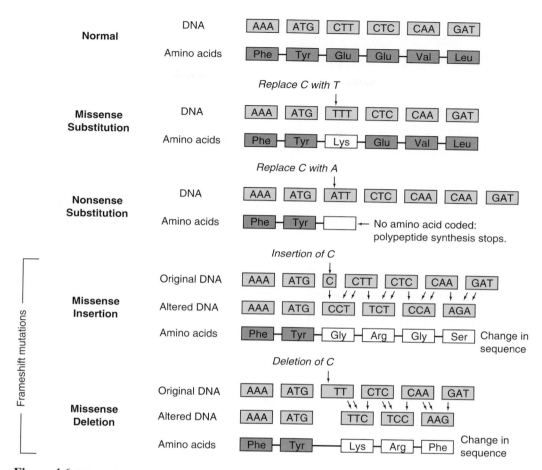

Figure 4.6 The molecular basis of point mutation. Each substitution, deletion, or insertion is a mutational event that alters the DNA sequence and, in turn, the polypeptide (protein) product.

500 amino acid units, the number of possible patterns (given 20 different amino acids) can be expressed by the number 1 followed by 1,100 zeros. How, then, does a cell form the particular amino acid patterns it requires out of the colossal number of patterns that are possible?

We may presume that the sequence of bases of the DNA molecule is in some way the master pattern, or code, for the sequence of the amino acids composing polypeptides. In 1954, the British physicist George Gamow suggested that each

amino acid is dictated by one sequence of three bases in the DNA molecule. As an example, the sequence cytosine-thymine-thymine (CTT) in the DNA molecule might designate that a particular amino acid (glutamic acid) be incorporated in the formation of a protein molecule, such as hemoglobin (fig. 4.7). Thus, the DNA code is to be found in *triplets*—that is, three bases taken together code one amino acid. It should be noted that only one of the two strands of the DNA molecule serves as the template for the genetic code. Biochemical

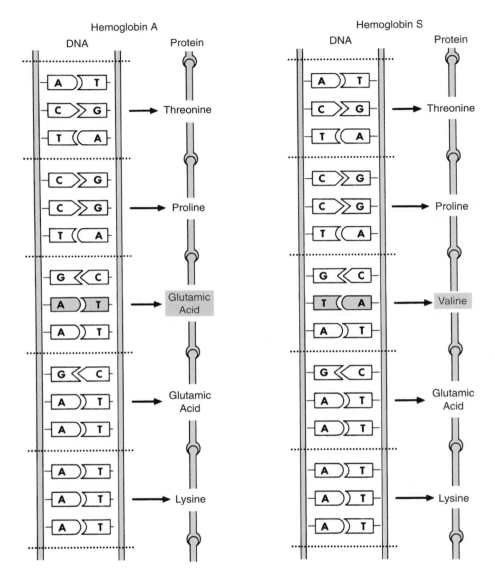

Figure 4.7 The abnormal hemoglobin that occurs in sickle-cell anemia (hemoglobin S) is the consequence of an alteration of a single base of the triplet of DNA that specifies a particular amino acid (glutamic acid) in normal hemoglobin (hemoglobin A). A simple base-pair switch, or mutation, in the DNA molecule results in the replacement of glutamic acid in hemoglobin A by another amino acid, valine, in hemoglobin S.

(Based on studies by Vernon M. Ingram.)

chaos would result if both complementary strands of DNA were to encode information.

The importance of the genetic code cannot be overstated. Let us assume the following sequence of bases in one of the strands of the DNA molecule: ... CGT ATC GTA AGC ..., and that these triplets specify four amino acids, designated **R, I, P,** and **E.** In other words, the following code exists: CGT=**R,** ATC=**I,** GTA=**P,** and AGC=**E.** This section of the DNA molecule thus specifies **RIPE,** and the message to make **RIPE** is passed on by this particular sequence from the nucleus to the cytoplasm of the cell. The message continues to flow out in the living cell, and **RIPE** will be made, copy after copy.

But what if a chance mishap occurs in one of the triplets? Let us say that T in the second triplet is substituted for A, so that the triplet reads TTC instead of ATC, which would specify an **O** instead of **I.** The word would now be **ROPE** instead of **RIPE.** This makes quite a difference, especially if the word is continually printed incorrectly throughout a novel. Just as a misprinted word can alter or destroy the meaning of a sentence, so an altered protein in the body fails to express its intended purpose. Sometimes the error is not tragic, but often the organism is debilitated by the misprint.

As a striking example, we may consider the case of individuals afflicted with sickle-cell anemia. The full story of sickle-cell anemia is narrated in chapter 8. For the moment, it suffices to say that the hemoglobin molecule in sickle-cell anemic patients is biochemically abnormal. The sole difference in chemical composition between normal and sickle-cell hemoglobin is the substitution of only one amino acid unit among several hundred. Specifically, at one site in a particular polypeptide chain of the sickle-cell hemoglobin molecule, the amino acid *glutamic acid* has been replaced by *valine.* The detrimental effect of sickle-cell anemia is thus traceable to an exceedingly slight alteration in the structure of the protein molecule. This, in turn, is associated with a simple base-pair switch, or mutation, in the DNA molecule in the chromosome (fig. 4.7).

The DNA molecule, like a tape recording, carries specific messages for the synthesis of a wide variety of proteins. We shall see presently that DNA does not directly form protein but works in an intriguing way through a secondary form of nucleic acid, *ribonucleic acid,* or RNA. At this point, it is important to recognize that a gene is a coded sequence in the DNA molecule. From a functional point of view, we may say that a gene is *a section of the DNA molecule (about 1,500 base pairs) involved in the determination of the amino acid sequence of a single polypeptide chain of a protein.*

TRANSCRIPTION OF DNA

It has been said that hemoglobin served as the Rosetta stone in revealing the hereditary language. It was the hemoglobin molecule that provided the first direct proof of the relationship of the gene to protein synthesis. Proteins are not synthesized directly on the DNA template. The genetic information (genetic code) of DNA is first transcribed to ribonucleic acid (RNA). The RNA molecule resembles the DNA molecule in structure except in three important respects. The pyrimidine base, *uracil,* is found in RNA, which replaces the thymine that is characteristic of DNA. Secondly, the sugar in RNA is *ribose,* which contains one more oxygen atom than does the deoxyribose sugar (fig. 4.4). Thirdly, RNA has only a *single* strand instead of two.

As seen in figure 4.8, one of the two strands of DNA (always the same one for a given gene) forms a complementary strand of RNA. The same rules of pairing hold as in replicating a copy of DNA, except that adenine attracts uracil instead of thymine. This RNA strand, which is responsible for carrying DNA's instructions out into the cytoplasm, is appropriately termed *messenger RNA,* or mRNA. The amino acids of a polypeptide chain are specified by the messenger RNA. In a large sense, then, the genetic code applies to RNA rather than to DNA itself.

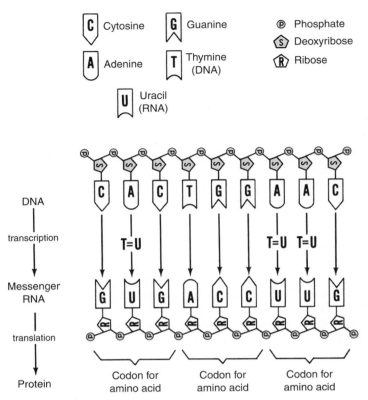

C Cytosine   G Guanine

A Adenine   T Thymine (DNA)

U Uracil (RNA)

P Phosphate
S Deoxyribose
R Ribose

DNA

transcription

Messenger RNA

translation

Protein

T=U   T=U T=U

Codon for amino acid   Codon for amino acid   Codon for amino acid

Figure 4.8 The language of life. The DNA molecule forms a single strand of messenger RNA that carries DNA's instructions out into the cell. The four bases of DNA (A, T, C, and G) are responsible for the 3-letter code words, or codons, of messenger RNA. Each of the twenty amino acids that make up the variety of body proteins is specified by at least one codon.

Scientists have broken the genetic code in its main elements. Each three-letter unit of the messenger RNA is called a *codon*. Each of the twenty main amino acids, ranging from alanine to valine, is specified by at least one codon. In fact, most amino acids have more than one codon. Serine, for example, has six codons; glycine has four; and lysine has two. Only methionine and tryptophan have one each. Because most amino acids are specified by more than one codon, the code is claimed to be *degenerate*. However, this kind of degeneracy is welcome since it ensures that the code works.

Details of the code are shown in table 4.4. This table of the genetic code represents for the biologist what the periodic table of the elements

represents for the chemist. It may be noted that three of the codons (UAA, UAG, and UGA) do not code for any of the 20 amino acids. They were originally, perhaps facetiously, described as "nonsense codons," but they are now known to function as *terminating codons* (or *stop codons*). They serve to signal the termination of a polypeptide chain.

One of the most impressive findings is that the genetic code is essentially universal. The same codon calls forth the same amino acid in organisms as widely separate phylogenetically as a bacterium *(E. coli)*, a flowering plant (wheat), an amphibian (the South African clawed frog), and mammals (guinea pig, mouse, and human). The essential universality of the genetic code supports the view that the code had its origin at least by the time bacteria evolved, 3 billion years ago. The genetic code may be said to be the oldest of languages. In this regard, the universality of the genetic code represent perhaps the best evidence for evolution.

The flow of genetic information is sketched in summary form in figures 4.9 and 4.10. The DNA molecule has the capacity to replicate as well as transcribe. The *transcription* of DNA into messenger RNA is the means by which genetic information is communicated from the nucleus to the cytoplasm. The transfer of information from messenger RNA to proteins is termed *translation*.

We may now take a closer look at sickle-cell hemoglobin in light of our knowledge of transcription and translation. This abnormal hemoglobin is caused by a mutation in one gene that codes a polypeptide chain, the so-called β-globin chain that is 146 amino acid residues in length. As we learned earlier, the mutation alters only one amino acid unit in the entire chain, changing a glutamic acid present in normal hemoglobin into another amino acid,

TABLE 4.4	The Three-Letter Codons of RNA and the Amino Acids Specified by the Codons*

AAU AAC }	Asparagine	CAU CAC }	Histidine	GQU GAC }	Aspartic acid	UAU UAC }	Tyrosine
AAA AAG }	Lysine	CAA CAG }	Glutamine	GAA GAG }	Glutamic acid	UAA UAG }	(Terminator)†
ACU ACC ACA ACG }	Threonine	CCU CCC CCA CCG }	Proline	GCU GCC GCA GCG }	Alanine	UCU UCC UCA UCG }	Serine
AGU AGC }	Serine	CGU CGC		GGU GGC }	Glycine	UGU UGC }	Cysteine
AGA AGG }	Arginine	CGA CGG }	Arginine	GGA GGG }		UGA UGG	(Terminator)† Tryptophan
AUU AUC AUA }	Isoleucine	CUU CUC CUA }	Leucine	GUU GUC GUA }	Valine	UUU UUC }	Phenylalanine
AUG	Methionine	CUG }		GUG }		UUA UUG }	Leucine

* The genetic code is a triplet RNA code that specifies the 20 essential amino acids, three stop or *termination sequences* (UAA, UAG, and UGA) and one start sequence (AUG which codes for the amino acid methionine). Most amino acids are coded for by more than one triplet sequence, called a codon. The start codon initiates *translation* and termination or "stop" codons signal the end of the formation of a polypeptide chain.
† Terminating codons signal the end of the formation of a polypeptide chain.

valine. The substitution of valine for glutamic acid can be accounted for by an alteration of a single base change of the triplet of DNA that specifies the amino acid, as depicted in figure 4.11. If the DNA triplet responsible for transcription in normal hemoglobin is CTT, then a single base change to CAT would alter the codon of messenger RNA to GUA. Accordingly, the amino acid specified by the codon GUA would be valine. The replacement of the glutamic acid by the valine causes a change in the conformation of the hemoglobin molecule. This, in turn, causes a distortion in the morphology ("sickling") of the red blood cell.

FREQUENCY OF MUTATIONS IN HUMAN DISORDERS

Most mutations present in an individual are inherited from the previous generation. The rate at which new disease-causing mutations arise in humans varies, and in our present state of knowledge the mutation rate for any single human gene must be considered as a rough estimate. It is likely that all human offspring contain at least two to five newly mutated genes capable of having an adverse effect.

Different species have different mutation rates, as do different proteins. This will be important to our later discussion of molecular clocks (chapter 15). For example, a large-size gene with many bases has a greater likelihood of experiencing a mutational change than does a small-size gene, just as in a manuscript a long sentence is more likely to contain a misspelling than in a short sentence.

It may be noted that the mutation rate for a recognizable human disorder is higher than the estimate, stated earlier in this chapter, of the frequency of uncorrected base changes in the DNA molecule. This reflects the fact that not all single-base changes (mutations) within a given gene have adverse effects or cause overt disease. Indeed,

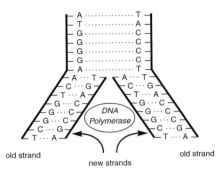

old strand   new strands   old strand

Replication

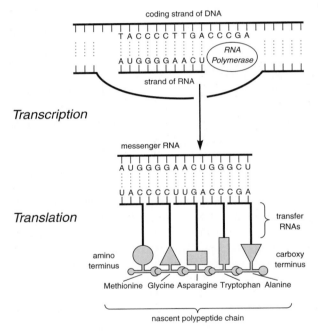

coding strand of DNA

T A C C C C T T G A C C C G A

RNA Polymerase

A U G G G G A A C U

strand of RNA

Transcription

messenger RNA

A U G G G G A A C U G G G C U

U A C C C C U U G A C C C G A

Translation

transfer RNAs

amino terminus

carboxy terminus

Methionine Glycine Asparagine Tryptophan Alanine

nascent polypeptide chain

Figure 4.9 Stepwise flow of information from DNA to RNA to protein. DNA not only copies itself by the act of *replication* with the aid of an enzyme (DNA polymerase), but expresses its information by generating a single-stranded RNA whose nucleotide sequence is coded by one of the two strands of the DNA duplex. The *transcription* of DNA into RNA is accomplished with the aid of an RNA polymerase. In the act of *translation,* the nucleotide sequence of the RNA transcript (messenger RNA) directs the sequence of amino acids comprising a protein. Transcription of DNA into RNA and translation of messenger RNA into protein are compartmentalized activities, the former event occurring in the nucleus and the latter in the cytoplasm.

several hundred different base changes are known to occur in the genes that code for hemoglobin, but many of them are inconsequential to the properties or activity of the hemoglobin molecule and are, accordingly, innocuous. There are, however, approximately 400 hemoglobin variants that have been identified as having a clinical impact. Sickle-cell anemia, earlier discussed, is a prominent example of a single base change resulting in a marked impairment of the hemoglobin molecule.

The mutation responsible for *achondroplasia* is notable in that it constitutes one of the most mutable single-base changes in the human genome. Achondroplasia is transmitted as an autosomal dominant trait. Affected individuals are small and disproportionate, with relatively long trunks and short arms and legs. Achondroplasia is a *congenital defect,* a defect present at birth. With an estimated prevalence of 1 per 40,000 live births, it is one of the more common Mendelian disorders.

Many achondroplastic dwarfs are stillborn or die in infancy; those surviving to adulthood produce fewer offspring than normal. About 80 percent of the children born with this condition in one generation will not replace themselves in the next generation. Given the high mortality and low fecundity, the frequency of the disorder would steadily decrease from one generation to the next, if it were not for the high mutability of the gene involved. In fact, more than 90 percent of affected individuals have newly arisen mutations. Interestingly, an increased paternal age at the time of conception has been observed. That is to say, *de novo* mutations occur more often in older fathers than in older mothers. Indeed, in several other dominantly inherited disorders, normal parents more often produce an affected child when the father is older at the time of conception.

Achondroplasia is like sickle-cell anemia in that it is caused by a single, specific amino acid substitution. However, this dominant disorder differs markedly from sickle-cell anemia in that

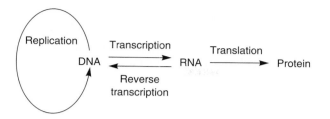

Figure 4.10 Simplified representation of the flow of genetic information showing the relationship between DNA, RNA and protein synthesis. DNA is *transcribed* into RNA and then RNA is *translated* into proteins. Some RNA viruses, called retroviruses, can convert RNA into DNA using the enzyme *reverse transcriptase.*

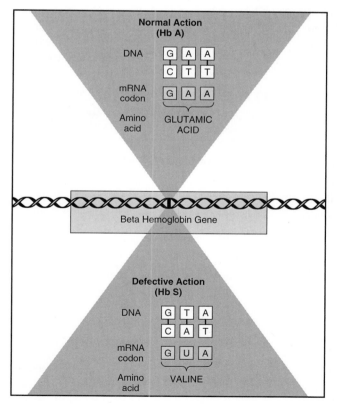

Figure 4.11 Derivation of abnormal hemoglobin S from normal hemoglobin A by a single nucleotide change in the triplet of DNA bases that normally codes for glutamic acid.

new mutations account for 90 percent of the cases. The mutation rate for the achondroplastic gene has been placed as high as about 4 mutations in every 100,000 gametes produced in a generation. This particular site is said to be a mutational "hot spot."

EVOLUTIONARY CONSEQUENCES OF MUTATIONS

By far, most of the gene mutations arising today in organisms are not likely to be beneficial. Existing populations of organisms are products of a long evolutionary past. The genes that are now normal to the members of a population represent the most favorable mutations selectively accumulated over eons of time. The chance that a new mutant gene will be more advantageous than an already established favorable gene is slim. Nonetheless, if the environment were to change, the mutant gene might prove to be beneficial in the new environmental situation. The microscopic water flea, *Daphnia,* thrives at a temperature of 20°C and cannot survive when the temperature rises to 27°C. A mutant strain of this water flea is known that requires temperatures between 25°C and 30°C and cannot live at 20°C. Thus, at high temperatures, the mutant gene is essential to the survival of the water fleas. This little episode reveals an important point: *A mutation that is inferior in the environment in which it arose may be superior in another environment.* Skeptics might contend that this proverbial declaration has no relevancy to complex higher organisms, including humans. Interestingly, the maxim derives overwhelming support from the varied frequencies of the gene for sickle-cell anemia in human populations. The obviously harmful sickling gene may actually confer an advantage to its carriers in certain geographical localities (see chapter 8).

The process of mutation furnishes the genetic variants that are the raw materials of evolution. Ideally, mutations should arise only when advantageous, and only when needed. This, of course, is fanciful thinking. Mutations occur irrespective of their usefulness or uselessness. The mutations responsible for achondroplasia and retinoblastoma (malignant eye tumors) in humans are certainly not beneficial. But novel heritable characters arise repeatedly as a consequence of mutation. Only one mutation in several thousands might be advantageous, but this one mutation might be important, if not necessary, to the continued success of a population. The harsh price of evolutionary potentialities for a population is the continual occurrence and elimination of mutant genes with detrimental effects. Thus, in evolutionary terms, a population, if it is to continue to evolve, must depend on the occasional errors that occur in the copying process of its genetic material.

CHROMOSOMAL ABERRATIONS AND PREGNANCY LOSS

Our discussion thus far has concentrated on changes at the molecular level within the gene, the so-called *point* (or *gene*) mutations. Grosser alterations involving large parts of the chromosomes have been termed *chromosomal aberrations*. Some of the chromosomal aberrations are numerical, affecting the amount of chromosomal material. A cell, for example, may have three sets of chromosomes (the *triploid* state) rather than the customary two chromosomal sets (the standard *diploid* state). Another example of a numerical abnormality is when there is an extra chromosome, or trisomy $(2n+1)$. The most common human trisomy is Down syndrome (or trisomy 21), where the individual has an extra chromosome 21. In general, missing a chromosome is less compatible with survival than having an extra chromosome. Other gross chromosomal deviations are classified as structural, as represented by the transfer of chromosomal segments between chromosomes *(translocation)*, the reversal of the order of a chromosome by turning a part of it upside down *(inversion)* or the loss of a part of a chromosome *(deletion)*. Once considered exceptionally rare and relatively unimportant, chromosomal aberrations now loom as a prominent cause of the prenatal loss of human embryos.

It may be disconcerting to learn that in humans there is a large natural (spontaneous) loss of embryos in early pregnancy, a level that far exceeds the incidence in most mammalian species. Of spontaneous abortions that occur before pregnancy has been diagnosed, most are due to gross chromosomal aberrations.

The data are too imprecise to provide a true rate of pregnancy loss in the human female, but available information does permit a mathematical estimation that appears to reliably approximate natural events. A table of intrauterine death (table 4.5) explores what happens to 100 eggs

TABLE 4.5	**Table of Intrauterine Deaths per 100 Ova Exposed to the Risk of Fertilization**

Weeks after ovulation	Survivors[a]	Failures[b]
	100	16[c]
0	84[d]	15
1	69[e]	27
2[f]	42	5.0
6	37	2.9
10	34.1	1.7
14	32.4	0.5
18	31.9	0.3
22	31.6	0.1
26	31.5	0.1
30	31.4	0.1
34	31.3	0.1
38	31.2	0.2
Live births	31	

[a] Pregnancies still in progress.
[b] Spontaneous abortions.
[c] Not fertilized.
[d] Number fertilized.
[e] Number implanted.
[f] Expected times of menses.

produced by women who are reproducing naturally. Sixteen of the 100 eggs will fail to be fertilized, even under optimal conditions. Of the 84 eggs that are fertilized, 15 will fail to implant. Of the 69 embryos implanted at the end of the first week, 27 of these characteristically will find the uterine lining inhospitable. Thus, up to the second week after ovulation, only 42 eggs of the original 100 will still be viable. Stated another way, by the time pregnancy is recognizable, more than half of the eggs have been lost. At 8 weeks' gestation, when the embryo is termed a fetus, about 65 eggs of the original 100 will have failed to survive. The incidence of spontaneous abortion during the fetal period is very low. By the end of gestation, the probability of a live birth is only 0.31, or 31 percent. For ease of comprehension, the numerical information in table 4.5 is pictorially represented in figure 4.12, in which the baseline is an original 20 eggs rather than 100 eggs.

With the rapid advances in chromosome methdology, it became possible to examine the human chromosome complement in the cells of spontaneous abortuses. To most scientists, the results of the analysis were astonishing. Several thorough investigations have revealed that 50 to 70 percent of first-trimester spontaneous abortuses are chromosomally abnormal. The data convincingly demonstrate that chromosomal aberrations are the major etiologic agents responsible for naturally occurring abortions.

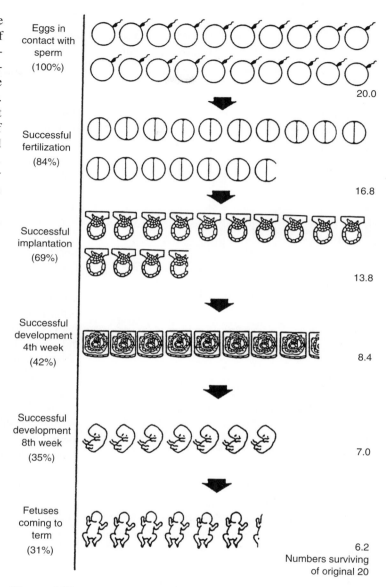

Eggs in contact with sperm (100%)

20.0

Successful fertilization (84%)

16.8

Successful implantation (69%)

13.8

Successful development 4th week (42%)

8.4

Successful development 8th week (35%)

7.0

Fetuses coming to term (31%)

6.2
Numbers surviving of original 20

Figure 4.12 The fate of 20 eggs produced by women who are reproducing naturally. Under conditions optimal for fertilization and development, only 6.2 eggs (31%) of the original 20 develop successfully to term.

In one study, an examination of the incidence of chromosomal anomalies in human abortuses found that among 1,097 specimens between the second and seventh weeks of gestation, 724, or

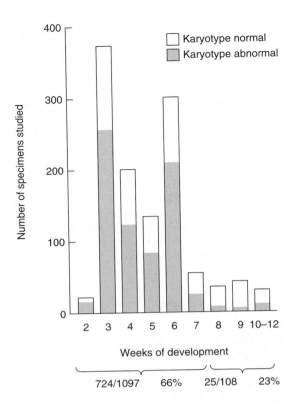

Figure 4.13 **Frequency** of chromosomal abnormalities in spontaneous abortuses in relation to weeks of development.

66 percent, had an abnormal karyotype (fig. 4.13). The incidence of chromosomal abnormalities fell to 23 percent among the 108 abortuses between 8 and 12 weeks of age. Most types of chromosomal aberrations are incompatible with survival to an advanced stage of pregnancy. It has been conservatively estimated that 1 in 5 of all human conceptuses carries a chromosome abnormality. At birth, the incidence declines dramatically to about 1 in 200. In other words, natural selection acts early in pregnancy before much energy is expended toward their development to eliminate organisms that are not likely to survive. Gross chromosomal abnormalities is thus the primary cause of pregnancy loss of conceptuses that would not survive outside the womb.

There is evidence that there is an optimal interval in humans between the delivery of one child and the conception of the next—that is, an interval associated with the greatest probability of a normal infant being born at full term. It can be argued that it is in the best interests of both the parents and offspring to expand the birth interval so that the parents can expend a greater effort on each child to better ensure the survival and ultimate fitness of each child. The investment of energy per child is likely to be less in situations in which successive births are very close to one another. Conversely, each child is likely to receive greater attention when the time period between births is extended. Research supports the notion that both short birth intervals (less than 6 months) and a long birth intervals (over 120 months) are associated with an increased risk of a variety of unfavorable birth outcomes. This has lead to the suggestion that the optimal birth-to-conception interval for safeguarding against adverse birth outcomes is approximately 2 years. This approximation fits with the concept that a prolonged birth interval may have been especially important during the nomadic hunting and gathering period in human evolution when having multiple children who could not walk during periods of nomadic migration was highly disadvantageous. Under this view, there may have been selective pressures on early humans to adapt an optimal birth interval to provide the greatest chance of infant survival. Whatever the case may be, it is certain that the length of the birth interval and the incidence of gross chromosomal aberration at birth have been the subject of selective pressure during the evolution of our species.

GENETIC EQUILIBRIUM

The opening chapter introduced us to natural populations of bullfrogs that conspicuously contained at one time several hundred multilegged variants. One of the hypotheses offered was that the multilegged anomaly is an inherited condition, transmitted by a detrimental recessive gene. The multilegged bullfrogs disappeared in nature as dramatically as they appeared. They unquestionably failed to reproduce and leave descendants. Now, let us imagine that the multilegged frogs were as reproductively fit as their normal kin. Would the multilegged trait eventually still be eliminated from the population?

A comparable question was posed to the English geneticist R. C. Punnett at the turn of the twentieth century. He was asked to explain the prevalence of blue eyes in humans in view of the acknowledged fact that the blue-eyed condition was a recessive characteristic. Would it not be the case that the dominant brown-eyed trait would in time supplant the blue-eyed state in the human population? The answer was not self-evident, and Punnett sought out his colleague Godfrey H. Hardy, the astute mathematician at Cambridge University. Hardy had only a passing interest in genetics, but the problem intrigued him as a mathematical one. The solution, which we shall consider below, ranks as one of the most fundamental theorems of genetics and evolution. As fate would have it, Hardy's formula was arrived at independently in the same year (1908) by a physician, Wihelm Weinberg, and the well-known equation presently bears both their names.

MENDELIAN INHERITANCE

We may recall that the genetic constitutions, or *genotypes*, of the normal and multilegged bullfrogs have been designated as *AA* (normal), *Aa* (normal but a carrier), and *aa* (multilegged). The kinds and proportions of offspring that can arise from matings involving the three genotypes are illustrated in figure 5.1. Six different types of matings are possible. The mating *AA* × *AA* gives rise solely to normal homozygous offspring, *AA*. Two kinds of progeny, *AA* and *Aa* in equal proportions, result from the cross of a homozygous normal parent

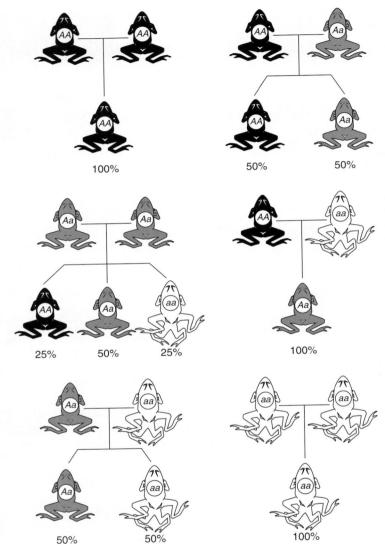

Figure 5.1 Six possible mating types with respect to one pair of genes, and the kinds and percentages of offspring from each type of mating. Normal frogs are either homozygous dominant *(AA)* or heterozygous *(Aa)*; multilegged frogs are recessive *(aa)*. The sex of the parent is not denoted; in crosses of unlike genotypes *(such as AA × Aa)*, either genotype may be the male or the female.

from the mating *AA × aa*. Both heterozygous *(Aa)* and recessive *(aa)* progeny, in equal numbers, arise from the cross of a heterozygous parent *(Aa)* and a recessive parent *(aa)*. Lastly, two recessive parents *(aa × aa)* produce only recessive offspring *(aa)*.

These principles of Mendelian inheritance merely inform us that certain kinds of offspring can be expected from certain types of matings. If we are interested in following the course of a population from one generation to the next, then additional factors enter the scene.

RANDOM MATING

One important factor that influences the genetic composition of a population is the system of mating among individuals. The simplest scheme of breeding activity in a population is referred to as *random mating,* wherein any one individual has an equal chance of pairing with any other individual. Random mating does not mean promiscuity; it simply means that those who choose each other as mating partners do not do so on the basis of similarity or dissimilarity in a given trait or gene.

The absence of preferential mating in a population has interesting consequences. Let us suppose that random mating prevails in our particular population of bullfrogs. This assemblage of frogs is ordinarily very large, numbering several thousand individuals. For ease of presentation, however, the size of the population is reduced to 48 males and 48 females. Moreover, for each sex,

(AA) and a heterozygous parent *(Aa)*. The mating of two heterozygotes *(Aa × Aa)* produces *AA, Aa,* and *aa* offspring in the classical Mendelian ratio of 1:2:1. Only heterozygous offspring *(Aa)* emerge

we may simplify the mathematical computations by assuming that 36 are normal (12 *AA* and 24 *Aa*) and 12 are multilegged *(aa)*. Accordingly, one-quarter of the individuals of each sex are homozygous dominant, one-half are heterozygous, and one-quarter are recessive. Now, if mating occurs at random, will the incidence of multilegged frogs decrease, increase, or remain the same in the next generation?

The problem may be approached by determining how often a given type of mating occurs. Here we will bring into play the multiplication rule of probability: *The chance that two independent events will occur together is the product of their chances of occurring separately.* The proportion of *AA* males in our arbitrary bullfrog population is 1/4. We may also say the chance that a male bullfrog is *AA* is 1/4. Likewise the probability that a female bullfrog is *AA* is 1/4. Consequently, the chances that an *AA* male will "occur together," or mate, with an *AA* female are 1/16 (1/4 × 1/4). The computations for all types of matings can be facilitated by coupling the males and females in a multiplication table, as shown in table 5.1.

Table 5.1 shows that there are nine combinations of mated pairs and that some types occur more frequently than others. It may be helpful to express the frequencies in terms of actual numbers. Thus, for a total of 48 matings, 3 (=1/16 × 48) would be *AA*♀ × *AA*♂, 6 (=1/16 X 48) would be *AA*♀ × *Aa*♂, 12 (= 4/16 × 48) would be *Aa*♀ × *Aa*♂, and so forth. The numbers of each type of mating are listed in table 5.2.

Our next step is to ascertain the kinds and proportions of offspring from each mating. We shall assume that each mated pair yields the same number of offspring—for simplicity, four offspring. (This is an inordinately small number, as a single female bullfrog can deposit well over 10,000 eggs.) We also take for granted that the genotypes of the four progeny from each mating are those that are theoretically possible in Mendelian inheritance (see fig. 5.1). For example, if the parents are *Aa* × *Aa*, their offspring will be 1 *AA*, 2 *Aa*, and 1 *aa*. In another instance, if the parents are *AA* × *Aa*, then the offspring will be 2 *AA* and 2 *Aa*. The outcome of all crosses is shown in table 5.2. It is important to note that the actual numbers of offspring recorded in table 5.2 are related to the frequencies of the different types of matings. For example, the mating of an *AA* female with an *Aa* male occurs six times; hence, the numbers of offspring are increased sixfold (from 2 each of *AA* and *Aa* to 12 each of the two genotypes).

An examination of table 5.2 reveals that the kinds and proportions of individuals in the new generation of offspring are exactly the same as in the

TABLE 5.1	Random Mating of Individuals

Female (♀)	Male (♂)		
	1/4 *AA*	**2/4 *Aa***	**1/4 *aa***
1/4 *AA*	1/16 *AA* × *AA*	2/16 *AA* × *Aa*	1/16 *AA* × *aa*
2/4 *Aa*	2/16 *Aa* × *AA*	4/16 *Aa* × *Aa*	2/16 *Aa* × *aa*
1/4 *aa*	1/16 *aa* × *AA*	2/16 *aa* × *Aa*	1/16 *aa* × *aa*

TABLE 5.2	First Generation of Offspring

Type of Mating (Female × Male)	Number of Each Type of Mating*	Number of Offspring		
		AA	*Aa*	*aa*
AA × *AA*	3	12		
AA × *Aa*	6	12	12	
AA × *aa*	3		12	
Aa × *AA*	6	12	12	
Aa × *Aa*	12	12	24	12
Aa × *aa*	6		12	12
aa × *AA*	3		12	
aa × *Aa*	6		12	12
aa × *aa*	3	—	—	12
		48	96	48
		(25%)	(50%)	(25%)

* Based on a total of 48 matings.

parental generation. There has been no change in the ratio of normal frogs (75 percent *AA* and *Aa*) to multilegged frogs (25 percent *aa*). In fact, the proportions of phenotypes (and genotypes) will remain the same in all successive generations, provided that the system of random mating is continued.

GENE FREQUENCIES

There is a less tedious method of arriving at the same conclusion. Rather than figure out all the matings that can possibly occur, we need only to consider the genes (alleles) that are transmitted by the eggs and sperm of the parents. Let us assume that each parent produces only 10 gametes. The 12 homozygous dominant males *(AA)* of our arbitrary initial population can contribute 120 sperm cells to the next generation, each sperm containing one *A* allele. The 24 heterozygous males *(Aa)* can transmit 240 gametes, 120 of them with *A* and 120 with *a*. The remaining 12 recessive males *(aa)* can furnish 120 gametes, each with an *a* allele. The total pool of alleles provided by all males will be 240 *A* and 240 *a*, or 50 percent of each kind. Expressed as a decimal fraction, the frequency of allele *A* is 0.5; of *a*, 0.5.

Since the females in our population have the same genetic constitutions as the males, their gametic contribution to the next generation will also be 0.5 *A* and 0.5 *a*. The eggs and sperm can now be united at random in a genetical checkerboard (fig 5.2).

It should be evident from figure 5.2 that the distribution of genotypes among the offspring is 0.25 *AA* : 0.50 *Aa* : 0.25 *aa*. The random union of eggs and sperm yields the same result as the random mating of parents (refer to table 5.2). Thus, using two different approaches, we have answered the question posed in the introductory remarks to this chapter. *If the multilegged frogs are equally as fertile as the normal frogs and leave equal numbers of offspring in each generation, then these anomalous frogs will persist in the population with the same frequency from one generation to the next.*

HARDY-WEINBERG EQUILIBRIUM

A population in which the proportions of genotypes remain unchanged from generation to generation is in *equilibrium*. The fact that a system of random mating leads to a condition of equilibrium was uncovered independently by G. H. Hardy and W. Weinberg, and has come to be widely known as the *Hardy-Weinberg equilibrium*. This theorem states that the proportions of *AA*, *Aa*, and *aa* genotypes, as well as the proportions of *A* and *a* alleles, will remain constant from generation to generation, provided that the bearers of the three genotypes have equal opportunities of producing offspring in a large, randomly mating population.

The above statement can be translated into a simple mathematical expression. If we let *p* be the frequency of allele *A* in the population, and *q* equal to the frequency of the alternative allele, *a*, then the distribution of the genotypes in the next generation will be $p^2 AA : 2pq\ Aa : q^2\ aa$. This relationship may be verified by the use, once again, of a genetical checkerboard (fig. 5.3). Mathematically inclined readers will recognize that $p^2 : 2pq : q^2$ is the algebraic expansion of the binomial $(p + q)^2$. The frequencies of the three genotypes (0.25 *AA* : 0.50 *Aa* : 0.25 *aa*) in our bullfrog population under

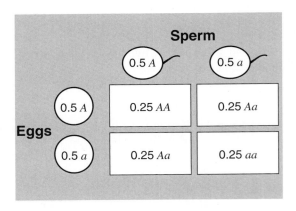

Figure 5.2 Random union of eggs and sperm yields the same outcome as the random mating of parents (refer to table 5.2).

the system of random mating is the expanded binomial $(0.5 + 0.5)^2$.

We may consider another arbitrary population in the equilibrium state. Suppose that the population consists of 16 *AA*, 48 *Aa*, and 36 *aa* individuals. We may assume, as before, that 10 gametes are contributed by each individual to the next generation. All the gametes transmitted by the 16 *AA* parents (numerically, 160) will contain the *A* allele, and half the gametes (240) provided by the *Aa* parents will bear the *A* allele. Thus, of the 1,000 total gametes in the population, 400 will carry the *A* allele. Accordingly, the frequency of allele *A* is 0.4 (designated *p*). In like manner, it can be shown that the frequency of allele *a* is 0.6 (*q*). Substituting the numerical values for *p* and *q* in the Hardy-Weinberg formula, we have:

p^2 *AA* : 2pq *Aa* : q^2aa
$(0.4)^2$ *AA* : 2(0.4) (0.6) *Aa* : $(0.6)^2$ *aa*
0.16 *AA* : 0.48 *Aa* : 0.36 *aa*.

Hence, the proportions of the three genotypes are the same as those of the preceding generation. Note that the frequency of the "A" allele (*p*) and the frequency of the "a" allele (*q*) are in a relationship such that *p + q* = 1.0. Likewise, the binomial expansion of this expression $(p + q)^2 = p^2 + 2pq + q^2 = 1$. This simply means that the total of all gene or genotypic frequencies equals 100%.

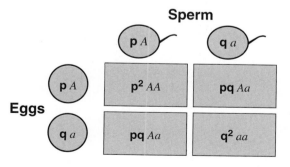

Figure 5.3 **The distribution of genotypes** in the next generation is p^2 *AA* : 2pq : *Aa* : q^2aa (Hardy-Weinberg formula).

It should also be clear that a recessive trait, such as blue eyes in humans, will not become rare just because it is governed by a recessive gene. Nor can the dominant brown-eyed condition become widespread simply by virtue of its dominance. Whether a given gene is common or rare is controlled by other factors, particularly natural selection.

ESTIMATING THE FREQUENCY OF HETEROZYGOTES

It comes as a surprise to many readers to discover that the heterozygotes of a rare recessive abnormality are rather common instead of being comparatively rare. Recessive albinism may be used as an illustration. The frequency of albinos is about 1/20,000 in human populations. When the frequency of the homozygous recessive (q^2) is known, the frequency of the recessive allele (*q*) can be calculated, as follows:

q^2 = 1/20,000 = 0.00005
$q = \sqrt{0.00005}$ = 0.007
     = about 1/140
     (frequency of recessive allele).

The heterozygotes are represented by *2pq* in the Hardy-Weinberg formula. Accordingly, the frequency of heterozygous carriers of albinism can be calculated as follows:

$q = 0.007$
$p = 1 - 0.007 = 0.993.$
$\therefore 2pq = 2(0.993 \times 0.007) = 0.014$
     = about 1/70 (frequency of heterozygote).

Thus, although 1 person in 20,000 is an albino (recessive homozygote), about 1 person in 70 is a heterozygous carrier. There are 280 times as many carriers as affected individuals! It bears emphasizing that the rarity of a recessive disorder does not signify a comparable rarity of heterozygous carriers. In fact, when the frequency of the recessive gene is extremely low, nearly all the recessive genes are in the heterozygous state.

IMPLICATIONS

The Hardy-Weinberg theorem is *entirely theoretical.* The set of underlying assumptions can scarcely be fulfilled in any natural population. We implicitly assume the absence of recurring mutations (no mutation), the absence of any degree of preferential matings (random mating), the absence of differential mortality or fertility (no selection), the absence of immigration or emigration of individuals (no gene flow), and the absence of fluctuations in gene frequencies due to sheer chance (theoretically, an infinitely large population). But therein lies the significance of the Hardy-Weinberg theorem. *In revealing the conditions under which evolutionary change cannot occur, it brings to light the possible forces that could cause a change in the genetic composition of a population.* The Hardy-Weinberg theorem thus depicts a static situation which is not the state of nature.

An understanding of Hardy-Weinberg equilibrium provides a basis for recognizing the forces that permit evolutionary change. The more obvious factors that prevent a natural population from attaining stationary equilibrium are: (1) mutation, (2) non-random mating, (3) natural selection, (4) chance events in small populations (e.g., genetic drift) and (5) migration. These factors or forces may profoundly modify the gene frequencies in natural populations. In essence, the Hardy-Weinberg theorem represents the cornerstone of population genetic studies, since deviations from the Hardy-Weinberg expectations direct attention to the evolutionary forces that upset the theoretical expectations. Evolution can be plausibly defined as the disturbance of, or shift in, the Hardy-Weinberg equilibrium.

CONCEPT OF SELECTION

We have remarked that the multilegged anomaly which arose in a local bullfrog population in Mississippi in 1958 has not been detected since its initial occurrence. One of our suppositions was that the multilegged trait is governed by a recessive mutant gene. We might presume that the mutant gene responsible for the abnormality has disappeared entirely from the population. But can a detrimental mutant gene be completely eradicated from natural populations of organisms, even in the face of the severest form of selection? Most persons are frankly puzzled when they are informed that the answer is no. Yet our knowledge of the properties of mutation and selection expressly permits a firm negative reply.

SELECTION AGAINST RECESSIVE DEFECTS

In the preceding chapter, in our consideration of the Hardy-Weinberg equilibrium, we assumed that the mutant multilegged frogs *(aa)* were as reproductively fit as their normal kin *(AA* and *Aa)* and left equal numbers of living offspring each generation.

Now, however, let us presume that all multilegged individuals fail to reach sexual maturity generation after generation. Will the incidence of the multilegged trait decline to a vanishing point?

We may start with the same distribution of individuals in the initial generation as postulated in chapter 5, namely, 24 *AA,* 48 *Aa,* and 24 *aa,* with the sexes equally represented. Since the multilegged frogs *(aa)* are unable to participate in breeding, the parents of the next generation comprise only the 24 *AA* and 48 *Aa* individuals. The heterozygous types are twice as numerous as the homozygous dominants; accordingly, two-thirds of the total breeding members of the population are *Aa* and one-third are *AA.* We may once again employ a genetic checkerboard (table 6.1) to ascertain the different types of matings and their relative frequencies.

The frequencies of the different matings shown in table 6.1 may be expressed as whole numbers rather than fractions. Given a total of 36 matings, 4 (= 1/9 × 36) would be *AA* ♀ × *AA* ♂, 8 (= 2/9 × 36) would be *AA* ♀ × *Aa* ♂, and 16 (= 4/9 × 36) would be *Aa* ♀ × *Aa* ♂. These numbers are recorded in table 6.2.

Our next task is to determine the outcome of each type of cross. We shall assume that each mated pair contributes an equal number of progeny to the next generation (say, four offspring). As revealed in table 6.2, the offspring are distributed according to Mendelian ratios, and the actual numbers of offspring reflect the frequencies of the different kinds of matings. For example, a single AA ♀ $\times Aa$ ♂ mating yields 4 offspring in the Mendelian ratio of 2 AA : 2 Aa. There are, however, eight matings of this kind; the numbers of offspring are correspondingly increased to 16 AA and 16 Aa.

Even though all the multilegged frogs fail to reproduce, the detrimental recessive genes are still transmitted to the first generation. The emergence of multilegged frogs in the first generation stems from the matings of two heterozygous frogs. However, as seen from table 6.2, the frequency of the multilegged trait (aa) decreases from 25 percent to 11.11 percent in a single generation.

The effects of complete selection against the multilegged frogs in subsequent generations can be determined by the foregoing method of calculation, but the lengthy tabulations can be wearisome. At this point we may apply a formula that will establish in a few steps the frequency of the recessive allele after any number of generations of complete selection:

$$q_n = \frac{q_o}{1 + nq_o} .$$

In the above expression, q_o represents the initial or original frequency of the recessive allele, and q_n is the frequency after n generations. Thus, with the initial value $q_o = 0.5$, the frequency of the recessive allele after two generations ($n = 2$) will be:

$$q_2 = \frac{q_o}{1 + 2q_o} = \frac{0.5}{1 + 2(0.5)} = \frac{0.5}{2.0} = 0.25.$$

If the frequency of the recessive allele (a) is q, then the frequency of the recessive individual (aa) is q^2. Accordingly, the frequency of the recessive homozygote is $(0.25)^2$, or 0.0625 (6.25 percent). In the second generation, therefore, the incidence of the multilegged trait drops to 6.25 percent.

If we perform comparable calculations through several generations, we emerge with a comprehensive picture that is tabulated in table 6.3 and portrayed in figure 6.1. In the third generation, the frequency of the recessive homozygote declines to 4.0 percent. Progress in terms of the elimination of the multilegged trait is initially rapid but becomes slower as selection is continued over many successive generations. About 20 generations are required to depress the incidence of the multilegged trait to 2 in 1,000 individuals (0.20 percent). Ten additional generations are necessary to effect a reduction to 1 in 1,000 individuals (0.10 percent). Thus, as a recessive trait becomes

TABLE 6.1	Matings and Relative Frequencies	

Female (♀)	Male (♂)	
	1/3 AA	2/3 Aa
1/3 AA	1/9 $AA \times AA$	2/9 $AA \times Aa$
2/3 Aa	2/9 $Aa \times AA$	4/9 $Aa \times Aa$

TABLE 6.2	First Generation of Offspring After Complete Selection			

Type of Mating (Female × Male)	Number of Each Type of Mating*	Number of Offspring		
		AA	Aa	aa
$AA \times AA$	4	16		
$AA \times Aa$	8	16	16	
$Aa \times AA$	8	16	16	
$Aa \times Aa$	16	16	32	16
		64	64	16
		(44.44%)	(44.44%)	(11.11%)

* Based on a total of 36 matings.

TABLE 6.3	Effects of Complete Selection Against a Recessive Trait

Generations	Recessive Allele Frequency	Recessive Homozygotes %	Heterozygotes %	Dominant Heterozygotes %
0	0.500	25.00	50.00	25.00
1	0.333	11.11	44.44	44.44
2	0.250	6.25	37.50	56.25
3	0.200	4.00	32.00	64.00
4	0.167	2.78	27.78	69.44
8	0.100	1.00	18.00	81.00
10	0.083	0.69	15.28	84.03
20	0.045	0.20	8.68	91.12
30	0.031	0.10	6.05	93.85
40	0.024	0.06	4.64	95.30
50	0.020	0.04	3.77	96.19
100	0.010	0.01	1.94	98.05

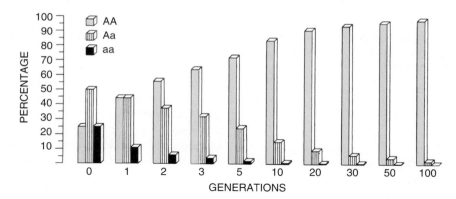

Figure 6.1   Effects of complete selection against recessive homozygotes *(aa)* occurring initially ("0" generation) at a frequency of 25 percent. The effectiveness of selection in reducing the incidence of the recessive trait decreases with successive generations. The frequency of recessive homozygotes drops markedly from 25 percent to 6.25 percent in two generations. However, 8 generations are required to reduce the incidence of the recessive trait to 1.0 percent, 30 generations are needed to achieve a reduction to 0.1 percent, and approximately 100 generations to depress the frequency to 0.01 percent.

rarer, selection against it becomes less effective. The reason is quite simple: very few recessive homozygotes are exposed to the action of selection. The now rare recessive allele *(a)* is carried mainly by heterozygous individuals *(Aa),* where it is sheltered from selection by its normal dominant partner *(A).* By contrast, rare lethal dominant alleles are eliminated in a single generation (fig. 6.2) and their subsequent frequency is a function of their rate of spontaneous mutation.

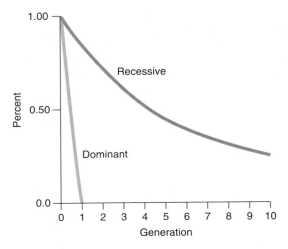

Figure 6.2 Total selection against rare lethal alleles. This graph shows the rate at which rare lethal alleles are eliminated by natural selection. Lethal dominant alleles are immediately exposed to natural selection and eliminated within a single generation after they appear. Rare lethal recessive alleles are reduced slowly, protected from selection by their dominant partner. In both instances, the graph assumes an initial allelic frequency of 1%, complete (100%) selection, and no new mutations.

SIGNIFICANCE OF THE HETEROZYGOTE

When the frequency of a detrimental recessive gene becomes very low, most affected offspring *(aa)* will come from matings of two heterozygous carriers *(Aa)*. For example, in the human population, the vast majority of newly arising albino individuals *(aa)* in a given generation (more than 99 percent of them) will come from normally pigmented heterozygous parents. Considerations of this kind led us to postulate in chapter 1 that the multilegged frogs that appeared suddenly in the natural population were derived from normal-legged heterozygous parents (refer to fig. 1.5).

Detrimental recessive genes in a population are unquestionably harbored mostly in the heterozygous state. As shown in table 6.4, the frequency of heterozygous carriers is many times greater than the frequency of homozygous individuals afflicted with a trait. Thus, an extremely rare disorder like alkaptonuria (blackening of urine) occurs in 1 in one million persons. This detrimental gene, however, is carried in the hidden state by 1 out of 500 persons. There are 2,000 times as many genetic carriers of alkaptonuria as there are individuals affected with this defect. For another recessive trait, cystic fibrosis, 1 out of 2,500 individuals is affected with this homozygous trait. One of 25 persons is a carrier of cystic fibrosis. In modern genetic counseling programs, an important consideration has been the development of simple, inexpensive means of detecting heterozygous carriers of inherited disorders.

INTERPLAY OF MUTATION AND SELECTION

Ideally, if the process of complete selection against the recessive homozygote were to continue for several hundred more generations, the detrimental recessive gene would be completely eliminated and the population would consist uniformly of normal homozygotes *(AA)*. But, *in reality,* the steadily diminishing supply of deleterious recessive genes is continually being replenished by recurrent mutations from normal *(A)* to abnormal *(a)*. Mutations from *A* to *a,* which inevitably occur from time to time, were not taken into account in our determinations. Mutations, of course, cannot be ignored.

All genes undergo mutations at some definable rate. If a certain proportion of *A* alleles are converted into *a* alleles in each generation, the population will at all times carry a certain amount of the recessive mutant gene *(a)* despite selection against it. Without any sophisticated calculations, it can be shown that a point will be reached at which the number of the variant recessive alleles eliminated by selection just balances the number of the same variant recessive alleles produced by mutation. An analogy shown in figure 6.3*A* will help in visualizing this circumstance. The

TABLE 6.4	Frequencies of Recessive Homozygotes and Heterozygous Carriers

Frequency of Homozygotes (aa)	Frequency of Heterozygous Carriers (Aa)	Ratio of Carriers to Homozygotes
1 in 500 (Sickle-Cell Anemia)[a]	1 in 10	50:1
1 in 2,500 (Cystic Fibrosis)	1 in 25	100:1
1 in 6,000 (Tay-Sachs Disease)[b]	1 in 40	150:1
1 in 20,000 (Albinism)	1 in 70	286:1
1 in 25,000 (Phenylketonuria)	1 in 80	313:1
1 in 50,000 (Acatalasia)[c]	1 in 110	455:1
1 in 1,000,000 (Alkaptonuria)	1 in 500	2,000:1

[a]Based on incidence among African Americans. Sickle-cell anemia is later described (chapter 8) as a co-dominant trait.
[b]In the United States, the disease occurs once in 6,000 Jewish births and once in 500,000 non-Jewish and Cajun American births.
[c]Based on prevalence rate among the Japanese.

water level in the beaker remains constant when the rate at which water enters the opening of the beaker equals the rate at which it leaves the hole in the side of the beaker. In other words, a state of equilibrium is reached when the rate at which the recessive gene is replenished by mutation equals the rate at which it is lost by selection. It should be clear that it is not mutation alone that governs the incidence of deleterious recessives in a population. The generally low frequency of harmful recessive genes stems from the dual action of mutation and selection. The mutation process tends to increase the number of detrimental recessives; the selection mechanism is the counteracting agent.

What would be the consequences of an increase in the mutation rate? Humans today live in an environment in which high-energy radiation

and copious chemical compounds promote a higher incidence of mutations. We may return to our analogy (fig 6.3B). The increased rate of mutation may be envisioned as an increased input of water. The water level in the beaker will rise and water will escape more rapidly through the hole in the side of the beaker. Similarly, mutant genes will be found more frequently in a population, and they will be eliminated at a faster rate from the population. As before, a balance will be restored eventually between mutation and selection, but now the population has a larger store of deleterious genes and a larger number of afflicted individuals arising each generation.

The supply of defective genes in the human population has already increased through the greater medical control of recessive disorders. The outstanding advances in modern medicine have served to prolong the lives of individuals who might otherwise not have survived to reproductive age. This may be compared to partially plugging the hole in the side of the beaker (fig. 6.3C). The water level in the beaker will obviously rise, as will the amount of deleterious genes in a population.

PARTIAL SELECTION

We have treated above the severest form of selection against recessive individuals. Complete, or 100 percent, selection against a recessive homozygote is often termed *lethal* selection, and the mutant gene is designated as a *lethal* gene. A lethal gene does not necessarily result in the death of the individual but does effectively prevent that individual from reproducing or leaving offspring. Not all mutant genes are lethal; in fact, the majority of them have less drastic effects on viability or fertility. A mildly handicapped recessive homozygote may reproduce but may be inferior in fertility to the normal individual. When the reproductive capacity of the recessive homozygote is only half as great as the normal type, he or she is said to be *semi-sterile,* and the mutant gene is classified as *semi-lethal.* A *subvital*

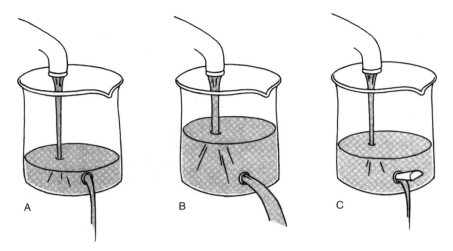

A   B   C

Figure 6.3 Interplay of detrimental mutant genes (water from faucet) and their elimination by selection (water escaping through hole) in a population (beaker) containing a pool of the harmful genes (water in beaker). *A.* State of genetic equilibrium (constant water level in beaker) is reached when the rates at which water enters and leaves the beaker are equal. *B.* The effect of an increase in the mutation rate (increased flow of faucet water) as might be expected from the continued widespread use of ionizing radiation. A new equilibrium (new constant water level) is established, but the frequency of the detrimental gene in the population is higher (higher water level in beaker). *C.* The effect of reducing selection pressure (decreased exit of water) as a consequence of improving the reproductive fitness of genetically defective individuals by modern medical practices. The mutation rate (inflow of water) is the same as in *A.* The inevitable result is a greater incidence of the harmful mutant genes in the population (higher level of water in the beaker).

recessive gene is one that, in double dose, impairs an individual to the extent that his or her reproductive fitness is less than 100 percent but more than 50 percent of normal proficiency.

The action of selection varies correspondingly with the degree of detrimental effect of the recessive gene. Figure 6.4 shows the results of different intensities of selection in a population that initially contains 1.0 percent recessive homozygotes. With complete (lethal) selection, a reduction in the incidence of the recessive trait from 1.0 percent to 0.25 percent is accomplished in 10 generations. Twenty generations of complete selection reduces the incidence to 0.11 percent. When the recessive gene is semilethal (50 percent selection), 20 generations, or twice as many generations as under complete selection, are required to depress the frequency of the recessive homozygote to about

0.25 percent. Selection against a subvital gene (for example, 10 percent selection) results in a considerably slower rate of elimination of the recessive homozygotes. When the homozygote is at a very slight reproductive disadvantage (1.0 percent selection), only a small decline of 0.03 percent (from 1.0 percent to 0.97 percent) occurs after 20 generations. It is evident that mildly harmful recessive genes may remain in a population for a long time.

These considerations are shown in table 6.5 also. Here, the results are expressed in terms of the value *s*, or the *selection coefficient*. The selection coefficient is a measure of the contribution of one genotype relative to the contributions of the other genotypes. Thus, an *s* value of 1 means that an individual leaves no offspring; the recessive homozygote is lethal. An *s* value of 0.10 signifies that the *aa* homozygote contributes only

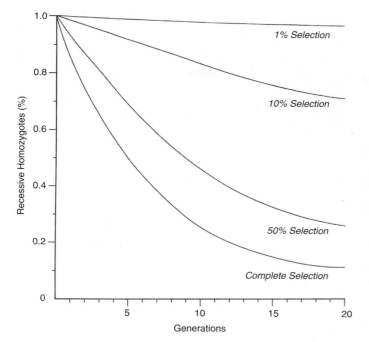

Figure 6.4 Different intensities of selection against recessive homozygotes occurring initially ("0" generation) at a frequency of 1.0 percent. The elimination of recessive individuals per generation proceeds at a slower pace as the strength of selection decreases.

90 offspring to the next generation as compared with *AA* and *Aa* individuals, each of whom contributes 100 offspring. It may be noticed, once again, that the rate of decline of the recessive homozygotes becomes slower as the selection coefficient decreases in value.

TABLE 6.5	Declines in Frequencies of Recessive Homozygotes Based on Different Values of the Selection Coefficient(s)

Generations	s = 1.0	s = 0.50	s = 0.10	s = 0.01
0	1.00*	1.00	1.00	1.00
10	0.25	0.46	0.84	0.98
20	0.11	0.26	0.71	0.97

* The initial frequency of the recessive homozygote in all cases is 1 percent, as in figure 6.4.

SELECTION AGAINST DOMINANT DEFECTS

If complete selection acts against an abnormal trait caused by a dominant allele *(A)* so that none of the *AA* or *Aa* individuals leave any progeny, than all the *A* genes are at once eliminated (fig. 6.2). In the absence of recurrent mutation, all subsequent generations will consist exclusively of homozygous recessive *(aa)* individuals.

However, we must contend again with ever-occurring mutations and the effects of partial selection. The late geneticist Curt Stern provides us with a simple, clear model of this situation. Imagine a population of 500,000 individuals, all of whom are initially homozygous recessive *(aa)*. Thus, no detrimental dominant genes *(A)* are present and the population as a whole contains 1,000,000 recessive genes *(a)*. In the first generation, 10 dominant mutant genes arise as a result of the recessive gene's mutating to the dominant state at a rate of 1 in 100,000 genes. We shall now assume that the dominant mutant gene is semilethal; in other words, only 5 of the newly arisen dominant genes are transmitted to the next, or second, generation. For ease of discussion, this is pictorially shown in figure 6.5 and represented also in table 6.6. It can be seen that the second generation would contain a total of 15 dominant genes—5 brought forth from the first generation and 10 new ones added by mutation. In the third generation, the 5 dominant genes carried over from the first generation would be reduced to 2.5, the 10 dominant alleles of the second generation would be depressed to 5, and 10 new abnormal alleles would arise anew by mutation. The total number of dominant genes would increase slightly with each subsequent generation, until a point is reached (about 12 generations) where the rate

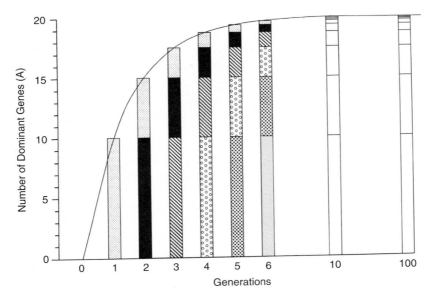

Figure 6.5 Establisment of a constant level of a semilethal dominant gene *(A)* in a population over the course of several generations. The fixed number of new dominant genes introduced each generation through mutations from *a* to *A* eventually exactly balances the number of dominant genes selectively eliminated each generation. In this particular case, an equilibrium is reached (after about 12 generations) when the total number of dominant genes is approximately 20.

TABLE 6.6	**Equilibrium Frequency of a Dominant Gene** **Conditions: 1. Size of Population: 500,000 Individuals** **2. Mutation Rate (*a ⟶ A*): 1 in 100,000** **3. Selection Coefficient (s) : 0.5 (semilethal)**

		Dominant Gene *A*		
Generation	**Recessive Gene *a***	**Left Over from Former Generations**	**Newly Mutated**	**Total**
0	1,000,000	—	—	—
1	1,000,000*	—	10	10
2	1,000,000	5	10	15
3	1,000,000	5 + 2.5	10	17.5
4	1,000,000	5 + 2.5 + 1.25	10	18.75
5	1,000,000	5 + 2.5 + 1.25 + 0.625	10	19.375
∞	1,000,000	5 + 2.5 + 1.25 + 0.625 + 0.3125 +⋯	10	20.00

*The total number of recessive genes should be reduced by the total number of dominant genes each generation, but this minor correction would be inconsequential.

of elimination of the abnormal dominant gene balances the rate of mutation. In other words, the inflow of new dominant alleles by mutation in each generation is balanced by the outflow or elimination of the dominant genes in each generation by selection.

The equilibrium frequency of a detrimental dominant gene in a population can be altered by changing the rate of loss of the gene in question. In humans, *retinoblastoma,* or cancer of the eye in newborn babies, has until recently been a fatal condition caused by a dominant mutant gene. With modern medical treatment, approximately 95 percent of the afflicted infants can be saved. The effect of increasing the reproductive fitness of the survivors is to raise the frequency of the detrimental dominant gene in the human population. The accumulation of deleterious genes in the human gene pool has been a matter of awareness and concern.

CONCEALED VARIABILITY IN NATURAL POPULATIONS

From what we have already learned, we should expect to find in natural populations a large number of deleterious recessive genes concealed in the heterozygous state. It may seem that this expectation is based more on theoretical deduction than on actual demonstration. This is not entirely the case. Penetrating studies by a number of investigators of several species of the fruit fly *Drosophila* have unmistakably indicated an enormous store of recessive mutant genes harbored by individuals in nature. We may take as an illustrative example the kinds and incidence of recessive genes detected in *Drosophila pseudobscura* from California populations. The following data are derived from the studies of the late geneticist Theodosius Dobzhansky.

Flies were collected from nature, and a series of elaborate crosses were performed in the laboratory to yield offspring in which one pair of chromosomes carried an identical set of genes. The formerly hidden recessive genes in a given pair of chromosomes were thus all exposed in the homozygous state. All kinds of recessive genes were uncovered in different chromosomes, as exemplified by those unmasked in one particular chromosome, known simply as "the second." About 33 percent of the second chromosomes harbored one or more recessive genes that proved to be lethal or semilethal to flies carrying the second chromosome in duplicate. An astonishing number of second chromosomes—93 percent—contained genes that produced subvital or mildly incapacitating effects when present in the homozygous state. Other unmasked recessive genes resulted in sterility of the flies or severely retarded the developmental rates of the flies. All these flies were normal in appearance when originally taken from nature. It is apparent that very few, if any, outwardly normal flies in natural populations are free of hidden detrimental recessive genes.

GENETIC LOAD IN HUMAN POPULATIONS

The study of the concealed variability, or *genetic load,* in humans cannot be approached, for obvious reasons, by the experimental breeding techniques used with fruit flies. Estimates of the genetic load in the human population have been based principally on the incidence of defective offspring from matings of close relatives (*consanguineous* matings). It can be safely stated that every human individual contains at least one newly mutated gene. It can also be accepted that any crop of gametes contains, in addition to one or more mutations of recent origin, at least 10 mutant genes that arose in the individuals of preceding generations and that have accumulated in the population. The average person is said to harbor four concealed lethal genes, each of which, if homozygous, is capable of causing death between birth and maturity. The most conservative estimates place the incidence of deformities to detrimental mutant genes in the vicinity of 2 per 1,000 births. *It is evident that humans are not exempt from their share of detrimental genes.*

GENERAL EFFECTS OF SELECTION

In our discussion thus far, we have seen that the selection process favors individuals who are best adapted to new situations or to new ecological opportunities. Such selection is said to be *directional* because the norm for the population is shifted with time in one direction (fig. 6.6A). The selection for melanic varieties of moths in industrial areas (see chapter 7) exemplifies directional

selection. The curve of distribution shifts to the right (fig. 6.6A) as the darker varieties supplant the formerly abundant light-colored moths.

The selection process may take other forms. *Stabilizing selection* acts to reduce the array of gene complexes that can be expressed in a population. As seen in figure 6.6B, the shape of the curve tends to narrow through the continual elimination of the less-adapted individuals at the extremes of the distribution curve. Stated differently, the

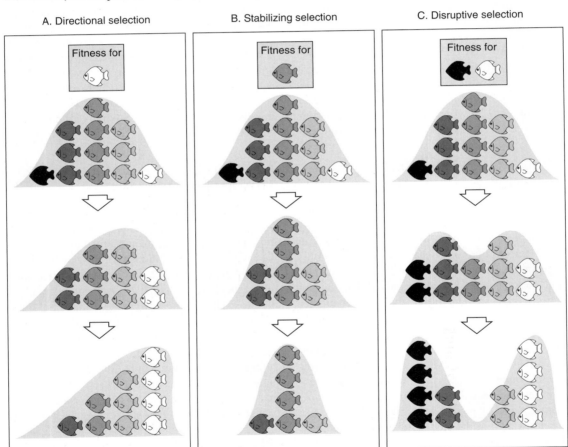

Figure 6.6 Schematic representation of three types of selection and their effects. Each curve represents the normal distribution of the trait in a population. From top to bottom, the lower curves show the expected distribution after the impact of selection. *(A) Directional selection.* The adaptive norm changes as less adapted genotypes are replaced by better adapted genotypes. *(B) Stabilizing selection.* The intermediate values for a given trait are favored (preserved) and the shape of the curve narrows through the elimination of the carriers of the extreme values. *(C) Disruptive selection.* Two adaptive norms are generated when the population exists in a heterogeneous environment.

intermediate values for a given trait are favored over the extreme values. The birth weight of newborns provides an instructive example of a human characteristic that has been subjected to stabilizing selection (fig. 6.7). The optimum birth weight is 7.3 pounds; newborn infants less than 5.5 pounds and greater than 10 pounds have the highest probability of mortality. Given the strong stabilizing influence of weeding out the extremes, the optimum birth weight is associated with the lowest mortality. The curve for mortality is virtually the complement of the curve for survival.

Disruptive selection is the most unusual of the three types of selection (fig. 6.6*C*). This form of selection occurs when the extreme values have the highest fitness and the intermediate values are relatively disadvantageous in terms of reproductive effectiveness. It is, essentially, selection for diversification with respect to a trait. The shell patterns of limpets (marine molluscs) form a continuum ranging from pure white to dark tan. Limpets typically dwell in one of two distinct habitats, attaching to either white gooseneck barnacles or tan-colored rocks. As might be expected, the light-colored limpets seek the protection of the white barnacles, whereas the tan limpets live by choice almost exclusively on the dark rocks. Limpets of intermediate shell patterns are conspicuous and are intensely selected against by predatory shore birds. If this disruptive type of selection (favoring the extremes) were to be accompanied by the sexual isolation of the two types of limpets, two new species could arise. The mode of origin of species is treated in detail in a subsequent chapter.

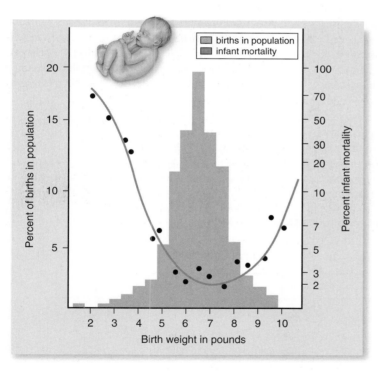

Figure 6.7 **The distribution of birth weights of human newborns** and the mortality of the various birth-weight classes. The histogram shows the proportions of the population falling into the various birth-weight classes. The mean birth weight is 7.1 pounds. The curve of mortality in relation to birth weight reveals that the lowest mortality is associated with the optimum birth weight (7.3 pounds). Recent medical advances have reduced mortality rates for small and larger babies.
(Based on data by Karn and Penrose, 1951.)

NATURAL SELECTION AND PREGNANCY LOSS

We remarked earlier (chapter 4) that there is a large natural loss of human embryos in pregnancy. The important consideration now is the high efficiency with which nature eliminates chromosomally abnormal embryos during the course of pregnancy. For every 1,000 chromosomal abnormalities that are present in embryos in the uterus, only 6 are expected to survive to the point of a live birth. Thus 99.4 percent of the chromosomal abnormalities are eliminated naturally through spontaneous abortion.

We may direct our attention to the 0.6 percent of newborn infants affected with a chromosomal abnormality. As seen in figure 6.8, some members of this affected group comprise the 45 XO (Turner Syndrome) and 47 XXY (Klinefelter Syndrome) conditions. These individuals are effectively eliminated from the reproductive pool by their sterility. In contrast, semisterile carriers of balanced translocation tend to be normal phenotypically but are at risk of having abnormal children. The risk, however, is reduced appreciably by the segregation of unbalanced chromosome complements in gametes that are likely to be inviable.

From the perspective of natural selection, the XXX females and the XYY males represent an instructive group. Both types of individuals are usually fertile, and their children are chromosomally normal (XX daughters and XY sons). Some mechanism operates during meiosis to eliminate the extra X chromosome from the egg, or the extra Y chromosome from the sperm. Thus, natural selection has modified the meiotic divisions to ensure that only normal haploid gametes are produced by XXX females and XXY males.

The salient feature of the foregoing considerations is the efficacy with which natural selection eliminates chromosomally abnormal conceptuses—largely through pregnancy loss and by sterility or special forms of meiosis that eliminate chromosome complements with extra chromosomes.

Nature has created a great barrier to the perpetuation of chromosomally abnormal offspring. Natural selection is not perfect, however. Some chromosomally abnormal fetuses escape nature's screening mechanism and survive to term.

SEXUAL SELECTION

Thus far we have focused on Darwin's concept of natural selection. In Darwin's seminal 1859 work, *The Origin of Species* the notion of *sexual selection* was first conceived. Later, in 1871, Darwin refined his ideas and devoted much of *The Descent of Man and Selection in Relation to Sex* to this subject. Darwin proposed sexual selection as the mechanism to explain the accentuation of apparently nonessential features to increase an individual's chance of reproducing. In *The Origin of Species,* Darwin described sexual selection as the "struggle between the individuals of one sex, generally the males, for the possession of the other sex". In *The Desent of Man,* Darwin further noted:

> The sexual struggle is of two kinds: in the one it is between the individuals of the same sex, generally the males, in order to drive away or kill their rivals, the females remaining passive; while in the other, the struggle is likewise between the

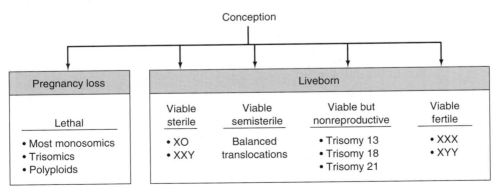

Figure 6.8 Natural selection operates in human populations to ensure the reduced survival and minimal reproductive potential of individuals with major chromosomal abnormalities.
Source: Data from Chandley, 1981.

individuals of the same sex, in order to excite or charm those of the opposite sex, generally the females, which no longer remain passive, but select the more agreeable partners.

In other words, for Darwin, sexual selection is a two-sided coin, male–male competition (also called *intrasexual selection*) resulting in horns, antlers, and other armaments for males to fight and compete for sexual access to passive females, and female choice (also called *intersexual selection*) in which characteristics such as a male bird's elaborate plumage attract the attention of observing females, resulting in the most attractive male suitor being selected as mate by the proactive female. Victorian England was not quite ready for the notion of female choice and largely ignored Darwin's offensive suggestions.

In intrasexual selection (male–male competition), selection favors those characteristics that make males better able to compete with one another for access to females. Females have no choice in the mating process; they are the sexual possession of whichever male wins the ritualized or actual fights between rival males. Male success is typically a function of overall size and the enlargement of the anatomical features used for fighting or display, such as teeth, antlers, and tusks. In intersexual selection (female choice), nature selects for traits that make males more attractive to the female, who then chooses which male or males have the most attractive traits. These sexy traits often translate into desirable genes that benefit the female and her offspring.

A female's control over reproduction may not end with copulation. One extreme form of female choice, called "cryptic female choice," is the notion that females of some species may be able to select which male sperm fertilizes her eggs. The concept of cryptic female choice posits that there are postcopulatory mechanisms by which females influence which of several sexual partners fathers her offspring, such as by choosing which size sperm or which sperm packet she uses to fertilize her eggs. While highly controversial, cryptic female choice has been described in a variety of species and may be a common aspect of reproduction.

Sexual selection is about Darwinian *fitness* in the strictest possible way. The measure of Darwinian fitness is the production of fertile offspring who themselves reproduce. Survival is a necessary prerequisite for mating, but in sexual selection, who actually mates is paramount. Sexual selection differs from natural selection in that it relates to selection occurring within a species *(intraspecific selection)* rather than selection occurring between species *(interspecific selection)*. Thus, while natural selection typically involves some type of interaction (such as competition for resources) between different species or between a species and the forces of nature, in sexual selection the interactions are framed completely within one species and revolve around reproduction rather than survival.

Sometimes sexual selection appears to be in opposition to natural selection. Darwin was keenly aware of this phenomenon and used the peacock as a prime example. In 1860 Darwin wrote his American friend, botanist Asa Gray, saying, "The sight of a feather in a peacock's tail, whenever I gaze at it, makes me sick." Darwin wondered whether the expenditure of effort and energy necessary to grow and maintain the most elaborate plumage to attract admiring pea hens might endanger the adorned males and thus ultimately make possessing these qualities disadvantageous to general fitness.

The contentious *handicap principle,* first articulated by Israeli zoologist Amotz Zahavi in the 1970s and since then by others, has been evoked to explain this apparent paradox. The basic tenet of the handicap principle is that sending false information is disadvantageous (i.e., that it is in the best interest of all concerned that information communicated between organisms is honest). According to Zahavi,

> An individual with a well developed sexually selected character [such as a peacock's flashy tail] is an individual which has survived a test. . . . Females which selected males with the most developed characters can be sure that they have selected from among the best genotypes of the male population.

We can reconcile elaborate sexual displays that put its owner at risk if those traits are in fact reliable indicators of fitness. In the case of the peacock, by maintaining such an elaborate display the male peacock demonstrates that he must be relatively free of infection, poor nutrition, parasites, and truly possess other real-life indictors of fitness. The handicap hypothesis argues that sexual selection is simply another form of natural selection in which sexiness equates with not only reproductive prowess, but to genes that are truly best adapted to the environment. An alternative resolution to the peacock dilemma is that reproductive success trumps survival if the end result is leaving more offspring, the defining principle of Darwinian fitness. In this way "survival of the fittest" does not necessarily select for the biggest, strongest, or even the most intelligent, but for the sexiest, even if that phenotype lasts only long enough to win the war of reproduction and pass more genes to the following generation.

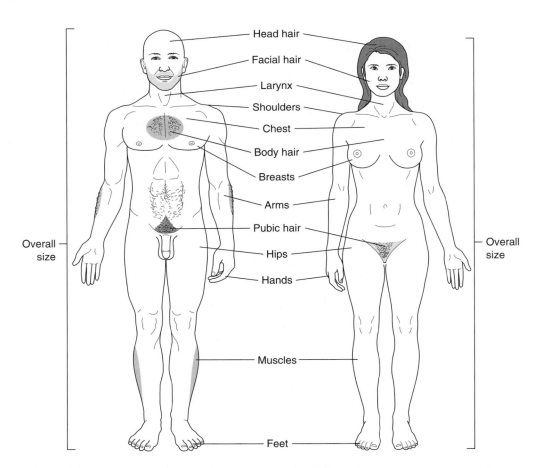

Figure 6.9 Generalized human secondary sex characteristics: Although there is a great deal of variation in secondary sex characteristics between the sexes and among different populations, there are many generalizable differences associated with mature adult males versus mature adult females. Table 6.7 lists some of the most common variations.

Phenotypic variation between the sexes of the same species is known as *sexual dimorphism,* from the Greek (di = two; morphism = form or morphology). Secondary sex characteristics evolved to provide reproductive advantages to their possessors by making them more attractive or better able to win competition between rivals for mates. Most sexually dimorphic traits arise or are associated with sexual maturation and are called *secondary sexual characteristics* (as compared to primary sexual characteristics, such as ovaries and testes). The appearance of secondary sexual characteristics is determined by sex hormones. In many species, especially long-lived species such as humans and turtles, it takes many years or even decades before secondary sex characteristics are apparent. Secondary sex characters are advertisements of sexuality and sexiness and hence are pivotal to mating and reproductive success. There are many well-known examples of secondary sex characteristics, including the ornamental feathers in birds, the presence or absence of horns, tusks, and antlers, the lion's mane, variation in body size (smaller vs. larger), different coloration, and so on. In humans, sexual dimorphism is initiated by the hormonal changes that coincide with puberty. Although human secondary sex characteristics show tremendous normal variation between individuals and across populations, there are many generalizable differences between mature adult males and mature adult females, such as modest body size differences, variation in breast development, voice tone, larynx growth, hormonal dominance (androgenic vs. estrogenic), distribution and amount of hair, distribution of sweat glands, fat distribution, and so on, as illustrated in figure 6.9 and listed in table 6.7.

TABLE 6.7	Human Secondary Sex Characteristics	
Characteristic	**Males**	**Females**
Head Hair	More prone to balding with age	Less prone to balding with age
Facial Hair	Robust, prominent throughout life	Faint, increases with age
Overall Size	Taller and heavier	Smaller and lighter
Larynx (Adam's apple)	Larger	Smaller
Shoulder	Broader, squarer	Rounded, sloping
Chest	Larger in all dimensions	Smaller in all dimensions
Body Hair	Prominent, especially chest, abdomen and arms	Sparse
Breasts	Unenlarged breasts	Enlarged breasts and nipples
Muscles	Larger, more obvious	Smaller, largely hidden
Arms	Longer, thicker	Shorter, thinner
Pubic Hair	Triangle-shaped	V-shaped
Hips	Narrower	Wider, rounded
Hands & Feet	Larger, stouter	Smaller, narrower
Hormones	Androgenic	Estrogenic
Voice Change at Puberty	Pronounced change, deeper	None
Fat Deposits	Mainly abdomen and waist	Buttocks, thighs, hips
Body Shape	Apple	Pear
Skin Texture	Coarser	Smoother

7

SELECTION IN ACTION

We have seen that many generations of persistent selection are required to drastically reduce the frequency of an unfavorable mutant gene in a population. Likewise, it generally takes an inordinately long period of time for a new favorable mutant gene to replace its allele throughout a large population. Yet, we have encountered situations in nature in which a favorable mutation has spread through a population in a comparatively short span of years. We shall look at some outstanding examples in which we have actually observed evolution in progress.

INDUSTRIAL MELANISM

One of the most spectacular evolutionary changes witnessed by us has been the emergence and predominance in modern times of dark, or *melanic,* varieties of moths in the industrial areas of England and continental Europe. Slightly more than a century ago dark-colored moths were exceptional. The typical moth in the early 1800s had a light color pattern, which blended with the light coloration of tree trunks on which the moths alighted. But then the industrial revolution intervened to alter materially the character of the countryside. As soot and other industrial wastes poured over rural areas, the vegetation became increasingly coated and darkened by black smoke particles. In areas heavily contaminated with soot, the formerly abundant light-colored moths were gradually supplanted by the darker varieties. This dramatic change in the coloration of moths has been termed *industrial melanism.* At least 70 species of moths in England have been so affected by the human disturbance of the environment.

Beginning in the late 1930s and 1940s several scientists, particulary E. B. Ford and H. B. D. Kettlewell at the University of Oxford, analyzed the phenomenon of industrial melanism. Kettlewell photographed the light and dark forms of the peppered moth, *Biston betularia,* against two different backgrounds (fig. 7.1). The light variety is concealed and the dark form is clearly visible when the moths rest on a light lichen-coated trunk of an oak tree in an unpolluted rural district. Against a sooty black oak trunk, the light form is conspicuous and the dark form is well camouflaged. Records of the

Figure 7.1 Dark and light forms of the peppered moth *(Biston betularia)* clinging to a soot-blackened oak tree in Birmingham, England *(right)* and to a light, lichen-coated oak tree in an unpolluted region *(left).* Arrow points to a barely visible light peppered moth.

Experimental breeding tests have demonstrated that the two varieties differ principally by a single gene, with the dark variant dominant to the light one. The dominant mutant allele was initially disadvantageous. However, as an indirect consequence of industrialization, the mutant allele became favored by natural selection and spread rapidly in populations in a comparatively short period of time. In unpolluted or nonindustrial areas in western England and northern Scotland, the dominant mutant allele does not confer an advantage on its bearers and the light recessive moth remains the prevalent type.

One of the many impressive features of Kettlewell's studies lies in the unequivocal identification of the selecting agent (bird predation). Selection, we may recall, has been defined as differential reproduction. The act of selection in itself does not reveal the factors or agencies that enable one genotype to leave more offspring than another. We may demonstrate the existence of selection, yet remain baffled as to the precise causative agent of selection. We might have reasonably suspected that predatory birds were directly responsible for the differential success of the melanic forms in survival and reproduction, but Kettlewell's laboriously accumulated data provided that all-important, often elusive ingredient: *verifiable evidence.*

If the environment of the peppered moth were to become altered again, natural selection would be expected to favor the light variety again. In the 1950s, the British Parliament passed the Clean Air Act, which decreed, among other things, that factories must switch from soft high-sulfur (sooty) coal to less smoky fuels. The enforcement of this enlightened smoke-abatement law has led to a marked reduction in the amount of soot in the atmosphere. In the 1970s, University of Manchester biologist L. M. Cook and his colleagues reported a small but significant increase in the frequency of

dark form of the peppered moth date back to 1848, when its occurrence was reported at Manchester in England. At that time, the dark form comprised less than 1 percent of the population. By 1898, only 50 years later, the dark form had come to dominate the Manchester locale, having attained a remarkably high frequency of occurrence estimated at 95 percent. In fact, the incidence of the melanic type has reached 90 percent or more in most British industrial areas.

The rapid spread of the dark variety of moth is certainly understandable. The dark variants are protectively colored in the smoke-polluted industrial regions. They more easily escape detection by predators, namely, insect-eating birds. Actual films taken by Kettlewell and Niko Tinbergen revealed that birds prey on the moths in a selective manner. That is to say, in woodlands polluted by industry, predatory birds more often capture the conspicuous light-colored moths. In a single day, the numbers of light forms in an industrial area may be pared by as much as one-half by bird predation.

the light-colored peppered moth in the Manchester area. This is further substantiation of the action and efficacy of natural selection. In response to recent criticism of Kettlewell's work, University of Cambridge geneticist Michael E. N. Majerus replicated this classic experiment, concluding that "differential bird predation here is a major factor responsible for the decline of *carbonaria* (the dark colored moth) frequency in Cambridge between 2001 and 2007," thus vindicating this enduring example of natural selection in action.

CANCER CELL HETEROGENEITY

Until recently, there has been a tendency to focus almost exclusively on environmental factors or *carcinogens* in the search for the causes of cancer. It cannot be denied that environmental pollutants, such as hydrocarbons, are involved in the etiology of malignancy. But, there is also the realization that carcinogens damage DNA, and that the roots of cancer reside in the genes. Current information indicates that the development of malignancy is a multi-step process involving sequential changes at the DNA level. The process begins when a cell's DNA is altered by some form of mutation which disrupts the harmonious checks and balances that regulate normal cell growth.

Human tumors are *clonal*—that is, they are derived from a single progenitor cell. However, as the tumor grows, the constituent cells become extremely heterogeneous. The heterogeneity represents the emergence of subpopulations, or subclones, of variant cell types. These subclones differ in chromosome numbers, invasiveness, growth rate, responsiveness to blood-borne antibodies, and ability to *metastasize*, or invade other areas of the body (fig. 7.2).

The different cell types do not survive fortuitously. Rather, they have been selected for their favorable attributes in perpetuating the existence of the cancerous growth, much to the detriment of the host. Evidently, the process of selection is not confined only to populations of organisms—selection operates on the cellular level as well!

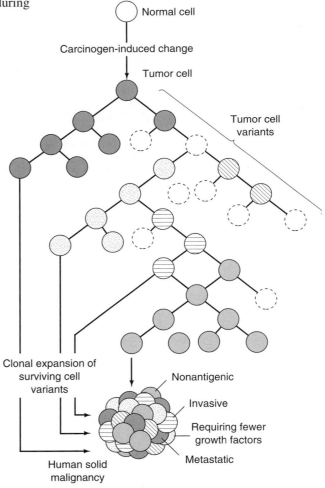

Figure 7.2 Tumor cell heterogeneity. New subpopulations arise continually from the descendants of the original malignant cell. Some subpopulations persist; those that survive and expand differ from each other in certain properties but all are adept at evading host defenses.

The subpopulations of cells in a solid tumor are continually changing. New variant subpopulations arise unceasingly, each population attempting to gain a growth advantage over the earlier population. The numbers of chromosomes become altered with time; in fact, chromosomal changes are almost a rule in cancer lineages. Most subpopulations are eliminated because of metabolic disadvantage or immunologic destruction. Those that prevail, or are selectively favored, have the greatest resistance to anticancer drugs.

The selection process is vividly revealed by the tumor's ability for angiogenesis and metastasis. As a solid tumor grows beyond the diameter of 2 millimeters, its nutritional requirements cannot be satisfied solely by the process of diffusion. Subclones of cell types within the tumor actually elicit the growth of new blood vessels from surrounding host tissue. The new growth of blood vessels is termed *angiogenesis*. The tumor cells themselves synthesize and release the growth factors that stimulate the formation of microvessels. Thus, the progression of cancer is dependent on the selection of potent cells that actually synthesize and secrete their own factors for the growth of blood vessels from pre-existing vessels. Similarly, the metastasis of a tumor to a distant site awaits the emergence of cell types that can direct the tumor to particular organs in the body. The organ chosen is usually one that provides a highly favorable environment for growth of the tumor.

SELECTION FOR RESISTANCE

Penicillin, sulfonamides (sulfa drugs), streptomycin, and other modern antibiotic agents made front-page headlines when first introduced. These wonder drugs were exceptionally effective against certain disease-producing bacteria and contributed immeasurably to the saving of human lives in World War II. However, the effectiveness of these drugs has been reduced by the emergence of resistant strains of bacteria. Medical authorities regard the rise of resistant bacteria as the most serious development in the field of infectious diseases over the past decade. Bacteria now pass on their resistance to antibiotics faster than people spread the infectious bacteria.

Mutations have occurred in bacterial populations that enable the mutant bacterial cells to survive in the presence of the drug. Here again we notice that mutations furnish the source of evolutionary changes and that the fate of the mutant gene is governed by selection. In an environment free of antibiotics, organisms that are resistant to antibiotic are rare and generally go undetected. In an environment changed by the addition of a drug, the drug-resistant mutants are favored and supplant the previously normal bacterial strains.

It might be thought that the mutations conferring resistance are actually caused or induced by the drug. This is not the case. Drug-resistant mutations arise randomly in bacterial cells, irrespective of the presence or absence of the drug. An experiment devised by the geneticist Joshua Lederberg provides evidence that the drug acts as a selecting agent, permitting preexisting mutations to express themselves. As seen in figure 7.3, colonies of bacteria were grown on a streptomycin-free agar medium in a petri plate. When the agar surface of this plate was pressed gently on a piece of sterile velvet, some cells from each bacterial colony clung to the fine fibers of the velvet. The imprinted velvet could now be used to transfer the bacterial colonies onto a second agar plate. More than one replica of the original bacterial growth can be made by pressing several agar plates on the same area of velvet. This ingenious technique has been appropriately called *replica plating*.

In preparing the replicas, Lederberg used agar plates containing streptomycin. On these agar plates, only bacterial colonies resistant to streptomycin grew. In the case depicted in figure 7.3, one colony was resistant. Significantly, this one resistant colony was found in exactly the same position in all replica plates. If mutations arose in response to exposure to a drug, it is hardly to be expected that mutant bacterial colonies would arise in precisely the same site on each occasion.

Spontaneous origin of mutation

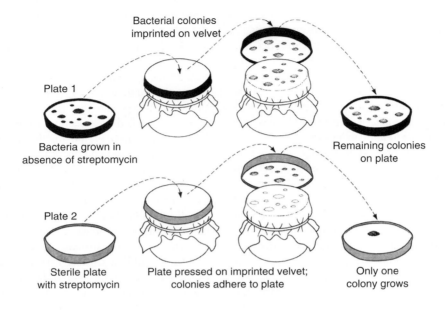

Bacterial colonies
imprinted on velvet

Plate 1

Bacteria grown in
absence of streptomycin

Remaining colonies
on plate

Plate 2

Sterile plate
with streptomycin

Plate pressed on imprinted velvet;
colonies adhere to plate

Only one
colony grows

Isolation and test of colonies

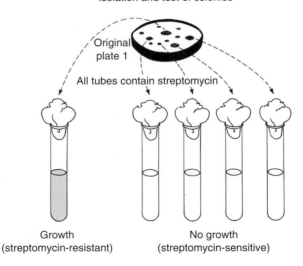

Original
plate 1

All tubes contain streptomycin

Growth
(streptomycin-resistant)

No growth
(streptomycin-sensitive)

Figure 7.3 Experiment by Joshua Lederberg revealing that drug-resistant mutations in bacterial cells had not been induced by the drug but had already been present prior to exposure of the bacteria to the drug.

In other words, a haphazard or random distribution of resistant bacterial colonies, without restraint or attention to location in the agar plate, would be expected if the mutations did not already exist in the original bacterial colonies.

Now, we can return to the original plate, as Lederberg did, and test samples of the original bacterial colonies in a test tube for sensitivity or resistance to streptomycin (bottom part of fig. 7.3). It is noteworthy that the bacterial colonies on the original plate had not been previously in contact with the drug. When these original colonies were isolated and tested for resistance to streptomycin, only one colony proved to be resistant. This one colony occupied a position on the original plate identical with the site of the resistant colony on the replica plates. The experiment demonstrated conclusively that the mutation had not been induced by streptomycin but was already present before exposure to the drug.

MULTIPLE ANTIBIOTIC RESISTANCE

The emergence of antibiotic-resistant bacteria appears deceptively simple. As described above, a spontaneous mutation conferring resistance permits the selective multiplication of mutant bacteria in the presence of a particular antibiotic. The mutation of a gene from sensitivity to resistance would be expected to occur about once in several million divisions, and cells containing the mutant gene would be expected to be resistant to only one particular antibiotic. It would therefore come as an unforeseen surprise to encounter bacteria that *instantaneously* become resistant to *several* antibiotics.

In 1955, a Japanese woman curtailed her visit to Hong Kong because of an unmanageable case of bacillary dysentery, an intestinal inflammation caused by bacteria of the genus *Shigella*. Upon culture, her intestinal *Shigellae* proved to be unusual. They were found to be resistant simultaneously to four antibiotics commonly used to control bacterial infections—namely, sulfanilamide, streptomycin, chloramphenicol, and tetracycline. Since the late 1950s, there have been innumerable outbreaks of intractable dysentery in Japan, all reflecting the rapid increase of multiply drug-resistant *Shigella dysenteriae*. Today, many more strains of bacteria are resistant to multiple antibiotics and their incidence is increasing at an alarming rate.

A bacterial cell contains a main, large circular strand of DNA and smaller circular pieces of DNA called *plasmids* (fig. 7.4). Plasmids do not have the same sequences of nucleic acid bases as the main ("chromosomal") DNA and replicate

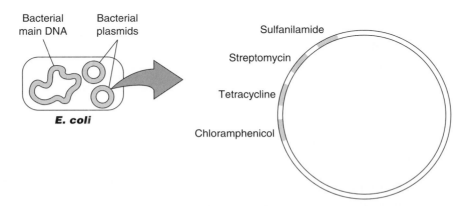

Figure 7.4 A simplified map of an R plasmid, showing four genetic determinants of resistance to four different antibiotics.

independently of the main DNA. A single bacterial cell can harbor as many as 25 plasmids, and each plasmid can carry from 4 to 250 genes, depending on the size of the ring. Plasmids may be viewed as accessory "chromosomes" serving varied functions. One type of plasmid, dubbed the *resistance (R) plasmid,* is clearly adapted to providing the bacterium with substantial resistance to antibiotics. As shown in figure 7.4, four resistance genes, each coding an enzyme that inactivates a specific antibiotic, are linked together on a single R plasmid.

The genetic determinants of an R factor plasmid can be transferred rapidly from one bacterium to another bacterium by cell-to-cell contact, in a process called *conjugation.* Ironically, R plasmids act like infectious agents as they expeditiously pass copies of themselves to other bacteria during conjugation. In a single conjugation event, several resistance genes can be instantly transferred. New bacterial strains evolved in a matter of months during epidemics of bacterial dysentery in Japan. In 1953, only 0.2 percent of the *Shigellae* isolated from Japanese patients were resistant to antibiotics. By 1965, the level had risen to 58 percent, with virtually all the pathogenic bacteria being resistant to the four aforementioned antibiotics.

Many of the patients who harbored drug-resistant *Shigellae* also quartered strains of the relatively harmless, common colon bacillus *Escherichia coli,* strains which were equally resistant to the four antibiotics. It is quite possible that the *Shigellae* obtained their R plasmids from *E. coli.* There is now experimental evidence that resistance factors can be transferred indiscriminately among different species—for example, *E. coli* can impart its R plasmids to *Shigella dysenteriae,* as well as to *Salmonella typhosa* (the causative agent of typhoid fever), *Klebsiella pneumoniae* (responsible for a virulent form of pneumonia), and *Pasturella pestus* (the bacterium involved in bubonic plague). The rise of pathogenic bacteria with multiple resistance to antibiotics has become a serious problem in human and veterinary medicine.

Remarkably, the resistance genes of a plasmid can become integrated into the bacterial "chromosome" (main DNA) and move from one plasmid to another. Resistance genes are found in highly mobile elements that molecular geneticists refer to as *transposable elements,* or *transposons.* As seen in figure 7.5, a transposon has an appealing architecture. The resistance gene (for tetracycline, in this example) is flanked on both sides by nucleotide sequences that are complementary to each other. These repeated sequences are called *insertion sequences,* since they are involved only in the insertion of a gene (resistance gene, in this case) into DNA molecules. The complementary sequences form the stem of a "lollipop" during the insertion of the gene, which resides in the head of the lollipop. Strange as it may seem, we identify the existence and position of a transposon by the lollipop configuration in a DNA preparation. Transposons can move readily between plasmids as well as integrate into bacterial or viral "chromosomes."

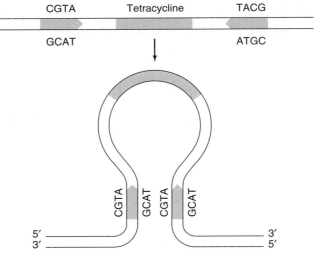

Figure 7.5 **The structure of a transposon** bearing a resistance gene for tetracycline and its conversion into a stem-loop configuration ("lollipop").

The transposable elements are not merely a curiosity of bacteria. Transposable elements have been discovered in yeast, the fruit fly, corn, and primates, including humans. Transposons are frequently spoken of as *jumping genes.* The pioneering studies on jumping genes in corn were performed in the 1940s by Barbara McClintock of Cold Spring Harbor Laboratory in New York. One cannot do justice to the analytical power of the innovative experiments conceived by her. It merits attention that on December 10, 1983, the Nobel Prize for Medicine or Physiology was awarded to Barbara McClintock for her discovery of unstable jumping genes. It took nearly 40 years for the genetic community to accept McClintock's heterodox genes, since the notion that genes were unstable both in their location and in their function was too disruptive to accept.

There are many mutiples of relatively short sequences of nucleotides in the human genome, which we may call simply *repetitive DNA.* The large number of repetitive nucleotide sequences in humans represents an awesome potential for the jumping of genes from one segment of the DNA molecule to another. The short repeats can promote their own movement from one location in the DNA molecule to another. These jumping repeats tend to inactivate genes in their new location or, at least, modify the expression of the genes in which they lodge. A stream of new information implicates transposable elements in several human genetic disorders, notably hemophilia A and type 1 neurofibromatosis. In its usual location, the repetitive piece of DNA lacks the capacity to direct the production of a protein. Its insertion, however, into a gene that does code for a protein leads to an abnormal protein product and the expression of a genetic defect. In large measure, jumping genes in humans act as mutagenic agents.

RECOMBINANT DNA

The painstaking studies of bacterial plasmids by a host of investigators have paved the way for the imaginative use of plasmids as carriers of foreign DNA. The plasmids can be manipulated in such a fashion that they can carry genes from very distantly related organisms, including frogs, rabbits, and humans. This sophisticated technology involves the use of a special class of enzymes called *restriction endonucleases.* These enzymes act as chemical scissors; they cut a DNA strand at precise points.

Restriction endonucleases occur widely within the true bacteria and the *extremophile* bacteria or *archaea* (See chapter 14) where they function as a primitive immune systems defending prokaryotic cells from foreign DNA originating in invading viruses. Over 3,500 different varieties of these enzymes have been identified and several hundred are commercially available. This makes restriction enzymes one of the largest group of enzymes that have the same basic function. This feature, along with their remarkable lack of sequence similarity, make them ideal subjects for evolutionary research. Each of these enzymes recognizes a particular short nucleotide sequence in any DNA exposed to it and snips the DNA at that point. If the sequence occurs more than once, as it generally does, the DNA molecules are cut in several fragments. The endonucleases were originally discovered in bacteria in 1970 by Werner Aber of the University of Basel (Switzerland) and by the team of Daniel Nathans and Hamilton D. Smith at Johns Hopkins University, for which achievement the three received the 1978 Nobel Prize for Medicine. Among the first restriction enzymes purified were *Eco*RI and *Eco*RV, from *Escherichia coli,* and *Hin*dII and *Hin*dIII, from *Haemophilus influenzae.* Some restriction enzymes and the base sequences they cleave are listed in table 7.1. As an example, *Eco*RI will cut a DNA molecule from any source at any point at which the nucleotide sequence GAATTC is encountered. Restriction enzymes get their names and number from the genus, species and strain in which they were found. For example, *Hin*dII was the second restriction enzyme found in *Haemophilus influenzae.* These enzymes are also *palindromes* meaning that they read the same way forward and in reverse, like the English phrase "never odd or even".

TABLE 7.1	Some Restriction Enzymes and the Base Sequences They Cleave

Bacterial source	Enzyme	Sequence*
Escherichia coli	*Eco*RV	5′ . . . G A T|A T C . . . 3′ 3′ . . . C T A|T A G . . . 5′
Escherichia coli	*Eco*RI	5′ . . . G|A A T T C . . . 3′ 3′ . . . C T T A A|G . . . 5′
Haemophilus haemolyticus	*Hha*I	5′ . . . G C G|C . . . 3′ 3′ . . . C|G C G . . . 5′
Providencia stuarti	*Pst*I	5′ . . . C T G C A|G . . . 3′ 3′ . . . G|A C G T C . . . 5′

* The upstream portion of a gene sequence is referred to as the 5′ end; the downstream portion, the 3′ end. The lines show the location in DNA strands where the restriction enzymes cleaves the DNA.

Since their discovery, scientists have debated whether restriction enzymes evolved only once or arose independently more than once. Molecular data supports the view that this enzyme group evolved only once and then diversified. This process is believed to have been hastened by various types of gene exchange between bacteria and archaea. The genes for restriction enzymes, like the genes for drug resistance, shuffle between bacterial strains. Natural selection for variety continually occurs. This results in never-ending cycles of adaptation by both the bacteria and viral variants. As viruses mutate and make new attempts to evade the bacteria's restriction enzymes, new generations of bacteria must either adapt (via the acquisition of new mutations) or fall victim to viral attack. This, in turn, selects for the next generation of viruses and the cycle continue like an arms race between warring armies.

Tens of thousands of human genes have already been isolated and cloned. An isolated human gene can be spliced into a plasmid that has been experimentally fragmented at specific points by a restriction endonuclease (fig. 7.6). The outcome is a new DNA circle, or *recombinant DNA molecule,* which contains both plasmid DNA and human DNA. This newly formed recombinant DNA molecule, when reintroduced into a bacterium, can replicate precisely and be passed on to daughter bacterial cells for many generations.

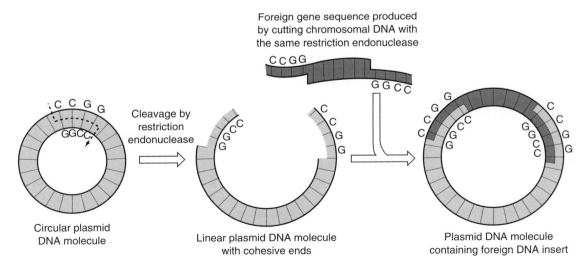

Figure 7.6 A recombinant DNA molecule, containing both plasmid (bacterial) DNA and human DNA. The circular plasmid DNA is cut by a restriction endonuclease that leaves cohesive single-stranded ends. A foreign (human) DNA sequence is generated by the same restriction endonuclease and, accordingly, has complementary cohesive ends that permit base-pair interactions.

Given that a bacterial cell can divide every 30 to 40 minutes, a single bacterium, under ideal growing conditions, can produce almost 70 billion new cells in 24 hours. At the same time, billions of copies of the product of the human gene become available!

This ingenious technique has permitted the large-scale transmission of genes that code for commercially indispensable molecules. As an admirable example, the insertion of the human insulin-producing gene into *Escherichia coli* has made possible the production of insulin in large quantities. Another notable example is the production of human growth hormone. In this instance, the genes coding for the polypeptide chains of the growth hormone were actually synthesized chemically in the laboratory. The synthetically created genes were incorporated into *E. coli* and efficient synthesis of human growth hormone was achieved (fig. 7.7). In addition to human insulin and growth hormone, many other valuable proteins have been produced by recombinant DNA technology.

PRUDENT USE OF ANTIBIOTICS

The increasing number of bacteria that have evolved resistance to antibiotics is one of the best examples of evolution via natural selection in our daily lives. When antibiotics were first discovered and later mass produced, they were exquisitely effective at rendering harmless a broad spectrum of formerly lethal bacteria. Most of us alive today do not remember the times before antibiotics when patients with bacterial infections were provided only palliative care because it was up to their own immune systems to effectively fight their bacterial infections. The lack of prudent use of antibiotics has resulted in a situation that is almost unimaginable. Where would we be without effective antibiotics? How would we deal with widespread pandemics? Certainly there would be some adaptation. Some of us carry with us the legacy of bacterial resistance. We are, after all, the descendants of those who survived the ravages of repeated plagues and other epidemics caused by bacterial disease. Some of us have strong and robust immune systems. And

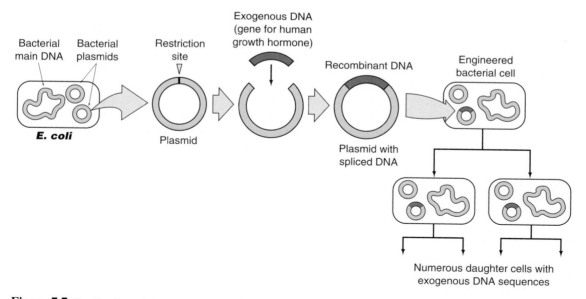

Figure 7.7 Production of virtually limitless quantities of an exogenous gene product (human growth factor, for example) is accomplished by inserting the exogenous gene in a bacterial plasmid.

some of us may, by chance, already have some resistance to new bacterial diseases. Nevertheless, we stand at the threshold of a world without effective antibiotics. Although some progress has been made, many human advocacy groups, both private and public, have clamored for a comprehensive program to monitor the use of antibiotics and forestall the seemingly endless emergence of antibacterial resistance. Leadership and change from government, the agricultural industry, and the medical profession has been slow.

The causes of increased bacterial resistance are varied and complex and include the widespread use of antibiotics—even when they are not needed. Physicians and patients must be educated to reduce this unnecessary use of these lifesaving drugs. Each time a bacterial population is exposed to an antibiotic, it is possible that one cell among billions carries a mutation allowing it to survive in the presence of that antibiotic. Over time, the likelihood increases that some bacterial cells will, by chance, contain a mutation that permits them to survive under antibiotic assault. Additional reasons for concern include the extensive use of antibiotics in the breeding of animals for food, the promiscuity of bacteria, and (as we have seen earlier in this chapter), bacteria's ability to exchange or swap DNA. Likewise, as the length of hospital stays has declined from weeks to hours, the likelihood of bacterial resistance has skyrocketed. Finally, increased travel, especially overseas travel, facilitates the potential rapid movement of resistant bacteria from one part of the world to another.

Prominent examples of antibiotic resistance are emerging new diseases, such as methicillin-resistant *Staphylococcus aureus,* better known as MRSA (also sometimes called multiple resistant *Staphylococcus aureus*), which is a virulent bacteria that has evolved resistance to many antibiotics, including the various penicillins (e.g., ampicillin, dicloxacillin, and methicillin) and the cephalosporins, a large group of broad-spectrum antibiotics. Once limited to hospital settings, MRSA is increasingly common throughout the environment and is a bellwether of microorganisms' growing resistance to our dwindling arsenal of antibiotics. Another nightmare from the past, tuberculosis (TB), is now increasingly being found in a multi-drug-resistant form. Antibacterial soaps are ubiquitous in our environment and may be a new danger on the horizon, selecting for an entirely new class of resistant bacteria. Whether this is a matter of concern is hotly debated. The recent installation of waterless handwash stations in most medical facilities, hospitals, and some grocery stores appears to be a helpful public health initiative, but it may be a mechanism to provide additional opportunities for mutated bacteria to surface.

Greater education efforts focused on patients and medical personnel has brought about increased awareness and some change in behavior, especially within the medical community. The same cannot be said of the agricultural industry, which continues the widespread use of subtherapeutic doses of antibiotics as growth enhancers. These low-dose uses are especially good at selecting for resistant bacterial stains.

EVOLUTIONARY IMPLICATIONS

The evolutionary implications of increasing bacterial resistance to antibiotics are instructive. As with all organisms, bacteria are constantly evolving and changing so as to better survive in their environment. Hence, under conditions of widespread use of antibiotics, mutated genes that provide resistance to antibiotics are relentlessly selected for by natural selection. Once mutated, only antibiotic resistant bacteria survive and multiply. Work to create new antibiotics is slow. This is partly because these drugs are relatively unprofitable to pharmaceutical companies. Compared to a new drug to lower hypertension or cholesterol, or for treating pain or depression, new antibiotics are relatively low-profit enterprises. Similarly, most research on creating new antibiotics is focused on highly dangerous pathogens such as smallpox, plague, and anthrax.

Most antibiotics are created from microscopic organisms that have been waging the war against pathogenic bacteria for longer than we have. These molds and other organisms have developed resistance to antibiotics over many eons of exposure.

Penicillin, the first of the antibiotics, was discovered, somewhat accidentally, by Scottish bacteriologist and later Nobel laureate Alexander Fleming. In 1928 Fleming arrived at his laboratory to find a zone of inhibition around an incubating container of virulent *Staphylococcus* where a mold (in the genus *Penicillium*) was growing. He quickly realized what he had inadvertently stumbled upon. He is famous for his quip that "chance favored the prepared mind" alluding to the fact that he appreciated what he found because of his training and background. It was not until late in World War II that penicillin became widely used. Today, just a little more than 65 years later, natural selection has had more than sufficient time to allow bacteria to evolve resistance to countless antibiotics. To emphasize this point, let us calculate the approximate number of bacterial generations that have occurred during the roughly 65 years that antibiotics have been in use. Assuming a bacterial generation time of 30 minutes, each year represents over a million bacterial generations (2 generations per

minute $\times$ 60 minutes per hour $\times$ 24 hours per day $\times$ 365 days in a year = 1,051,200). Sixty-five years is thus over 65 million bacterial generations or the human equivalent of (based on a 20 years generation time for humans) over a BILLION years.

During most of the history of human civilization, pathogenic bacterial diseases, such as plague, smallpox, and diphtheria, repeatedly ravaged human populations, often killing as much as 50 percent or more of the population each time they swept through an area. Even though many of our ancestors developed natural immunities to these diseases in the same way that modern bacteria have developed resistance to antibiotics, the selective environment is now different and many of us might be unable to ward off bacterial infections without the assistance of antibiotics. Regardless, the return to the days when bacterial diseases simply ran their course, and you either survived or didn't, is a world that is hard to imagine. That said, bacteria are a normal and essential component of the envelope that covers the inside and outside of our bodies. Most bacteria are not pathogenic— quite the opposite, they are necessary to our very existence. Humans and bacteria evolved together, and although we have sometimes been at war, we have frequently achieved a state of détente.

CHAPTER

8

BALANCED POLYMORPHISM

The concepts presented in Chapter 7 led us to believe that selection operates at all times to reduce the frequency of a detrimental gene to a low equilibrium level. This view is not entirely accurate. We are aware of genes with deleterious effects that occur at fairly high frequencies in natural populations. A striking instance is the high incidence in certain human populations of a mutant gene that causes a curious and life-threatening form of blood cell destruction, known as *sickle-cell anemia*. It might be presumed that this harmful gene is maintained at a high frequency by an exceptionally high mutation rate. There is, however, no evidence to indicate that the sickle-cell gene is unusually mutable. We now know that the maintenance of deleterious genes at unexpectedly high frequencies involves a unique, but not uncommon, selective mechanism, which results in a type of population structure known as *balanced polymorphism*.

SICKLE-CELL ANEMIA

This disease was discovered by the American physician James B. Herrick, who in 1904 made an office examination of an anemic African American male residing in Chicago. The patient's blood examined under the microscope showed the presence of numerous crescent-shaped erythrocytes (red blood cells). The peculiarly twisted appearance of the red blood cell is shown in figure 8.1. The patient was kept under observation for six years, during which time he displayed many of the distressing symptoms we now recognize as typical of the disease (fig. 8.2).

The bizarre-shaped red cells in the form of a sickle blade cannot easily negotiate the thin spaces of the capillaries and cause miniature "log jams." The clogging of small blood vessels can occur anywhere in the body, denying vital oxygen to the tissues and bringing about the painful crises and other debilitating symptoms (fig. 8.2) that are so characteristic of the disease. Life expectancy is reduced—half the victims succumb before the age of 20 and most do not survive beyond 50. The clinical picture, however, is quite variable, with some patients having only few crises and mild pains for many years, while others become severely disabled or die at an early age.

Sickle-cell anemia occurs predominantly among African Americans but is also found in

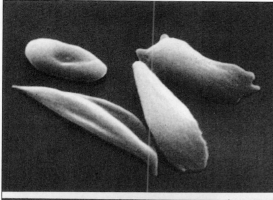

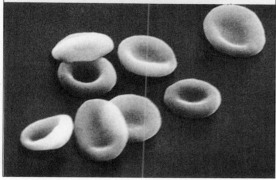

Figure 8.1 **Peculiarly twisted shape** of red blood cells of an individual suffering from sickle-cell anemia *(top),* contrasted with the spherical appearance of normal red blood cells *(bottom).*
(World Wide Photos.)

other populations such as Hispanics of Caribbean ancestry, persons of Mediterranean ancestry (e.g. Turkey, Greece and Italy), as well as people from parts of India and Saudi Arabia. The incidence at birth of the disabling sickle-cell anemia in the United States varies by ethnicity. In African Americans, the incidence is approximately 2 per 1000 live births and in Hispanic Americans it is approximately 1 in 35,000 live births. In the United States, nearly 70,000 people are afflicted with sickle cell disease. Roughly 2 million or 1 in 12 African Americans is a carrier of this disease. It was not until after World War II that the hereditary basis of sickle cell disease was elucidated. The irregularity was shown in the late 1940s

by James V. Neel of the University of Michigan to be inherited as a simple Mendelian character. The sickle-cell anemic patient inherits two variant (sickling) alleles, one from each parent (fig. 8.3). Individuals with one normal and one variant allele are generally healthy but are carriers—they are said to have *sickle-cell trait.* If two heterozygous carriers marry, the chances are one in four that a child will have sickle-cell anemia, and one in two that a child will be a carrier (fig. 8.3).

Since the homozygous state of the detrimental allele is required for the overt expression of the disease, sickle-cell anemia may be considered to be recessively determined. However, although heterozygous individuals are, on the whole, normal, even the red cells of the heterozygotes can undergo sickling under certain circumstances, producing clinical manifestations. As an instance, heterozygous carriers have been known to experience acute abdominal pains at high altitudes in unpressurized planes. The lowering of the oxygen tension is sufficient to induce sickling. The abdominal pains can be traced to the packing of sickled erythrocytes in the small capillaries of the spleen.

According to some authors, since the detrimental gene can express itself in the heterozygous state (by producing a positive sickling test under certain circumstances), the gene should be considered dominant. This reveals that "dominance" and "recessiveness" are somewhat arbitrary concepts that depend on one's point of view. Indeed, from a molecular standpoint, the relation between the normal and variant allele in this case may best be described as *codominant.* This means that at the molecular level neither allele masks the expression of the other. As we shall see the next section "Sickle-Cell Hemoglobin", the heterozygote produces both normal and abnormal hemoglobin. *There is no blending of inheritance at the molecular level.*

SICKLE-CELL HEMOGLOBIN

In 1949, the late distinguished chemist and Nobel laureate Linus Pauling and his co-workers made the important discovery that the detrimental sickling

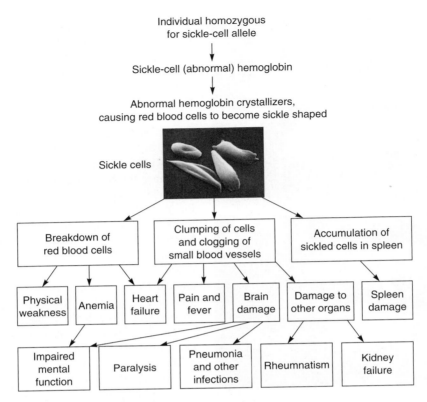

Figure 8.2 Sickle-cell anemia. In affected individulas, sickle-cell anemia causes a constellation of symptoms in many different organs of the body.

allele alters the configuration of the hemoglobin molecule. Pauling used the then relatively new technique of electrophoresis, which characterizes proteins according to the manner in which they move in an electric field. The hemoglobin molecule travels toward the positive pole. Hemoglobin from a sickled cell differs in speed of migration from normal hemoglobin; it moves more slowly than the normal molecule (fig. 8.4). The variant allele thus functions differently from the normal allele, and, in fact, acts independently of its normal partner allele. As a result, the heterozygote does not produce an intermediate product, but instead produces both kinds of hemoglobin in nearly equal quantities—the normal type (designated *hemoglobin A,* or *Hb A*) and the sickle-cell anemic variety *(Hb S).* The dual electrophoretic pattern of hemoglobin

from a heterozygous individual can actually be duplicated experimentally by mechanically mixing the hemoglobin taken from blood cells of a normal person and a sickle-cell anemic patient. The mixed solution separates in an electric field into the same two hemoglobin components as those characteristic of a heterozygous person with the sickle-cell trait.

Further chemical analysis revealed that sickle-cell hemoglobin differed only slightly in its chemical constitution from normal hemoglobin. As mentioned earlier (chapter 4), a glutamic acid residue at a specific site in normal hemoglobin is replaced by valine in sickle-cell hemoglobin (fig. 8.5). The substitution of valine for glutamic acid can be accounted for by an alteration of a single base change of the triplet of DNA that specifies the amino acid, as depicted in figure 8.6. If the

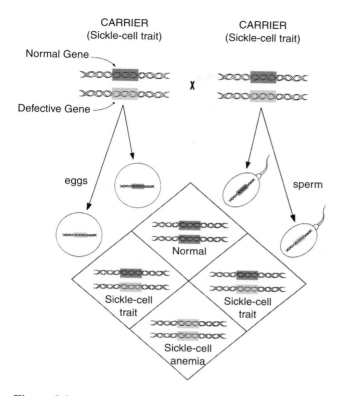

Figure 8.3 **Types of children** that can result from a marriage of two heterozygous carriers of sickle-cell anemia. Individuals homozygous for the variant gene suffer from sickle-cell anemia: the benign heterozygous state is referred to as the sickle-cell trait.

DNA triplet responsible for transcription in normal hemoglobin *(Hb A)* is CTT, then a single base change to CAT would alter the codon of messenger RNA to GUA. Accordingly, the amino acid specified by the codon GUA would be valine. The substitution of a glutamic acid by valine causes the abnormal hemoglobin molecule *(Hb S)* to aggregate or polymerize into strands. The strands are laid down to form cable-like fibers. As greater numbers of fibers accumulate, the large aggregates or polymers of linearly arranged fibers attain sufficient length and rigidity to distort the cell membrane into an odd, crescent shape.

THEORY OF BALANCED POLYMORPHISM

Natural selection confers survival and selective advantages to certain alleles under specific environmental conditions. Of the alleles known to cause human genetic diseases (table 8.1), several have been identified in which the heterozygotes exhibit resistance to specific infectious diseases (environmental conditions).

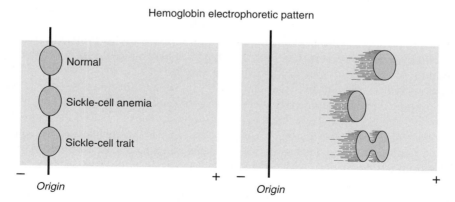

Figure 8.4 **Electrophoretic patterns of hemoglobins.** The hemoglobin of the heterozygous person with sickle-cell trait is not intermediate in character, but is composed instead of approximately equal proportions of the normal hemoglobin and the sickle-cell anemia variety. [Based on studies by Linus Pauling, in Pauling, Itano, Singer, and Wells, "Sickle-Cell Anemia: A Molecular Disease," *Science* 110 (25 November 1949): 543–548.]

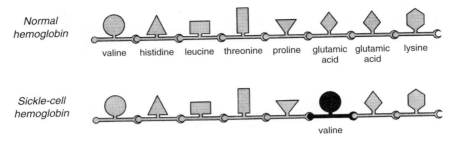

Normal hemoglobin

valine   histidine   leucine   threonine   proline   glutamic   glutamic   lysine
                                                       acid   acid

Sickle-cell hemoglobin

valine

Figure 8.5 Amino acid sequences in a small section of the normal hemoglobin molecule and of the sickle-cell hemoglobin. The substitution of a single amino acid, glutamic acid by valine, is responsible for the abnormal sickling of human red blood cells.

(Based on studies by Vernon Ingram.)

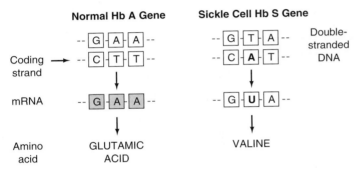

Normal Hb A Gene   **Sickle Cell Hb S Gene**

Coding strand

mRNA

Amino acid   GLUTAMIC ACID   VALINE

Double-stranded DNA

Figure 8.6 The CTT in the sixth triplet of the coding strand of the normal gene changes to CAT. This leads to a change from GAA to GUA in the sixth codon of the beta-globulin mRNA of sickle cells. This, in turn, results in the insertion of a valine in the sixth amino acid position of sickle-cell beta globulin, where a glumatic acid normally is present.

Sickle-cell anemia is a model of this phenomenon, which is called *balanced polymorphism.*

Until recently, most people with sickle-cell anemia did not survive to reproductive age, and thus it might be expected that the detrimental allele would pass rapidly from existence. Each failure of the homozygous anemic individual to transmit his or her alleles would result each time in the loss of two aberrant alleles from the population. And yet the sickle-cell allele reaches remarkably high frequencies in the tropical zone of Africa. In several African populations, 20 percent or more of the individuals have the sickle-cell trait, and frequencies as high as 40 percent have been reported for some African tribes. As mentioned earlier in this chapter, the sickle-cell trait is not confined to the African continent. What can account for the high incidence of the sickle-cell allele, particularly in light of its detrimental action?

For simplicity in the presentation of the population dynamics, we will consider the sickling allele to be recessive. Accordingly, we can symbolize the normal allele as *A* and its allele for sickling as *a*. The explanation for the high level of the deleterious sickle-cell allele is to be found in the possibility that the heterozygote *(Aa)* is superior in fitness to both homozygotes (*AA* and *aa*). In other words, selection favors the heterozygote, and both types of homozygotes are relatively at a disadvantage. Let us examine the theory behind this form of selection.

Figure 8.7 illustrates the theory. The classical case of selection discussed in chapters 6 and 7 is portrayed in the first part of the figure. In this case, when *AA* and *Aa* individuals are equal in reproductive fitness, and the *aa* genotype is completely selected against, the recessive allele will be eliminated. Barring recurring mutations, only *A* genes will ultimately be present in the population.

Let us assume in another situation (case II in fig. 8.7) that both homozygotes (*AA* as well as *aa*) are incapable of leaving surviving progeny. The only effective members in the population are the

TABLE 8.1	Balanced Polymorphism and Human Disease

Disease	Protects Against	Carrier Frequency	Target Population (S)	Mechanism
Sickle-Cell Anemia	Malaria	1/4–1/12 varies	Agricultural Africans Mediterraneans	Parasite cannot exist in abnormal RBC
G6PD	Malaria	1/5 1/3–1/20	African, Middle Eastern & S. Asian	Parasites unable to reproduce in abnormal RBC
Thalassemia	Malaria	1/30	Mediterranean Greece and Italy	Makes RBC unsuitable for lethal host parasite
PKU	Fungal infection of fetuses	1/23	European Caucasian	Inactivation of fungal toxin by high levels of phenylalanine
Tay-Sachs Disease	Tuberculosis	1/30	Ashkenazi Jews	Unknown
Cystic Fibrosis	Diarrheal disease (Cholera, Typhus)	1/23	European Caucasians	Water loss prevented by reduced chloride channels

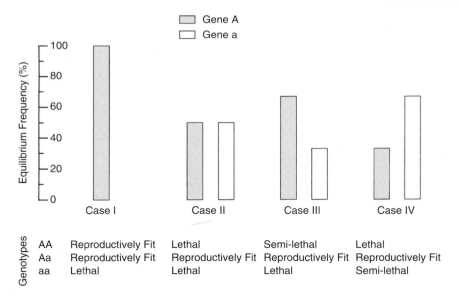

Figure 8.7 **Equilibrium frequencies** of two alleles *(A* and *a)* under different conditions of selection (ignoring mutation). In contrast to case I (complete selection and total elimination of *a*), the recessive gene can be retained at appreciable frequencies in a population when the heterozygote *(Aa)* is superior in reproductive fitness to *both* homozygotes (cases II, III, and IV, showing different relative fitnesses of the two homozygotes, *AA,* and *aa).*

heterozygotes *(Aa).* Obviously, the frequency of each allele, *A* and *a*, will remain at a constant level of 50 percent. The inviability of both homozygous types probably never exists in nature, but the scheme does reveal how a stable relation of two

alleles at high frequencies is possible. If, as in case III, the *AA* genotype leaves only half as many progeny as the heterozygote, and the recessive homozygote is once again inviable, it is apparent that more *A* alleles are transmitted to each new generation than *a* alleles. Eventually, however, the *A* allele will reach an equilibrium point at 0.67 percent. Note that although the recessive homozygote is lethal, the frequency of the recessive allele *(a)* is maintained at 0.33 percent. In the last illustrative example (case IV), the recessive homozygote is not as disadvantageous as the dominant homozygote, but both are less reproductively fit than the heterozygote. Here also, both alleles remain at relatively high frequencies in the population. Indeed, the recessive allele *(a)* will constitute 67 percent of the gene pool.

We have thus illustrated in simplified form the selective forces that serve to maintain two alleles at appreciable frequencies in a population. This phenomenon is known as *balanced polymorphism*. The loss of a deleterious recessive allele through deaths of the homozygotes is balanced by the gain resulting from the larger numbers of offspring produced by the favored heterozygotes. Balanced polymorphism results in a stable equilibrium of the two alleles; their equilibrium frequencies are determined by the relative fitness of the two homozygotes.

SUPERIORITY OF THE HETEROZYGOTES

The high frequency of the sickle-cell allele in certain African populations can be explained by assuming that the heterozygotes (individuals with the sickle-cell trait) have a selective advantage over the normal heterozygotes (fig. 8.8). What might be the nature of the advantage? Field work undertaken in Africa by the British geneticist Anthony Allison revealed that the incidence of the sickle-cell trait is high in regions where malignant subtertian malaria caused by the parasite *Plasmodium falciparum* is *hyperendemic*—that is, transmission of the infection occurs throughout most of the year. Thus, the population is almost constantly reinfected with malaria. Under such circumstances, relative immunity to falciparum malaria would be most beneficial.

Allison examined blood from African children and found that carriers of the sickle-cell trait were relatively resistant to infection with *Plasmodium falciparum*. The heterozygous carriers were infected less often with the parasite than the homozygous dominant nonsicklers. Moreover, among those heterozygotes that were infected, the incidence of severe, or fatal, attacks of malaria was strikingly low. The evidence is strong that the sickle-cell allele affords young children some degree of protection against malarial infection. Hence, in areas where malaria is common, children possessing the sickle-cell trait will tend to survive malaria, and are more likely to pass on their genes to the next generation. The heterozygotes *(Aa)* are thus superior in fitness to both homozygotes, which are likely to succumb from either anemia on the one hand *(aa)* or malaria on the other *(AA)*.

The spread of the sickling allele was greatly enhanced by the development of agriculture in Africa. The clearing of the forest in the preparation of ground for cultivation provided new breeding areas for the mosquito *Anopheles gambiae*, the *vector*, or carrier, of the malaria plasmodium. The spread of malaria has been responsible for the spread of the selective advantage of the sickle-cell allele, which in the heterozygous state imparts resistance to malaria. In essence, the selective advantage of the heterozygote tends to increase in direct proportion to the amount of malaria present in a given area. Hunting populations in Africa show a very low incidence of malaria and an equally low frequency of the sickling allele. The Pygmies of the Ituri Forest constitute a good example.

Communities in Africa with the greatest reliance on agriculture (rather than on hunting or animal husbandry) tend to have the highest frequencies of the sickle-cell trait. A high incidence of the sickle-cell trait in an intensely malarious environment has the consequence of reducing the number of individuals capable of being infected by the malarial parasite and, accordingly, of lowering the mortality from such infections. More human energy, or greater "manpower," is thus made available for raising and harvesting crops. Ironically, then, the sickle-cell trait carries with it the beneficent effect of enabling tribes to develop and

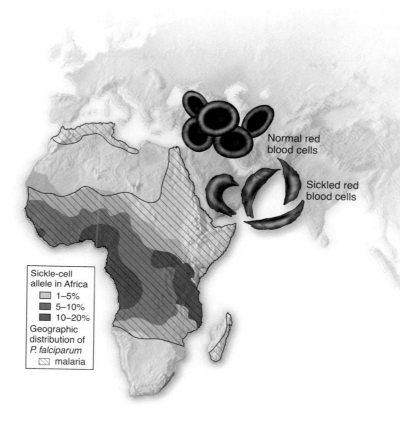

Figure 8.8 Balanced polymorphism and heterozygote advantage. This illustration compares of the frequency of sickle-cell anemia and the distribution of malaria *(Plasmodium falciparum)* in Africa. Carriers of the sickle-cell allele do not have impaired health but are resistant to the malaria parasite. This is an example of heterozygote superiority, also called balanced polymorphism.

maintain an agricultural culture rather than adhere to a hunting or pastoral existence. This is a curious but striking instance of the interplay of biological change and socioeconomic adaptation.

In a similar finding, evidence exists that persons whose red blood cells are deficient in the enzyme G6PD (glucose-6-phosphate-dehydrogenase) are less likely to be affected by malaria. Data also show a strong correlation between the incidence of malaria and the frequency of thalassemia (another type of anemia) in the Mediterranean peninsula and in Sardinia. In all these situations, it is as if

one inherits something that is bad (hemolytic disease) to afford protection against something that is worse (malaria). Malaria apparently has had a profound influence on human events.

RELAXED SELECTION

We should expect the frequency of the sickle-cell allele to be low in malaria-free areas, where the selective advantage of the heterozygote would be removed. We do find that the lowest frequencies of the sickle-cell allele occur consistently in regions free of malaria. The frequency of the sickle-cell allele has fallen to relatively low levels in the African American populations of the United States. The frequency of the sickle-cell trait among African Americans is currently about 9 percent, corresponding to an allele frequency of 0.045. This places the frequency of the recessive homozygote (sickle-cell anemia) at 0.002, or 2 per 1,000 individuals.

The relaxation of selection in this case is *eugenic;* that is, the changes occur in the desired direction. The disappearance of malaria disrupts the balanced polymorphic state, and the allele for sickle-cell anemia begins to decline at a slow rate. Table 8.2 presents computations of the gene frequencies of the heterozygotes and the recessive homozygotes, starting with the frequency of the carrier at 18 percent. It may be noted that 30 generations are required for the incidence of the heterozygous carrier to be depressed from 18.0 to 6.5 percent. At present, medical researchers are attempting to find a cure for sickle-cell anemia. If sickle-cell anemic individuals could be completely cured, then obviously the selective process would be thwarted and the sickle-cell allele would no longer decline in frequency.

TAY-SACHS DISEASE

Tay-Sachs disease is a very rare, recessively inherited condition that is untreatable and always fatal. The disease results from mutations on the HEXA gene on chromosome 15. The basic defect is the abnormal accumulation of a fatty substance (specifically, ganglioside GM_2) in the brain cells as a consequence of the absence of a particular enzyme (beta hexosaminidase A), which plays an essential role in the spinal cord and brain. The storage of massive amounts of the lipid in the brain leads to profound mental and motor deterioration. Affected children appear normal and healthy at birth. Within six months, however, motor weaknesses become obvious as the muscles twitch and the infant experiences periodic convulsions. The muscles deteriorate until the infant becomes completely helpless, unable to sit up or stand. At the age of one year, the child lies still in the crib. The child becomes mentally retarded, progressively blind, and finally paralyzed. The disease exacts its lethal toll by the age of three to four years. There are no known survivors and no cure.

A feature of special interest is that 9 out of 10 affected children are of Jewish heritage. It is especially common among the Ashkenazi Jews of northeastern European origin, particularly from provinces in Lithuania and Poland. In the United States, Tay-Sachs disease is about 100 times more prevalent among the Ashkenazi Jews than among other Jewish (Sephardi) groups and non-Jewish populations. It is estimated that 1 of 30 Jewish persons is a heterozygous carrier, whereas only about 1 of 300 non-Jewish persons is a heterozygous carrier. If the high incidence of heterozygotes is maintained by mutation alone, then an extraordinarily

TABLE 8.2	Selection Relaxation Against Sickle-Cell Anemia (in Nonmalarial Areas) Conditions: Initial fitness of recessive homozygote: 0.33 Initial gene frequency of sickling gene: 0.10 Initial frequency of heterozygote: 0.18

Generation	% Heterozygote (Sickle-Cell Trait)	% Recessive Homozygote (Sickle-Cell Anemia)
0	18.0	1.0
1	17.0	0.88
3	15.3	0.70
10	11.4	0.37
30	6.5	0.11
100	2.6	0.02

high mutation rate of the detrimental recessive allele would have to be postulated. However, some reproductive advantage for the heterozygote carrier would seem the most plausible explanation.

It has been suggested that the Ashkenazi Jews, who have lived for many generations in the urban ghettos in Poland and the Baltic states, have been exposed to different selective pressures than other Jewish groups (for example, those who have lived in countries around the Mediterranean and Near East). The densely populated urban ghettos may have experienced repeated outbreaks of infectious diseases. In 1972 the geneticist Ntinos Myrianthopoulos of the National Institutes of Health in Bethesda, Maryland, presented data that shows that pulmonary tuberculosis is virtually absent among grandparents of children afflicted with Tay-Sachs disease, although the incidence of Jewish tuberculosis patients from eastern Europe is relatively high. The findings suggest that the heterozygous carrier of Tay-Sachs disease is resistant to pulmonary tuberculosis in regions where this contagious disease is prevalent.

In terms of theoretical expectations, one can calculate (in the absence of genetic screening) that 50 children will be born with the disorder each year in the United States, of whom 45 will be of Ashkenazi Jewish origin. Yet the number of actual cases of affected newborns is far less than expected. Indeed, since genetic screening program began in the early 1970s the incidence of births of children afflicted with Tay-Sachs disease has been so dramatically reduced that less than a dozen cases are now known in North America, nearly all of whom are not of Jewish ancestry. Through genetic screening (below) this disease has ostensibly been eliminated in the Jewish population.

The dramatic reductions in births of affected infants reflects a voluntary, widespread adult screening program initiated in 1971 among the Jewish population with the objective of detecting carriers of the fatal recessive gene. The screening for heterozygotes has been followed by the intrauterine monitoring of the fetus in instances where both parents are heterozygous carriers. In most cases, when parents have discovered through screening and subsequent prenatal diagnosis that they have conceived an affected child, they have opted to terminate the pregnancy. The decision is generally not thought of as a choice between life and death for the affected fetus. The choice is between prenatal death and a lingering, painful, postnatal death. At issue is not the saving of the life of an affected child, but the prevention of undeniable suffering.

Interestingly, the same mutation found in Ashkenazi Jews has been discovered, also at an unusually high incidence (1/30), in the ethnically and linguistically isolated American Cajuns who live in south Louisiana. There has been diverse speculation as to why this same mutation occurs in such diverse populations as Ashkenazi Jews and Cajuns, but no definitive resolution to this phenomenon has emerged. What is known is that Cajun Tay-Sach's mutation can be traced to an apparently non-Jewish eighteenth-century couple from France.

Intriguingly, another Tay-Sachs mutation, with similar pathology and frequency (1/30), is also known in French Canadian families and has been traced to a single seventeenth-century family from southern Quebec. In both the Cajuns and the French Canadians, the strikingly high prevalence of the Tay-Sachs gene variants is the result of a *founder effect,* a kind of genetic drift that is discussed in chapter 9. No similar relationship to natural selection has thus far emerged for these two populations.

CYSTIC FIBROSIS

Cystic fibrosis (CF) is not a rare genetic condition; one child in 2,000 Caucasians is affected, making it the most common autosomal recessive disease in Caucasian Americans of European descent. The frequency of carriers vary across ethnicities, being highest (1 in 23) among Caucasians and rarer in Hispanic Americans (1 in 46), African Americans (1 in 66), and Asian Americans (1 in 150), making the overall frequency about 1 in 3,900 live births per year in the United States. Approximately 2,500 affected newborns may be expected each year in the United States.

Cystic fibrosis (CF) is a multisystem disease, affecting many parts of the body. Specifically, cystic fibrosis is identified by a triad of abnormal conditions—a highly elevated concentration of chloride in sweat, pancreatic insufficiency, and chronic obstructive pulmonary disease. Almost equally consistent is *azoospermia* (lack of sperm in the semen), seen in more than 95 percent of men with cystic fibrosis. Almost universally, the sperm ducts atrophy, or degenerate, as a consequence of prolonged blockage by thick mucus secretions. The female patient does not have a comparable anatomical abnormality. However, her fertility is below normal, as the thick desiccated cervical mucus tends to be inhospitable to the sperm cells. One in every three affected females is married, and among affected women who become pregnant, there is an increased incidence of premature birth and perinatal death.

The frequency of heterozygous carriers is many times greater than the frequency of affected homozygous individuals. Assuming that 1 of 25 persons is a carrier of CF, the chance of having the disease is 100 times greater than of being a carrier. Like the other diseases discussed in this chapter, CF's relatively high frequency in certain populations may be explained by the concept of balanced polymorphism. A reproductive advantage for the heterozygote *(Aa)* over the normal homozygote *(AA)* would account for the high frequency of the disabling recessive allele for cystic fibrosis in Caucasian populations. The hypothesis currently favored is that heterozygous carriers are more resistant to infantile gastroenteritis (inflammation of the stomach and intestines) than noncarriers. The basic defect expressed in cystic fibrosis is a failure of transport of chloride ions across cell membranes. Ironically, the same chloride ion channels are involved in another devastating disease with high mortality, cholera. This disease is marked by excessive diarrhea and vomiting, associated with the profuse passage of chloride ions

through channels in the cells lining of the intestinal tract into the lumen of the digestive tract. In a curious manner, it may be that the heterozygous carrier of cystic fibrosis is resistant to cholera in regions where this contagious, infectious disease is, or once was, prevalent.

In 1989 the gene responsible for cystic fibrosis was identified on the long arm of chromosome 7, caused by a mutation in a gene called the cystic fibrosis transmembrane conductance regulator (CFTR). The CFTR gene controls the chloride ion channel, which is needed in creating sweat, digestive juices, and mucus. Knowledge of the precise location of the mutant allele for cystic fibrosis has permitted the development of a diagnostic test for family members at risk. The thrust for prenatal screening may well be a two-edged sword. The survival of patients with cystic fibrosis has improved dramatically in recent decades. In the mid-1960s, affected children could not look forward to reaching the age of 10 years, despite the best efforts of the medical profession. Today this disorder still remains a burden to children, but a limited life span is no longer foreboding or inevitable. Presently, with increased clinical awareness and aggressive therapeutic measures, more than half the babies born with this disorder can expect to live at least to their late thirties (fig. 8.9), although this is highly variable, depending on the severity of the disease. Cystic fibrosis has graduated from the pediatric clinic and has become a disease of adults. The disorder, however, still retains its distressing aspects, and the greater longevity may be viewed as imposing greater rather than lesser burdens on the patients and their families.

SELECTION AGAINST THE HETEROZYGOTE

The mother-child blood incompatibility in humans, *erythroblastosis fetalis,* represents an interesting case of selection *against* the heterozygote. This unique hemolytic disorder is often referred to simply as *Rh disease* because it involves the Rh

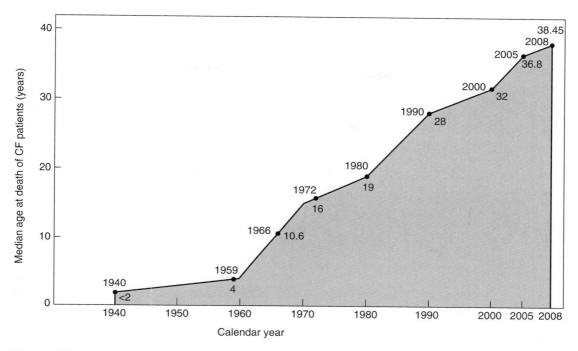

Figure 8.9 **Medical advances** in recent decades have enhanced the life span of infants affected with cystic fibrosis.

antigen on the red blood cell. Rh disease was once an alarming thorn to procreation. The conquest of this curious incompatibility between mother and child is one of the more notable accomplishments of modern medicine.

A relation between the Rh antigen and hemolytic disease was first postulated by Dr. Philip Levine and his colleague Dr. B. E. Stetson in 1939, when they discovered a then unknown antibody in the serum of a woman who had recently delivered a stillborn infant. The antibody was subsequently identified as anti-Rh, which is produced by the mother in response to, and directed against, the Rh antigen of the blood cells of her own fetus. Clinical records reveal that the mothers of affected newborns lack the Rh antigen (that is, are *Rh-negative*) whereas their husbands and affected infants possess the Rh antigen (that is, are *Rh-positive*).

The Rh antigen in the blood cell is controlled by a dominant allele, designated *R*. An Rh-positive person has the dominant allele, either in the homozygous *(RR)* or heterozygous *(Rr)* state. All Rh-negative individuals carry two recessive alleles *(rr)* and are incapable of producing the Rh antigen. The inheritance on the Rh antigen follows simple Mendelian principles (fig. 8.10). A mother who is Rh-negative *(rr)* need not fear having Rh-diseased offspring if her husband is likewise Rh-negative. If the husband is heterozygous *(Rr)*, half of the offspring will be Rh-negative *(rr)* and none of these will be afflicted. The other half will be Rh-positive *(Rr)*, just like the father, and are potential victims of hemolytic disease. If the Rh-positive father is homozygous *(RR)*, then all the children will be Rh-positive *(Rr)* and potential victims. In essence, an Rh-positive child carried by an Rh-negative mother is the setting for possible, though not inevitable, trouble.

The chain of events leading to hemolytic disease begins with the inheritance by the fetus of the dominant *R* allele of the father. The Rh antigens are produced in the red blood cells of the fetus.

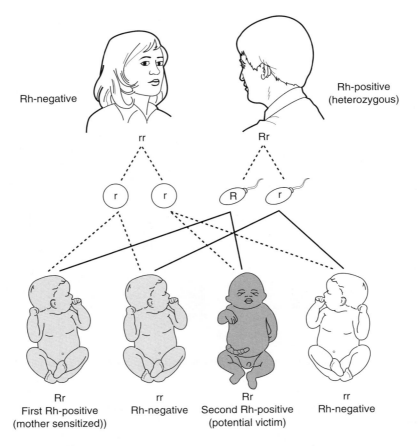

Rh-negative

Rh-positive (heterozygous)

rr

Rr

r   r   R   r

Rr
First Rh-positive
(mother sensitized))

rr
Rh-negative

Rr
Second Rh-positive
(potential victim)

rr
Rh-negative

Figure 8.10 Offspring that may arise from a marriage of an Rh-negative woman
and an Rh-positive man (in this case, heterozygous). The mother must be sensitized to the
Rh antigen before delivering a child affected with Rh disease.

The fetal red cells bearing the Rh antigens escape
through the placental barrier into the mother's circulation, and stimulate the production of antibodies (anti-R) against the Rh antigens on the fetal
red cells (fig. 8.11). The mother having produced
antibodies, is said to be immunized (or sensitized)
against her baby's blood cells. The maternal antibodies hardly ever attain a sufficient concentration during the first pregnancy to harm the fetus.
In fact, although fetal cells cross the placenta
throughout pregnancy, they enter the maternal
circulation in much larger numbers during delivery, when the placental vessels rupture. It is now

generally conceded that sensitization of the mother
takes place shortly *after* the delivery of the first
Rh-positive child. Accordingly, the firstborn is
rarely affected, unless the mother had previously
developed antibodies from having been transfused
with Rh-positive blood or has had a prior pregnancy terminating in an abortion.

The antibodies remain in the mother's system, and may linger for many months or years
(fig. 8.11). If the second baby is also Rh-positive,
the mother may send sufficient antibodies into the
fetus' bloodstream to destroy the fetal Rh-positive
red cells. The majority of affected fetuses survive

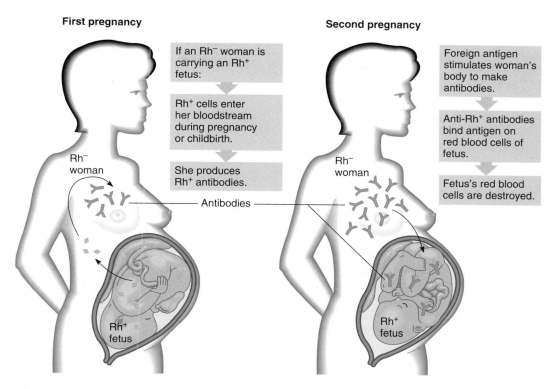

First pregnancy

If an Rh⁻ woman is carrying an Rh⁺ fetus:

Rh⁺ cells enter her bloodstream during pregnancy or childbirth.

She produces Rh⁺ antibodies.

Rh⁻ woman

Antibodies

Rh⁺ fetus

Second pregnancy

Foreign antigen stimulates woman's body to make antibodies.

Anti-Rh⁺ antibodies bind antigen on red blood cells of fetus.

Fetus's red blood cells are destroyed.

Rh⁻ woman

Rh⁺ fetus

Figure 8.11 Rh incompatibility. Fetal cells entering the pregnant woman's bloodstream can stimulate her immune system to make anti-Rh antibodies. If the fetus is Rh⁺ and she is Rh⁻. A drug called RhoGAM prevents attacks on subsequent fetuses.

for the usual gestation period but are born in critical condition from anemia. Severely anemic individuals are likely to be jaundiced and develop heart failure. A grave threat to the newborn infants is bilirubin, which is a product of red cell destruction. During pregnancy, fetal bilirubin is transported across the placenta and eliminated by the mother. From the time of birth, however, bilirubin accumulates as the affected infant fails to dispose of it. Bilirubin has been shown to be highly toxic to the soft brain tissues; the brain may be permanently damaged.

Another aspect of the hemolytic condition, responsible for its original name *(erythroblastosis fetalis),* is the presence of an extraordinarily large number of immature red cells (the erythroblasts) in the circulating blood. It is as if the liver and the spleen, in an attempt to combat the severe anemic

condition, produce vast numbers of "unfinished" red blood cells. Unattended, erythroblastosis fetalis leads to stillborn or neonatal death. Many of the erythroblastotic babies are saved by exchange transfusion of Rh-negative blood; others, however, die despite treatment. Exchange transfusions are essentially a flushing-out process, whereby the infant's blood is gradually diluted with Rh-negative donated blood until, at the end of the procedure, most of the infant's circulating blood is problem free. In severe cases, where it has been predicted that the fetus would die before it was mature enough for premature delivery, intrauterine transfusions have been used successfully.

As might have been anticipated, the Rh gene has turned out to be more complex than initially envisioned. There are several variant alleles, and there is a corresponding diversity of antigenic constitutions. This diversity need not concern us here. The most common antigen is the one that was first recognized, known more specifically now as Rh_0, or D. It is the presence of Rh_0 that is tested in ordinary clinical work.

Among Caucasians in the United States, the incidence of Rh-negative persons is approximately 16 percent. In certain European groups, such as the Basques in Spain, the frequency of Rh-negative individuals rises as high as 34 percent. Non-Caucasian populations are relatively free of Rh hemolytic disease. The incidence of Rh-negative persons among the full-blooded Native Americans, Eskimos, African blacks, Japanese, and Chinese is 1 percent or less. In contrast, the frequency of Rh-negative African Americans is high (9 percent), which reflects the historical consequences of intermarriages.

We can now appreciate the reasons for viewing Rh disease as an instance of selection *against* the heterozygote. The erythroblastotic infant is always the heterozygote *(Rr)*. Each death of an erythroblastotic infant *(Rr)* results in the elimination of one *R* and one *r* gene. In such a situation, where selection continually operates against the heterozygote, the rarer of the two genes should ultimately become lost (or decline to a low level to be maintained solely by mutation). In populations where the *R* gene is much more common that the *r* allele, we should be witnessing a gradual dwindling of the *r* gene.

No decline in the frequency of the *r* gene is evident, however. One counterbalancing factor is the tendency of parents who have lost infants from erythroblastosis to compensate for their losses by having relatively large numbers of children. Thus, if a father is heterozygous *(Rr)* and the mother is homozygous *(rr)*, there is an even chance that the infant will be *rr* and unaffected. Each unaffected child born restores two *r* genes lost by the death of two *Rr* erythroblastotic sibs. Accordingly, an excess of homozygous children *(rr)* counterbalances the *r* genes lost through erythroblastosis. This consideration alone overrides the selective force against the heterozygote.

THE CONTROL OF RH DISEASE

In the 1960s Drs. Vincent Freda and John Gorman at the Columbia-Presbyterian Medical Center in New York and Dr. William Pollack at the Ortho Research Foundation in New Jersey sought the means of suppressing the production of antibodies in Rh-negative mothers who had recently delivered an Rh-positive infant. Experiments performed 50 years earlier by the distinguished American bacteriologist, Dr. Theobald Smith, furnished an important clue to the solution. In 1909, Smith arrived at the general principle that passive immunity can prevent active immunity. That is, an antibody given passively by injection can inhibit the recipient from producing its own active antibody. After five years of experimentation and testing, Drs. Freda, Gorman, and Pollack successfully developed an immunosuppressant consisting of a blood fraction (gamma globulin) rich in Rh antibodies. Injected into the bloodstream of the Rh-negative mother no later than three days after the birth of her first Rh-positive child, the globulin-Rh antibody preparation (known as RhoGAM) suppresses the mother's antibody-making activity. Countless numbers of mothers have received the preparation; none have formed active antibodies. More impressive, many of them have delivered a second Rh-positive baby and none of the babies has been afflicted with Rh disease. The evidence is overwhelming that the immunosuppressant is very effective.

The immunosuppressant does not prevent hemolytic disease if maternal antibody is already present by earlier pregnancies or by previous transfusions of Rh-positive blood. For this reason the preparation is administered only to Rh-negative women who do not have anti-Rh in their sera at the time of delivery. With each delivery, opportunity for exposure to Rh-positive fetal cells is repeated. The protection given at the delivery of the first infant does not protect the mother from exposure to the antigen received at a subsequent pregnancy. Hence, the immunosuppressant must be given immediately following each pregnancy.

Since the management and prevention of Rh disease have advanced considerably, we have witnessed the end of Rh disease as a troublesome clinical problem. We can also anticipate that the recessive *r* allele will not disappear from the human gene pool, but rather endure. In other words, the *r* allele will be perpetuated by the now-surviving heterozygotes.

IMPLICATIONS OF BALANCED POLYMORPHISM

According to the classical concept of selection (discussed in chapter 6), a deleterious gene has a harmful effect when homozygous and virtually no expression in the heterozygous state. Deleterious genes will, by selection, be reduced in frequency to very low levels and will be maintained in the population by recurrent mutations. The fittest individuals are homozygous for the normal, or "wild type", allele at most loci. Stated another way, the fittest individuals carry relatively few deleterious genes in the heterozygous state.

In this chapter, we have considered examples of genes that impair fitness when homozygous and actually improve fitness in the heterozygous state. To what extent are individuals heterozygous at their gene loci? Estimates of the number of genetically *polymorphic* loci (i.e., the number of loci where two or more allelic forms occur) varies across species and between regions of the genome that code for proteins (genes) and those that do not code for genes. If the alternative alleles at these polymorphic loci are maintained by selection that favors the heterozygote, then low fitness would have to be assigned to an unusually large number of homozygous loci. In other words, an unreasonably high level of unfitness would prevail in the population because of the selective disadvantage of many alleles in the homozygous state.

The possibility exists that most of the alternative alleles at polymorphic loci are *selectively neutral*—that is, the different alleles at one locus confer neither selective advantage nor disadvantage to the individual. Much of the observed variation, then, would represent merely the accumulation of neutral mutations. The notion of selectively neutral alleles has been much debated. Are there mutant alleles whose effects on fitness are not at all different from the more frequent alleles that lead to a normal phenotype? And, can some of these neutral mutant alleles, purely by chance, reach frequencies as high as the frequencies that characterize the state of balanced polymorphism? We will attempt to resolve these vexing queries in subsequent chapters.

GENETIC DRIFT AND GENE FLOW

The peculiar multilegged condition of the bullfrog (see fig. 1.1) is a rarity in nature. Yet, we witnessed in a particular locality an exceptionally high incidence of this trait. We surmised that the deformity may have been caused by a recessive mutant gene. Since harmful recessive genes in a population tend to be carried mostly in the heterozygous state, the multilegged frogs probably arose from matings of heterozygous carriers. The probability that two or more carriers will actually meet is obviously greater in a small population than in a large breeding assemblage. In fact, the number of matings of carriers of a particular recessive allele in a population is mainly a function of the size of the population. Most populations are not infinitely large, and many fluctuate in size from time to time.

During a period when a population is small, chance matings and segregations could lead to an uncommonly high frequency of a given recessive gene. For example, it is not unthinkable that an unduly harsh winter sharply reduced the size of our particular bullfrog population. By sheer chance, an unusually large proportion of heterozygous carriers of the multilegged condition might have survived the winter's severity and prevailed as parents in the ensuing spring's breeding aggregation. In this manner, the "multilegged" gene, although not at all advantageous, would occur with an extraordinarily high incidence in the new generation of offspring. Such a fortuitous change in the genetic makeup of a population that may arise when the population becomes restricted in size is known as *genetic drift*.

ROLE OF GENETIC DRIFT

Examination of a natural situation that may be illustrative of genetic drift will lead us into a simplified mathematical consideration of the concept. Coleman Goin, a naturalist at the University of Florida, studied the distribution of pigment variants of a terrestrial frog, known impressively as *Eleutherodactylus planirostris*. We shall refer to the frog simply by its vernacular name, the greenhouse frog. This frog may possess either of two pigmentary patterns, mottled or striped (fig. 9.1). A unique feature of the greenhouse frog is the terrestrial development of its eggs. That is, the eggs need not be submerged completely in water, but

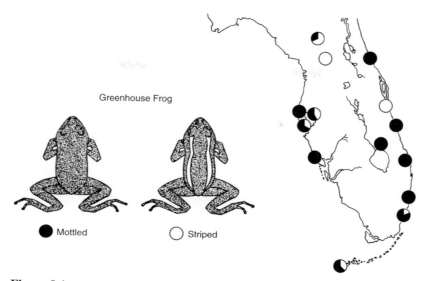

Greenhouse Frog

● Mottled   ○ Striped

Figure 9.1 **Distribtion of the greenhouse frog** in Florida, and the relative frequencies of the two pattern variants, *mottled* and *striped*. Immediately after their initial introduction, the populations were small and isolated and differed appreciably in the incidences of mottled and striped forms. The varied frequencies were likely not be due to natural selection but represented the outcome of chance fluctuations of genes, or genetic drift.
(Based on studies by Coleman Goin.)

can develop in moist Earth. This important quality has a considerable bearing on the dispersal of the frogs. Goin successfully reared eggs in flower pots two-thirds filled with beach sand and placed in a finger bowl of water. The analysis of a large number of progeny hatched from many different clutches of eggs revealed that the striped pattern is dominant to the mottled pattern.

The aptly named greenhouse frog is native to Cuba, the Cayman Islands and the northern Bahamas and later became established in Florida. Cuba was the center of dispersal from which the Florida populations is derived. In recent decades this tiny frog has become much more widespread throughout southern Florida, and disjunct colonies have been introduced into Georgia, Louisiana, Mississippi, Alabama, Missouri, Oklahoma, and even Hawaii. However, shortly after its introduction into Florida, the greenhouse frog population consisted of a series of small isolated colonies, as shown in figure 9.1. The proportions of the two patterns varied in different colonies in Florida. In several colonies, only the

mottled type occurred, while in other colonies only the striped type could be found. Most populations consisted of a mix of the two types of frog. What could account for the local preponderance of one or the other pattern, or even the absence of one of the contrasting patterns?

Goin conjectures that the greenhouse frog was introduced into Florida by means of clutches of eggs accidentally included, as stowaways, in shipments of tropical plants and landscaping material from Cuba, a distinct possibility in view of the terrestrial development of the eggs that undoubtedly accounts for its introduction into other U.S. states. Thus, a single clutch of introduced eggs could initiate a small colony that, in turn, would establish at the outset a given pattern or proportion of patterns. The presence of only mottled forms in a population may be due to the chance circumstance that only mottled eggs were introduced. Or perhaps both striped and mottled eggs were included in the shipment, but by sheer accident one type was lost in succeeding generations.

It should be understood that Goin did not prove that the unusual distribution and frequency of the two pigment patterns were due solely to chance. The demonstration of genetic drift in any natural population is an extremely difficult task. Genetic drift of the variant pigment patterns is, however, a reasonable explanation. Today the greenhouse frog has a much more extensive distribution, making for a much more complex state of affairs.

THEORY OF GENETIC DRIFT

The theory of genetic drift was systematically developed in the 1930s by the distinguished American geneticist Sewall Wright. In fact, the phenomenon of drift is frequently called the *Sewall Wright effect*. The Sewall Wright effect refers specifically to the random fluctuations (or drift) of gene frequencies from generation to generation in a population of small size. Because of the limited size of the breeding population, the *gene pool* of the new generation may not be at all representative of the parental gene pool from which it was drawn.

The essential features of the process of drift may be seen in the following modest mathematical treatment. Let us assume that the numerous isolated colonies in Florida were each settled by only two frogs, a male and a female, both of constitutions *Aa*. Let us further suppose that each mated pair produces only two offspring. The possible genotypes of the progeny, and the chance associations of the genotypes, are shown in table 9.1.

Several meaningful considerations emerge from table 9.1. For example, the chance that the first offspring from a cross of two heterozygous parents will be *AA* is 1/4. The second event is independent of the first; hence, the chance that the second offspring will be *AA* is also 1/4. The chance that *both* offspring will be *AA* is the product of the separate probabilities of the two independent events, 1/4 × 1/4, or 1/16.

We may now ask: What is the probability of producing two offspring, one *AA* and the other *Aa*, *in no particular order?* From table 9.1, we can see that the chance of obtaining an *AA* individual followed by an *Aa* individual is 2/16. Now, the wording of our question requires that we consider a second possibility, that of an *Aa* offspring followed by an *AA* offspring (also 2/16). These two probabilities must be added together to arrive at the chance of producing the two genotypes irrespective of the order of birth. Hence, in the case in question, the chance is 2/16 + 2/16, or 4/16. In like manner, it may be ascertained that the expectation of obtaining one *AA* and one *aa* offspring (in no given order) is 2/16 and that of producing one *Aa* and one *aa* (in any sequence) is 4/16.

The essential point is that any one of the above circumstances may occur in a given colony. We

TABLE 9.1	Chance Distribution of Offspring of Two Heterozygous Parents (*Aa* × *Aa*)			
Genotype of first offspring	Probability of first event	Genotype of second offspring	Probability of second event	Total Probability
AA	1/4	*AA*	1/4	Both offspring *AA*, 1/16
AA	1/4	*Aa*	2/4	*AA* followed by *Aa*, 2/16
AA	1/4	*aa*	1/4	*AA* followed by *aa*, 1/16
Aa	2/4	*AA*	1/4	*Aa* followed by *AA*, 2/16
Aa	2/4	*Aa*	2/4	Both offspring *Aa*, 4/16
Aa	2/4	*aa*	1/4	*Aa* followed by *aa*, 2/16
aa	1/4	*AA*	1/4	*aa* followed by *AA*, 1/16
aa	1/4	*Aa*	2/4	*aa* followed by *Aa*, 2/16
aa	1/4	*aa*	1/4	Both offspring *aa*, 1/16

may concentrate on one situation. The probability that a colony will have only two AA offspring is 1 in 16. Thus, by the simple play of chance, the parents initiating the colony might not leave an aa offspring. The a gene would be immediately lost in the population. Subsequent generations descended from the first-generation AA individuals would contain, barring mutation, only AA types. Chance alone can thus lead to an irreversible situation. A gene once lost could not readily establish itself again in the population. The decisive factor is the size of the population. When populations are small, striking changes can occur from one generation to the next. Some genes may be lost or reduced in frequency by sheer chance; others may be accidentally increased in frequency. Thus, the genetic architecture of a small population may change irrespective of the selective advantage or disadvantage of a trait. Indeed, a beneficial gene may be lost in a small population before natural selection has had the opportunity to act on it favorably.

FOUNDER EFFECT

When a few individuals or a small group migrate from a main population, only a limited portion of the parental gene pool is carried away. In the small migrant group, some genes may be absent or occur in such low frequency that they may be easily lost. The unique frequencies of genes that arise in populations derived from small bands of colonizers, or "founders," has been called the *founder effect*. This expression emphasizes the conditions or circumstances that foster the operation of genetic drift.

The American Indians afford a possible example of the loss of genes by the founder principle. North American Indian tribes, for the most part, surprisingly lack the allele (designated I^B) that governs type B blood. However, in Asia, the ancestral home of the American Indian, the I^B allele is widespread. The ancestral population that migrated across the Bering Strait to North America might well have been very small. Accordingly,

the possibility exists that none of the prehistoric immigrants happened to be of blood group B. It is also conceivable that a few individuals of the migrant band did carry the I^B allele but they failed to leave descendants.

The interpretation based on genetic drift should not be considered as definitive. The operation of natural selection cannot be flatly dismissed. Most Native Americans possess only blood group O, or stated another way, contain only the recessive blood allele i. With few exceptions, the North American Indian tribes have lost not only blood group allele I^B but also the allele that controls type A blood I^A. The loss of both alleles, I^A and I^B, by sheer chance, perhaps defies credibility. Indeed, many modern students of evolution are convinced that some strong selective force led to the rapid elimination of the I^A and I^B genes in the Native American populations. If this is true, it would offer an impressive example of the action of natural selection in modifying the frequencies of genes in a population.

Native Americans are also known to have a high frequency of albinism. The incidence of albinism among the Cuna Indians of the San Blas Province in Panama is about 1 in 200, which contrasts sharply with the 1 in 20,000 figure for European Caucasians. The Hopi of Arizona and the Zuni of New Mexico, like the Cuna, are also remarkable in their high numbers of albino individuals.

Because these Native American populations are small, one might suspect the operation of genetic drift, notwithstanding the adverse effects of the albino allele. It is difficult to imagine, however, that by chance alone this detrimental allele could reach a high frequency independently in several Native American populations. Charles M. Woolf, geneticist at Arizona State University, has suggested that the high incidence of albinism among the Hopi of Arizona reflects an inimitable form of selection that he terms *cultural selection*. Albinos have been highly regarded in the traditional Hopi society and actually have enjoyed appreciable success in sexual activity. The albino male has been admired, and some have become legendary for leaving large numbers of offspring. Woolf further notes that the fading of old customs among the Hopi is beginning to nullify any reproductive advantage held by albino males in past generations. The frequency of albinism may be expected to decline with the dissolution of the traditional Hopi way of life.

RELIGIOUS ISOLATES

The most likely situation to witness genetic drift is one in which the population is virtually a small, self-contained breeding unit, or *isolate,* in the midst of a larger population. This typifies the Dunkers, a very small religious sect in eastern Pennsylvania. The Dunkers are descendants of the Old German Baptist Brethren, who came to the United States in the early eighteenth century. In the 1950s, Bentley Glass, a professor at Johns Hopkins University, studied the community of Dunkers in Franklin County, Pennsylvania, which then numbered about 300 individuals. In each generation, the number of parents has remained stable at about 90. The Dunkers live on farms intermingled with the general population, but are genetically isolated by rigid marriage customs. The choice of mates is restricted to members within the religious group.

Glass, with his colleagues, compared the frequencies of certain traits for the Dunker community, the surrounding heterogeneous American population, and the population in western Germany from which the Dunker sect had emigrated centuries ago. Such a comparison of a small isolate with its large host and parent populations should reveal the effectiveness, if any, of genetic drift. In other words, if the small isolated population shows aberrant gene frequencies as compared to the large parent population, and if the other forces of evolution can be excluded, then the genetic differences can be ascribed to drift.

Analyses were made of the patterns of inheritance of three blood group systems—the ABO blood groups, the MN blood types, and the Rh blood types. In addition, data were accumulated on the incidences of four external traits—namely, the configuration of the ear lobes (which either may be attached to the side of the head or hang free), right- or left-handedness, the presence or absence of hair on the middle segments of the fingers (mid-digital hair), and "hitchhiker's thumb," technically termed distal hyperextensibility (fig. 9.2).

The frequencies of many of these traits are strikingly different in the Dunker community from those of the general United States and West German populations. Blood group A is much more frequent among the Dunkers; the O group is somewhat rarer in the Dunkers; and the frequencies of groups B and AB have dropped to exceptionally low levels in the Dunker community. In fact, the I^B allele has almost been lost in the isolate. Most of the carriers of the I^B allele were not born in the community but were converts who entered the isolate by marriage.

A noticeable change has also occurred in the incidences of the M and N blood types in the Dunker community. Type M has increased in frequency, and type N has dwindled in frequency as compared with the incidences of these blood types in either the general United States population or the West German population. Only in the Rh blood groups do the Dunkers conform closely to their surrounding large population.

In physical traits, equally striking differences are found. Briefly, the frequencies of mid-digital hair patterns, distal hyperextensibility of the thumb, and attached ear lobes are significantly lower in the Dunker isolate than in the surrounding American populations. The Dunkers do, however, agree well with other large populations in the incidence of left-handedness. It would thus appear that the peculiar constellation of gene frequencies in the Dunker community—some uncommonly high, others uniquely low, and still others, unchanged from the general large population—can be best attributed to chance fluctuations, or genetic drift.

There is no concurrence of opinion among evolutionists concerning the operation of genetic drift in natural populations, but few would deny

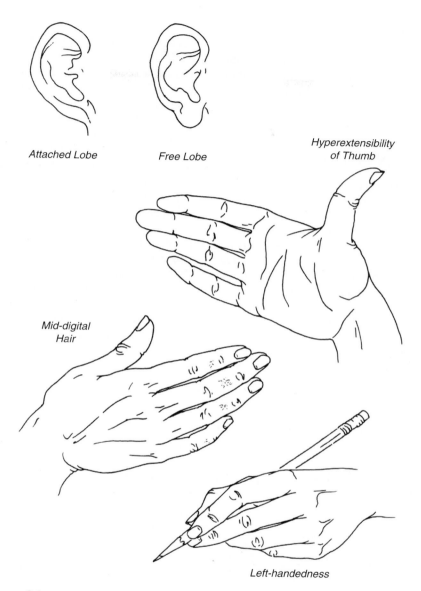

Attached Lobe   Free Lobe   Hyperextensibility
of Thumb

Mid-digital
Hair

Left-handedness

Figure 9.2 Inheritable physical traits—nature of ear lobes, "hitchhiker's thumb," mid-digital hair, and left-handedness—studied by Bentley Glass and his co-workers in members of the small religious community of Dunkers in Pennsylvania. The distinctive frequency of most of these traits in the Dunker population suggests the operation of genetic drift.

that small religious isolates have felt the effect of random sampling. It should be clear, however, that genetic drift becomes ineffectual when a small community increases in size. Fluctuations or shifts in gene frequencies in large populations are determined almost exclusively by selection.

AMISH OF PENNSYLVANIA

We have seen that gene frequencies in small religious isolates may differ significantly from the original large populations from which the isolates were derived. Another feature of small isolates is the occurrence of rare recessive traits in greater numbers than would be expected from random mating in a large population. This is witnessed among the Old Order Amish societies in the eastern United States.

The Amish sect is an offshoot of the Mennonite Church; both religious groups settled in the United States to escape persecution in Europe in past centuries. The Amish are plain dressed, rural-living people who cultivate the religious life apart from the world, resisting modern conveniences. Present-day communities were founded by waves of Amish immigration

that began about 1720 and continued until about 1850. The vast majority of Amish live in relatively isolated colonies in Pennsylvania, Ohio, and Indiana. Each community is descended from a small immigrant stock, as attested by the relatively few family names in a given community. Analyses have shown that eight names account for 80 percent of the Amish families in Lancaster County, Pennsylvania. Other Amish communities also have a high frequency of certain family names, as table 9.2 shows.

Marriages have been largely confined within members of the Amish sect, with a resulting high degree of *consanguinity*. Matings of close relatives have tended to promote the meeting of two normal, but carrier, parents. Four recessive disorders manifest themselves with uncommonly high frequencies, each in a different Amish group: the Ellis–van Creveld syndrome, pyruvate kinase-deficient hemolytic anemia, hemophilia B (Christmas disease), and a form of limb-girdle muscular dystrophy (Troyer syndrome).

We may consider in some detail the Ellis-van Creveld syndrome, which occurs in the Lancaster County population (see fig. 1.6). Dozens of affected persons have been identified in a small number of sibships, most of which have unaffected parents.

TABLE 9.2	Old Order Amish Family Names in Three American Communities*				
Lancaster Co., Pa.		**Holmes Co., Ohio**		**Mifflin Co., Pa.**	
Stolzfus**	23%	Miller	26%	Yoder	28%
King	12%	Yoder	17%	Peachey	19%
Fischer	12%	Troyer	11%	Hostetler	13%
Beiler	12%	Hershberger	5%	Byler	6%
Lapp	7%	Raber	5%	Zook	6%
Zook	6%	Schlabach	5%	Speicher	5%
Esh***	6%	Weaver	4%	Kanagy	4%
Glick	3%	Mast	4%	Swarey	4%
	81%		77%		85%
Totals:					
1,106 families, 1957		1,611 families, 1960		238 families, 1951	

*From data compiled by Victor A. McKusick of John Hopkins University.
**Including Stolzfoos.
***Including Esch.

Pedigree analysis has revealed that Samuel King and his wife, who immigrated in 1744, are ancestral to all parents of the sibships. Either Samuel King or his wife carried the recessive gene. None of their children were affected, but subsequent generations were. Evidently, previously concealed detrimental recessive genes are brought to light by the increased chances of two heterozygotes meeting in a small population.

CONSANGUINITY AND GENETIC DRIFT

Small populations not only provide opportunities for genetic drift but for consanguinity as well. As witnessed among the Amish, offspring afflicted with recessive disorders arise more often from unions of close relatives than from matings between unrelated persons. Yet the Amish sternly frown on first-cousin marriages. Nonetheless, although consanguineous marriages are not made by choice, the limited size of the Amish population restricts the availability of potential mates and virtually forces marriages of close relatives. Indeed, second-cousin and third-cousin marriages are rather common among the Amish. The relationship between genetic drift and inbreeding thus becomes clear: the closed nature of a small population creates a situation wherein few nonrelatives are present in the population. Because matings of genetically related individuals increase the probability of homozygosity in the progeny, a relatively large number of rare recessive homozygotes are expected, and do occur in *endogamous* populations.

The smaller the population and the longer it has been isolated, the greater the chance that most members of the population are related to each other through common ancestors. This is exemplified by the isolated island of Tristan da Cunha in the South Atlantic (fig. 9.3). This small island is of historical interest. In 1816, a British military

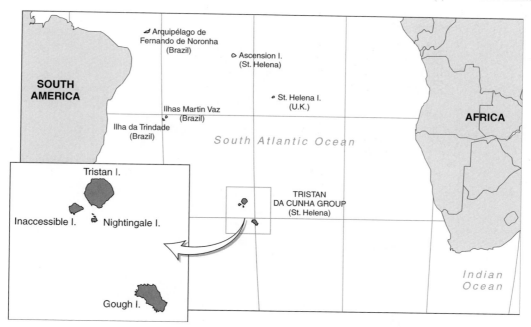

Figure 9.3 **Tristan da Cunha** and its near neighbors Inaccessible Island, Nightingale Island, and Gough Island make up the most remote inhabited archipelago in the world. These volcanic islands in South Atlantic Ocean are 1,750 miles (2,816 kilometers) from South Africa and 2,088 miles (3,360 kilometers) from South America.

garrison was established on Tristan for the sole purpose of safeguarding against the escape of Napoleon from the neighboring, though distant, St. Helena. Within a few months, it became glaringly apparent that Tristan was inconsequential to Napoleon's safekeeping. The battery of soldiers was hastily withdrawn, but a Scots corporal, William Glass, his family, and a few soldiers remained behind. From 1817 until 1908, women from St. Helena and shipwrecked sailors joined the community to bring the total to a mere 15 individuals. The population of roughly 270 traces their origin to the 15 early settlers.

The English geneticist D. F. Roberts has calculated that in Tristan the probability is high that any two young individuals contemplating marriage share numerous genes due to the common ancestry of their parents. Roberts stresses that the relatively high level of consanguinity on the island does not reflect a conscious preference for marriage between close relatives.

Rather, random mating prevails, but the marriages prove to be consanguineous because most of the potential mates are already relatives by common descent.

GENE FLOW

A rich archeological record reveals appreciable movement on the part of early human populations. Some migrations were sporadic, in small groups; others were more or less continual streams involving large numbers of people. Large-scale immigrations followed by interbreeding have the effect of introducing new genes to the host populations. The diffusion of genes into populations through migrations is referred to as *gene flow*.

The graded distribution of the I^B blood-group allele in Europe represents the historical consequence of invasions by Mongolians who pushed westward repeatedly between the sixth

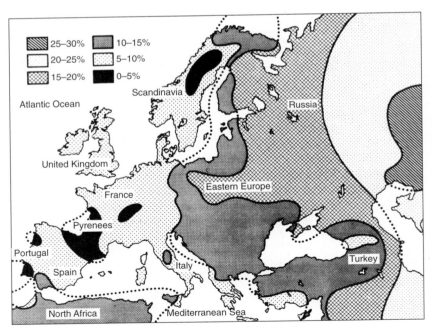

Figure 9.4 Gradient of frequencies of the I^B blood-group gene from central Asia to western Europe.

(Based on studies by A. E. Mourant.)

and sixteenth centuries (fig. 9.4). There is a high frequency of the I^B allele in central Asia. In Europe, the frequency of the I^B gene diminishes steadily from the borders of Asia to a low level of 5 percent or less in parts of Holland, France, Spain, and Portugal. The Basque peoples, who inhabit the region of the Pyrenees in Spain and France, have the lowest frequency of the I^B allele in Europe—below 3 percent. From a biological standpoint, the Basque community of long standing is a cohesive, endogamous mating unit. The exceptionally low incidence of the I^B gene among the Basques may be taken to indicate that there has been little intermarriage with surrounding populations. It is possible that a few centuries ago the I^B allele was completely absent from the self-contained Basque community.

The exchange of genes between populations may have dramatic consequences. Until recently, Rh disease was virtually unknown in China. Less than 80 years ago, all Chinese women were Rh-positive *(RR)*. However, intermarriage between immigrant Americans and the native Chinese has led to the introduction of the Rh-negative allele *(r)* in the Chinese population. No Rh disease would be witnessed in the immediate offspring of American men and Chinese women. By contrast, all marriages of Rh-negative American women *(rr)* and Rh-positive Chinese men *(RR)* would be of the incompatible type. All children by these Chinese fathers would be Rh-positive *(Rr)* and potential victims of hemolytic disease.

Whereas American immigrants introduce the Rh-negative allele *(r)* into Chinese populations where it formerly was not present, Chinese immigrants (all of whom are *RR*) introduce more Rh-positive genes *(R)* into the American populations, thus diluting the Rh-negative gene pool in the United States. Initially, the Rh-positive Chinese men *(RR)* married to Rh-negative American women *(rr)* would result in an increased incidence of Rh-diseased infants. In later generations, however, the frequency of Rh-negative women in the United States would be lower, inasmuch as women of mixed Chinese-American origin would be either *RR* or *Rr,* predominantly the former. Thus, in the

United States, the long-range effect of Chinese-American intermarriage is a reduction in the incidence of hemolytic disease of the newborn.

EUROPEAN GENES IN AFRICAN AMERICANS

Increasingly, people today question the need for and validity of the term *race* when applied to humans. This is due, in part, to the realization that there is more genetic diversity within so-called human races than between races. Likewise, studies continue to show that all people are closely related and have a common ancestor that is less than 200,000 years old (see chapter 17). Nonetheless, people still self-identify as being part of a particular racial group.

For nearly four centuries, *admixture,* or the interchange of genes among populations, between American Caucasians of European ancestry and African Americans has been commonplace and widespread. The forced migration of African American slaves to the New World can be traced to 1619, when slaves first arrived at Jamestown Virginia; Spanish expeditions likely first brought African slaves to this country as early as 1526 but no permanent settlements resulted from those forays to North America.

Over the past 300 plus years, the extent of Caucasian–African American admixture has been variable. The recent mobility of peoples from one part of the country to another has diminished the variability in admixture measures compared to earlier studies done in the 1950s and 1960s. With advances in DNA technology, the ability to measure this phenomenon has become more refined. Historically, geneticists and anthropologists used single blood groups, such as the Duffy blood factor gene, but now there are more informative genetic markers for measuring admixture.

During the peak of the Atlantic slave trade, the majority of the African slaves originated from the coastal regions of western and central Africa. Anthropological geneticist Mark D. Shriver of Pennsylvania State University and colleagues estimated the European genetic contribution to nine

U.S. African American populations using nine autosomal DNA markers. Shriver used highly variable DNA markers capable of differentiating major geographic and ethnic groups. These "population specific alleles" have great utility for studying population genetics, forensics, and epidemiology. Three of the nine genes studied are shown in table 9.3. They provide examples of the extraordinary variability in modern genetic markers and the varying amount of admixture between Europeans and African Americans. For example, in the FY-NULL*1 gene, there is

TABLE 9.3	Frequencies of Population-Specific Alleles in African, Jamacan, and American Populations and Estimates of Genetic Contribution to U.S. African American Populations from Europeans

Population	Allelic Frequency for APO*1	Allelic Frequency for FY-NULL*1	Allelic Frequency for GC-15	Mean Admixture
African				
Nigeria 1	0.409	0.000	0.081	*
Nigeria 2	0.480	0.000	0.085	*
Central African Republic	0.435	0.000	0.067	*
African Average	0.441	0.000	0.078	*
European				
England	0.934	1.000	0.622	*
Ireland	0.915	1.000	0.633	*
Germany	0.933	1.000	0.567	*
European Average	0.927	1.000	0.607	*
African American **1. Non-Southern**				
Detroit	0.533	0.133	0.177	16.3
Maywood, Ill.	0.520	0.185	0.200	18.8
New York	0.522	0.210	0.146	19.8
Philadelphia	0.499	0.153	0.142	12.9
Pittsburg	0.551	0.217	0.129	20.2
Baltimore	0.505	0.141	0.176	15.5
2. Southern				
Charleston, S.C.	0.500	0.112	0.133	11.6
New Orleans	0.593	0.200	0.215	22.5
Houston	0.525	0.188	0.313	16.9
Caribbean				
Jamaica	0.511	0.065	0.113	6.8
European Americans				
Detroit	0.935	0.990	0.564	*
Pittsburg	0.917	0.983	0.569	*
Louisiana (Cajuns)	0.935	0.989	0.628	*

Compiled from Parra, E., Marcini, A., Akey, J., Martinson, J., Batzer, M., Cooper, R., Forrester, T., Allison, D. B., Deka, R., Ferrell, R. E., and Shriver, M. D. (1998) Estimating African-American Admixture Proportions by use of Population-Specific Alleles *Am. J. Hum. Genet.* 63:1839-1851.

100 percent divergence between Africans and Europeans, making it an especially discerning tool for differentiating between populations. For the other genes, APO*1 and GC-15, the divergence is not quite as great but they do show significant divergence, allowing them to serve as effective tools for differentiating between populations.

By using a combination of DNA markers, called a *haplotype,* Shriver was easily able to differentiate populations and more powerfully measure the amount of genetic admixture in each population than would be possible with any single genetic marker. Overall, the amount of admixture in U.S. populations varies from the lowest, 11.6 percent in Charleston, South Carolina, to the highest of 22.5 percent in New Orleans. By comparison, the Caribbean island of Jamaica has a dramatically lower rate of admixture of 6.8 percent.

New Orleans has the highest level of admixture of the cities studied, and the history of slavery in Louisiana territory is markedly different from that of other regions of the American South. Louisiana, and especially New Orleans, was a bastion for African Americans who could escape harsh slavery if they could find their way there. Since early colonial days, African Americans and people of mixed racial ancestry lived freely among Europeans in New Orleans, and many were able to establish themselves as "free people of color" *(les gens de couleur libres).* Historically known as Creoles or Creoles of Color, this population had and still has great influence on the ethos of this culturally unique American city. Although some people take offense to the use of the term *Creole,* it is not considered a racially derogatory term in New Orleans and South Louisiana—quite the opposite. The Creole culture has contributed immensely to the culture and traditions of New Orleans, and native New Orleanians and tourists alike relish in the food, architecture, and other cultural features that "Creoles of Color" have brought to the Crescent City.

10

RACES AND SPECIES

Any large assemblage of a particular organism is generally not distributed equally nor uniformly throughout its territory or range in nature. A widespread group of plants or animals is typically subdivided into numerous local populations, each physically separated from the others to some extent. The environmental conditions in different parts of the range of an organism are not likely to be identical. We may thus expect that a given local population will consist of genetic types adapted to a specific set of prevailing environmental conditions. The degree to which each population maintains its genetic distinctness is governed by the extent to which *interbreeding* between the populations occurs. A free interchange of genes between populations tends to blur the differences between the populations. But what are the consequences when gene exchange between populations is greatly restricted or prevented? This chapter addresses itself to this question.

VARIATION BETWEEN POPULATIONS

Our first consideration is to demonstrate that heritable variations exist among the various breeding populations in different geographical localities of an organism. Jens Clausen, David Keck, and William Hiesey of the Carnegie Institution of Washington at Stanford, California, demonstrated that each of the populations of the yarrow plant *Achillea lanulosa* from different parts of California are each adapted to their respective habitat. As figure 10.1 shows, the variations in height of the plant are correlated with altitudinal differences. The shortest plants are from the highest altitudes, and the plants increase in height in a gradient fashion with decreasing altitude. The term *cline,* or character gradient, has been applied to such situations where a character varies more or less continuously with a gradual change in the environmental terrain.

The observation by itself that the yarrow plants are phenotypically dissimilar at different elevations does not indicate that they are genetically different. If the observed variations are claimed as local adaptations resulting from natural selection, then a hereditary basis for the differences in height should be demonstrated. It is often difficult to obtain data that discloses the hereditary nature of population differences. In this respect, the studies of Clausen and his co-workers are commendable.

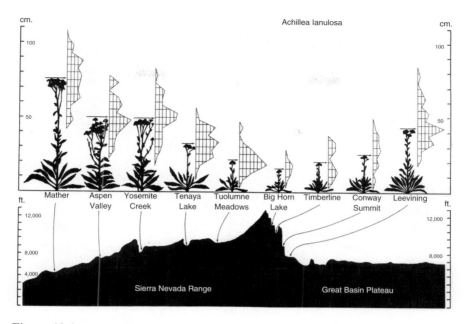

Figure 10.1 Clinal variation in the yarrow plant, *Achillea lanulosa*. The increase in height of the plant is more or less continuous with decreasing altitude. The plants shown here are representatives from different populations in the Sierra Nevada Mountains of California that were grown in a uniform garden at Stanford, California. Each plant illustrated is one of the average height for the given population; the graph adjacent to the plant reveals the distribution of heights within the population.

(From Clausen, Keck, and Hiesey, *Carnegie Institution of Washington Publication 581*, 1948.)

The plants shown in figure 10.1 had actually been grown together in a uniform experimental garden at Stanford, California. Transplanted from various localities, the plants developed differently from one another in the same experimental garden, revealing that each population had evolved its own distinctive complex of genes. At each environmental gradient, local populations are genetically adapted to narrow selective pressures.

SUBSPECIES

The variation pattern in organisms may be discrete, or discontinuous, particularly when the populations are separated from each other by pronounced physical barriers. This is exemplified by the varieties of the carpenter bee *(Xylocopa nobilis)* in the Celebes and neighboring islands of Indonesia (fig. 10.2). As shown by the studies of J. van der Vecht of the Museum of Natural History at Leiden in the Netherlands, there are three different varieties of carpenter bee on the mainland of Celebes and at least three kinds on the adjacent small islands. These geographical variants differ conspicuously in the coloration of the small, soft hairs that cover the surface of the body. The first abdominal segment is invariably clothed with bright yellow hairs. However, each variety has evolved a unique constellation of color on the other abdominal segments and also on the thorax.

The variations in the carpenter bees within and between islands are well defined and easily distinguishable. One may refer to populations with well-marked discontinuities as *race* or *subspecies*. Races are simply geographical aggregates of populations that differ in the incidence of genetic

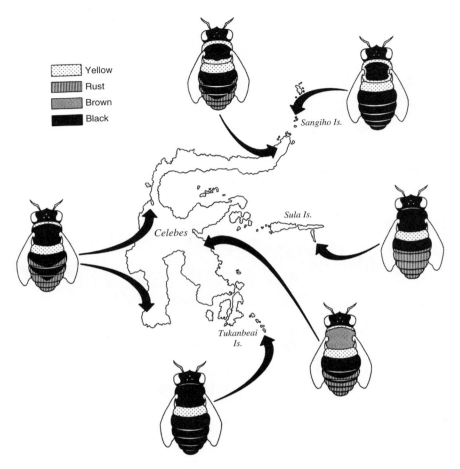

Figure 10.2 Geographic variation of color patterns in females of the carpenter bee, *Xylocopa nobilis,* in the Celebes and neighboring islands in Indonesia. Each geographic subspecies has evolved a distinctive constellation of colors.
(Based on studies by J. van der Vecht.)

traits. How genetically different two assemblages of populations must be to warrant racial designations is an open question.

Some of the problems inherent in delimiting races are exemplified by the different temperature-adapted populations of the North American leopard frog, *Lithobates pipiens.* The late John A. Moore, then at Columbia University and later at the University of California, tested the effects of temperature on the development of the embryos of

frogs from widely different localities. He wished to ascertain the limits of temperatures that the embryos can endure or tolerate. The findings on four different geographic populations in the eastern United States are shown in figure 10.3.

Embryos of northern *Lithobates pipiens* populations are more resistant to low temperatures and less tolerant of high temperatures than are embryos from southern populations. Embryos of populations from Vermont and New Jersey have

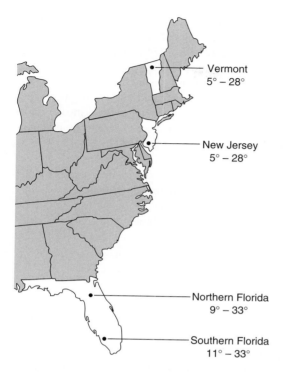

Figure 10.3 **Limits of temperature tolerance** of embryos of the leopard frog *(Lithobates pipiens)* from different geographical populations. Embryos of northern populations are more resistant to low temperatures and less tolerant of high temperatures than are embryos from southern populations.
(Based on studies by John A. Moore.)

comparable ranges of temperature tolerances. These northern embryos can resist temperatures as low as 5 °C. Embryos from Florida differ markedly from those of northern populations. Embryos from southern Florida (latitude 27 °N) can tolerate temperatures as high as 33 °C, but are very susceptible to low temperatures. Hence, northern and southern populations have become adapted to different environments in their respective territories.

Since Moore's classic 1940s experiments on leopard frog embryonic temperature tolerance, other more telling attributes have been studied in this broad-ranging amphibian. As it turns out, the diversity of mating calls across thermal and geographic clines reduces or eliminates interbreeding.

Using a combination of physical, biochemical, genetic, and behavioral characters, scientists have been able to deduce a more unified picture of the relationship among the various American leopard frogs. Today scientists recognize that what were once thought to be northern and southern races or subspecies of leopard frogs are distinct species: the southern leopard frog *(Lithobates utricularia)* and the northern leopard frog *(Lithobates pipiens)*. Males of the northern species call to their waiting females with "a deep rattling snore interspersed with clucking grunts," lasting about a second, while the southern species invites its mate with "short chuckle-like guttural trill." Mating calls are excellent examples of prezygotic (ethological/behavioral) isolating mechanisms, which are discussed in the section 'Reproductive Isolating Mechanisms' later in this chapter. Using similar criteria, a plains leopard frog *(Lithobates blairi)* and a Rio Grande leopard frog *(Lithobates berlandieri)* have also been recognized as distinct species within what used to be considered a large species complex with several geographic races or subspecies (fig. 10.4). This example demonstrates how new information leads to the re-evaluation of hypotheses and the formulation of new models, as discussed in chapter 1 regarding hypothetico-deductive reasoning and the scientific method.

Conceptually, subspecies may be best thought of as units of organization below the species level. In other words, subspecies may be considered as stages in the transformation of populations into species. But what constitutes a species? Up to this point, we have assiduously avoided the use of the term *species*. A discussion of the process leading to the formation of species will facilitate understanding of the term itself.

THE SPECIES CONCEPT

In recognition of the centrality of the *species* concept to modern biology, Harvard evolutionist and two-time Pulitzer Prize winner Edward O. Wilson declared the species "the atomic unit" of biology. Darwin likewise recognized the paramount importance of the species concept but, like present-day

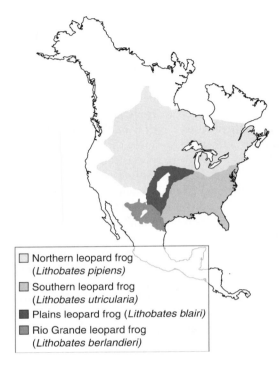

Figure 10.4 Leopard frogs. Based on Moore's 1940 breeding experiments examining defective embryos, leopard frogs were initially considered to be subspecies of a single, highly variable species. Modern classification based on male mating call variation (a prezygotic isolating mechanism) and genetic data now recognizes several reproductively isolated species.

scientists, found that no single definition is satisfactory to all situations observed in nature. Darwin said it this way:

> No one definition has satisfied all naturalists; yet every naturalist knows vaguely what he means when he speaks of a species.

Darwin's solution was to trust the experts:

> In determining whether a form should be ranked as a species or a variety, the opinion of naturalists having sound judgment and wide experience seems the only guide to follow.

In other words, for Darwin the best remedy to the species dilemma was to leave the definition of what is or is not a species to the experts who study

that group of organisms. Although this is not a scientifically rigorous definition, this is largely how today's scientists deal with the process of naming organisms. Ultimately, any new species, or name change of an existing species, or other species designation becomes accepted and valid when other experts agree with each other and, most importantly, start using a particular species name in their professional communications.

Today, there are many definitions of a species used in different contexts to describe particular observations. The most traditional definition of a species, the *biological species concept,* was articulated by one of evolutionary biology's giants, Harvard's German-born Ernst Mayr, who defined a species as

> a group of actually or potentially interbreeding natural populations, which are reproductively isolated from other such groups.

Mayr's species concept emphasizes reproductive isolation (see Reproductive Isolating Mechanisms) and posits that, over time and via natural selection, populations accumulate sufficient differences that if or when these populations are again proximate for interbreeding, they can no longer do so.

FORMATION OF SPECIES

Let us imagine a large assemblage of land snails subdivided in three geographical aggregations or races, *A, B,* and *C,* each adapted to local environmental conditions (fig. 10.5). There are initially no gross barriers separating the populations from each other, and where *A* meets *B* and *B* meets *C,* interbreeding occurs. Zones of intermediate individuals are thus established between the races, and the width of these zones depends on the extent to which the respective populations intermingle. It is important to realize that races are fully capable of exchanging genes with one another.

We may now visualize (fig. 10.5) some striking feature, such as a great river, forging its way through the territory and effectively isolating the land snails of race *C* from those of *B.* These two assemblages may be spatially separated from each other for an indefinitely long period of time, affording an

Geographical variation

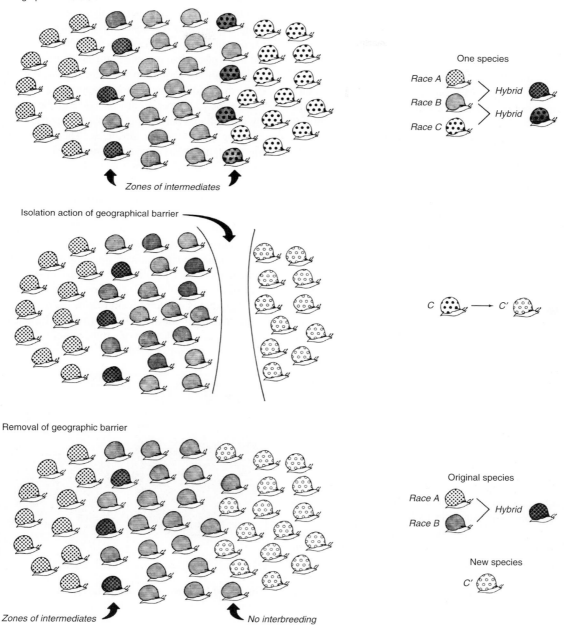

One species

Race A

Race B   Hybrid

Race C   Hybrid

Zones of intermediates

Isolation action of geographical barrier

$C \longrightarrow C'$

Removal of geographic barrier

Original species

Race A

Race B   Hybrid

New species

C'

Zones of intermediates   No interbreeding

Figure 10.5 Model for the process of geographic speciation. Members of populations (or race) *C* had diverged genetically during geographical isolation in ways that have made them reproductively incompatible with race *B* when they met again. Race *C* has thus transformed into a new species, *C′*.

opportunity for race C to pursue its own independent evolutionary course. Two populations that are geographically separated, like B and C in our model, are said to be *allopatric.* (Technically speaking, A and B are also allopatric since, for the most part, they occupy different geographical areas.)

After eons of time, the river may dry up and the hollow bed may eventually become filled in with land. Now, if the members of populations B and C were to extend their ranges and meet again, one of two things might happen. The snails of the two populations might freely interbreed and establish once again a zone of intermediate individuals. Or, the two populations might no longer be able to interchange genes because they have become so different genetically that they cannot. If the two assemblages can exist side by side without interbreeding, then the two groups have reached the evolutionary status of separate species. *A species is a breeding community that preserves its genetic identity by its failure to exchange genes with other such breeding communitites.* In our pictorial model (fig. 10.5), race C has become transformed into a new species, C'. Two species *(A-B and C')* have now arisen where formerly only one existed. It should be noted that races A and B are treated as members of a single species since no barriers to gene exchange exist between them.

NOMENCLATURE

The scientific names that the taxonomist would apply to our populations of land snails deserve special comment. The technical name of a species consists of two words, in Latin or in latinized form. An acceptable designation of the original species of land snails depicted in figure 10.5 would be *Helix typicus.* The first word is the name of a comprehensive group, the *genus,* to which land snails belong; the second word is a name unique to the species. The taxonomist would be obliged to create a different latinized second name for the newly derived species of the land snail, the C' population in figure 10.5. This new species might well be called *Helix varians.* The name of

the genus remains the same since the two species are closely related. The genus, therefore, denotes a group of interrelated species. Taxonomists choose a given latinized species name for a variety of reasons, and more often than not, the latinized name does not connote much information about the organism itself. Thus, the student should not imagine that the key features of each species are encoded in the name, any more than a person's given name is particularly revealing of his or her attributes. The names are important, however, in revealing relationships.

The binomial ("two-named") system of nomenclature, universally accepted, was devised by the Swedish naturalist Carolus Linnaeus (born Karl von Linné) in his monumental work, *Systema Naturae,* first published in 1735 (fig. 10.6). Convention dictates that the first letter of the generic name be capitalized and that the specific name begin with a small letter. It is also customary to print the scientific name of a species in italics, or in a type that is different from that of the accompanying text. A modern refinement of the Linnaean system is the introduction of a third italicized name,

(a) (b)

Figure 10.6 Carolus Linnaeus. (a) Portrait of Carolus Linnaeus (1707–1778), Swedish-born physician, botanist, and zoologist who is best known for creating the modern binomial (two-name) system of classification. (b) Cover page of the 1758 tenth edition of *Systema Naturae* (System of Nature), Linnaeus's pathbreaking volume on the classification of organisms.

Courtesy of the National Library of Medicine.

which signifies the subspecies. Geographical races are recognized taxonomically as subspecies. Thus, it would be appropriate to designate races *A* and *B* (fig. 10.5) as *Helix typicus elegans* and *Helix typicus eminens,* respectively. Such a species composed of two (or more) subspecies is said to be *polytypic.* A *monotypic* species is one that is not differentiated into two or more geographical races or subspecies. *Helix varians* would be a monotypic species.

REPRODUCTIVE ISOLATING MECHANISMS

We have seen that two populations (or subspecies), while spatially separated from each other, may accumulate sufficient genetic differences in isolation that they would no longer be able to interchange genes if they came into contact with one another. When the geographical barrier (or other barrier to gene flow) persists, it is difficult to judge the extent to which the two *allopatric* populations have diverged genetically from each other. Only when the two populations come together again does it become apparent whether or not they have changed in ways that would make them reproductively incompatible. Two populations that come to occupy the same territory are called *sympatric.* The agencies that prevent interbreeding between sympatric species are known as *reproductive isolating mechanisms.*

Reproductive isolating mechanisms are of different types, and one or more of the types may be found separating two species. The various types may be grouped into two broad categories. One category includes the prezygotic (or premating) mechanisms, which serve to prevent the formation of hybrid zygotes. The other category encompasses the postzygotic (or postmating) mechanisms, which act to reduce the viability or fertility of hybrid zygotes. The specific types of isolating mechanisms under these two groupings can be listed in table 10.1.

Two related species may live in the same general area but differ in their ecological requirements. The scarlet oak *(Quercus coccinea)* of eastern North America grows in moist or swampy soils, whereas the black oak *(Quercus velutina)* is adapted to drier soils (fig. 10.7). The two kinds of oak are thus effectively separated by different *ecological* or *habitat* preferences. Two sympatric species may also retain their distinctness by breeding at different times of the year *(seasonal isolation).* Evidently,

TABLE 10.1	Reproductive Isolating Mechanisms

I. Prezygotic (premating) mechanism: Mechanisms that reduce the formation of hybrid zygotes.

1. *Habitat (ecological) isolation:* Ecological niche separates coexisting populations.

2. *Seasonal (temporal) isolation:* Species breeds at different times; mating is restricted by season.

3. *Sexual (ethological) isolation:* Species engage in species-specific courtship and/or mating rituals. Sexual attraction is species-specific.

4. *Mechanical Isolation:* Prevention of copulation or transfer of pollen by physical means such as genitals not matching.

5. *Gametic isolation:* Incompatibility between male and female gametes.

II. Postzygotic (postmating) Isolating mechanisms: Mechanisms that reduce the viability or fertility of hybrid zygotes.

1. *Hybrid inviability:* Fertilized hybrid eggs (zygotes) have reduced or absent viability.

2. *Hybrid sterility:* Hybrid progeny (of one sex or the other or both) are viable but cannot produce functional gametes (i.e., they are sterile).

3. *Hybrid breakdown:* Hybrids are fertile and viable, but less so in successive generations.

Based on Dobzhansky, Theodosius (1970). *Genetics of the Evolutionary Process.* Columbia University Press, New York.

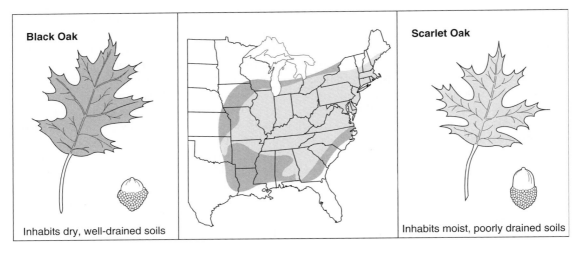

Figure 10.7 Ecological isolation is exemplified by the different habitat requirements of two species of oak. Although both species occur in the eastern United States, the scarlet oak *(Quercus coccinea)* is adapted to the moist bottom lands, whereas the black oak *(Quercus velutina)* is adapted to the dry upland soils.

cross-fertilization is not feasible between two species of frogs that release their gametes on different months even in the same pond, or between two species of pine that shed their pollen in different periods. The breeding seasons of two species may overlap, but interbreeding may not occur because of the lack of mutual attraction between the sexes of the two species *(sexual isolation)*. Among birds, for example, elaborate courtship rituals play important roles in species recognition and the avoidance of interspecific matings (chapter 6).

In many insects, interbreeding of species is hindered by differences in the structure of the reproductive apparatus *(mechanical isolation)*. Copulation is not possible because the genitalia of one species is physically incompatible with the genitalia of the other species. In fact, several closely related species of insects can often be accurately classified by their distinctive genitalia. In some instances, the male of one species may inseminate the female of the other species, but the sperm cells may be inviable in the reproductive tract of the female. This form of *gametic isolation* is not unique to animals; in plants, such as the Jimson weed *(Datura),* the sperm-bearing pollen tube of one species encounters a hostile environment in the flower tissue of the other species and is unable to reach the egg.

Cross-fertilization between two species may be successful, but the hybrid embryos may be abnormal or fail to reach sexual maturity *(hybrid inviability)*. For example, two species of the chicory plant, *Crepis tectorum* and *Crepis capillaris,* can be crossed, but the hybrid seedlings die in early development. Crosses between the bullfrog, *Lithobates catesbeiana,* and the green frog, *Lithobates clamitans,* result in inviable embryos. In certain hybrid crosses, such as between female Fowler's toad *(Bufo=Anaxyrus fowleri)* and male Gulf Coast toad *(Bufo=Incilius nebulifer—*formerly *valliceps),* the hybrids may survive but are completely sterile *(hybrid sterility)*. The familiar example of hybrid sterility is the mule, the offspring of a male ass and a female horse. In some situations, the F_1 hybrids appear to be vigorous and fertile, but the viability of a subsequent generation is very reduced. Such a case of *hybrid breakdown* has been described in several species of cotton—*Gossypium hirsutum* (Upland cotton) and *Gossypium barbadense* (Pima cotton). These species produce normal fertile F_1 hybrids but the majority of the F_2 hybrid cotton seedlings fail to germinate.

In essence, two populations can remain distinct, and be designated as species, when gene exchange between them is prevented or limited by one or more reproductive isolating mechanisms. More often than not, we are unable to obtain direct evidence for the presence or absence of interbreeding in nature between two groups. The degree of reproductive isolation is then indirectly gauged by the extent to which the members of two populations differ in morphological, genetic, physiological, and behavioral characteristics. Two populations that are morphologically very dissimilar are likely to be distinct species. It should be understood, however, that the level of morphological differentiation cannot be used as a reliable criterion of a species. For example, two species of fruit fly, *Drosophila pseudoobscura* and *Drosophila persimilis,* are reproductively isolated but are almost indistinguishable on morphological grounds.

In recent years, biologists have increasingly relied on molecular genetic traits to distinguish one species from another. As a result of the application of these newer techniques, many relationships between species and other taxonomic units (e.g., genera) have been elucidated. As unbiased as molecular data may seem, it is not without its critics. This harkens back to our earlier definition of a species as whatever the experts in the field says it is! After all, what we call a particular organism is simply meant to define our best current approximation of its evolutionary affinities. In this regard, biological classification has become more fluid than it was in the past as the newer molecular genetic data is melded with the morphological, physiological, ecological and behavioral attributes.

ORIGIN OF ISOLATING MECHANISMS

How do reproductive isolating mechanisms arise? In the 1940s, John A. Moore undertook a series of instructive evolutionary studies on the leopard frog, *Lithobates pipiens.* The leopard frog is widely distributed in North America, ranging from northern Canada through the United States and Mexico into the lower reaches of Central America. Moore obtained leopard frogs from different geographical populations and crossed them in the laboratory. When frogs from the northeastern United States (Vermont) were crossed with their southerly distributed lowland relatives in eastern Mexico (Axtla

in San Luis Potosi), the hybrid embryos failed to develop normally. Thus, of this species, the members geographically most distant from one another have diverged genetically to the extent that when cross-mated in the laboratory they are incapable of producing viable hybrids.

It must be admitted that the possibility of a Vermont frog crossing with a Mexican frog in nature is extremely remote. It took a biologist to bring these two frogs together. Yet, it is just this point that emphasizes that an isolating mechanism, such as hybrid inviability, does not develop for the effect itself; it is simply the natural consequence of sufficient genetic differences having accumulated in two populations during a long geographical separation. The late Hermann J. Muller of Indiana University was among the first to suggest that isolating mechanisms originate as a by-product of genetic divergence of allopatric populations. The genetic changes that arise to adapt one population to particular environmental conditions may also be instrumental in reproductively isolating that population from other populations that are themselves developing adaptive gene complexes. Indeed, the embryos of Vermont leopard frogs differ considerably in their range of temperature tolerance from embryos of eastern Mexican frogs. It might well be, then, that the embryonic defects in hybrids between these northern and southern frogs are associated with the different temperature adaptations of the parental eggs.

If the Vermont and Mexican leopard frogs were ever to meet in nature, any intercross between them would lead to the formation of inviable hybrids. This would represent a wastage of reproductive potential of the parental frogs. Theodosius Dobzhansky advanced the engaging hypothesis that under such conditions, natural selection would promote the establishment of isolating mechanisms that would guard against the production of abnormal hybrids. In frogs, a normal mating or a mismating in a mixed population depends principally on the discrimination of the female. The reproductive potential is obviously lower for an undiscriminating female than for a female who leaves normal offspring. If the tendency to mismate is heritable, then the genes responsible for this tendency will eventually be lost or sharply reduced in frequency by elimination of the indiscriminate females, an elimination effectively accomplished by the inviability of their offspring. Thus, the continual propagation of females that most resist the attentions of "foreign" males will lead eventually to a situation in which mismatings do not occur and abnormal hybrids are no longer produced.

Karl Koopman, an able student of Theodosius Dobzhansky, tested the thesis that natural selection tends to strengthen, or make complete, the reproductive isolation between two species coexisting in the same territory. Koopman used for experimentation two species of fruit fly, *Drosophila pseudoobscura* and *Drosophila persimilis*. In nature, sexual selection (chapter 6) between these two *sympatric* species is strong, and interspecific matings do not occur. However, in a mixed population in the laboratory, particularly at low temperatures, mismatings do take place. Koopman accordingly brought together members of both species in an experimental cage and purposely kept the cage at a low temperature (16 °C). Hybrid flies were produced and were viable, but Koopman in effect made them inviable by painstakingly removing them from the breeding cage when each new generation emerged. Over a period of several generations, the production of hybrid flies dwindled markedly and mismatings in the population cage were substantially curtailed. This is a dramatic demonstration of the efficacy of selection in strengthening reproductive isolation between two *sympatric* species.

HUMANS: A SINGLE VARIABLE SPECIES

There is only one present-day species of human, *Homo sapiens*. Different populations of humans can interbreed successfully and, in fact, do. The extensive commingling of populations renders it difficult, if not impossible, to establish discrete racial categories in humans. Races, as we have seen, are geographically defined aggregates of local populations. The populations of humankind are no longer sharply separated geographically from one another. Multiple migrations of peoples and innumerable intermarriages have tended to blur the genetic contrasts between populations. The boundaries of human races, if they can be delimited at all, are at best fuzzy, ever shifting with time.

Recent DNA studies demonstrate that there is much more variation within the so-called human racial groups than between them. This tells us that the superficial differences we see are trivial in terms of speciation and evolution. Today scientists agree that observed human genotypic and phenotypic variations are examples of *clinal variation,* which was discussed earlier and that race is a social construct that is not biologically meaningful for humans.

The term *race* is regrettably one of the most abused words on the English vocabulary. The biologist views a race as synonymous with a geographical subspecies; a race or subspecies is a genetically distinguishable subgrouping of a species. It is exceedingly important to recognize that a race is *not* a community based on language, literature, religion, nationality, or customs. There are Aryan languages, but there is no Aryan race. Aryans are people of diverse genetic makeups who speak a common language (Indo-European). *Aryan* is therefore nothing more than a linguistic designation. In like manner, there is a Jewish religion, but not a Jewish race. And there is an Italian nation, but not an Italian race. A race is a reproductive community of individuals occupying a definite region, and in one and the same geographical region may be found Aryans, Jews, and Italians. Every human population today consists of a multitude of diverse genotypes. A pure population or race, in which all members are genetically alike, is nonexistent.

MICROEVOLUTION AND MACROEVOLUTION

Evolutionary changes in populations ordinarily are visualized as gradual, built upon many small genetic variations that arise and are passed on from generation to generation. The shifting gene frequencies in local populations may be thought of as *microevolution.* The progressive replacement of light-colored moths by dark moths in industrial regions of England exemplifies the microevolutionary processes. Most population geneticists subscribe to the view that the same microevolutionary processes have been involved in the major transformations of organisms over long spans of geologic time *(macroevolution).* The traditional outlook is that small variations gradually accumulate in evolving lineages over periods of millions of years. If we were to recover a complete set of fossil specimens of a lineage, we would expect to find a graded series of forms changing continuously from the antecedent species to the descendant species. The temporally intermediate forms would be intermediate morphologically. Under this perspective, our inability to find transitional fossil forms between the ancestral and descendant populations would represent merely the imperfect nature of the fossil record.

PUNCTUATED EQUILIBRIUM

The late Stephen Jay Gould of Harvard University and Niles Eldredge of the American Museum of Natural History have questioned the conventional view that evolutionary changes in the distant past are principally the outcome of the gradual accumulation of slight inherited variations. They advocate that most evolutionary changes have consisted of rapid bursts of speciation alternating with long periods in which the individual species remain virtually unmodified. Gould and Eldredge maintain that most lineages display such limited morphological changes for long intervals of geologic time as to remain in stasis, or "equilibrium." Conspicuous or prominent evolutionary changes are concentrated in those brief periods ("punctuations") when the lineages actually split or branch. This is the tenet of *punctuated equilibria.*

A single line of descent, or lineage, may persist for long reaches of geologic time. As small changes accumulate over periods of millions of years within one lineage, the descendant populations may eventually be recognized as a species distinct from the antecedent populations. The persistent accumulation of small changes within a lineage has been

termed *phyletic gradualism,* and the transformation of a lineage over time has been termed *anagenesis.* As depicted in figure 10.8, a new species (B) arises from the slow and steady transformation of a large antecedent population (A). If only transformation (i.e., *anagenesis*) occurred, then life would cease as single lineages became extinct. Hence, a new species (C in fig. 10.8) can also arise by the splitting or branching of a lineage. The splitting of one phyletic lineage into two or more lines is termed *cladogenesis.* But here again the splitting is conceived of as proceeding slowly and gradually, with the two branching lineages progressively diverging, without significant reduction in population size. Thus, paleontologists tend to view lineage splitting in terms of gradual morphological divergence.

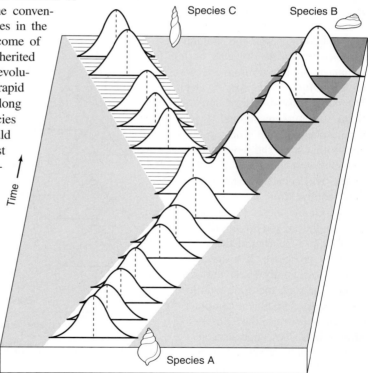

Phyletic Gradualism

Figure 10.8 **According to phyletic gradualism,** a lineage consists of a graded series of intermediate forms connecting ancestral and descendant organisms. Species A gradually transforms into species B and may split into a slowly evolving species C. Redrawn by Phil Mattes, in Volpe, E. P., *Biology and Human Concerns,* 4th ed., 1993, Dubuque, IA:WCA. fig. 41.1, p. 443.

Viewed as a slow process, lineage splitting becomes reduced to a special case of the phyletic model.

Gould and Eldredge maintain that phyletic gradualism is much too slow to produce the major events of evolution. In fact, the morphological changes in successive populations of lineage usually are directionless and involve only minor temporal variations. The theory of punctuated equilibria advocates that the prominent episodes of evolution in the history of life are associated with the splitting of lineages, but it does not see the splitting as slow and steady. The new species arises through rapid evolution when a small local population becomes isolated at the margin of the geographic range of the parent species (fig. 10.9). Indeed, the successful branching of a small isolated population from the periphery of the parental range virtually assures the rapid origin of a new species. In geologic time, the branching is sudden—thousands of years or less, compared to the millions of years of longevity of the species itself. The punctuational change does not entail any unconventional evolutionary phenomenon. The expectation is that once the new species is established, little evolutionary change will occur in that species over geologic time.

Given the thesis of punctuated equilibria, the fossil record is a faithful rendering of the evolutionary processes. If a species remains essentially static during its long lineage, then one would not anticipate finding a continuous series of transitional fossils. Moreover, if a descendant species arises in small, peripheral populations, then it would be rare to find evidence of preservation of a small population evolving in a relatively brief period of time. Thus, the so-called gaps in the fossil

record may not be gaps at all. The notion of "gaps" presupposes the existence of "fill." In this case, the fill would be intermediate forms. But none exist; hence neither do gaps. Before his death, Gould argued that there is no dichotomy between punctuated equilibrium and phyletic gradualism because evolutionary time is measured on a geologic time scale and therefore what is fast (punctuated) may take thousands or even tens of thousands of years whereas what is slow (gradualism) may take hundreds of thousand to millions of years to occur.

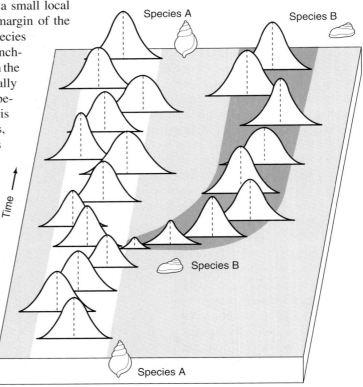

Punctuated Equilibria

Figure 10.9 **The theory of punctuated equilibria** proposes that a small isolate from the parental population (species A) evolves rapidly into a new entity (species B), which undergoes a long period of stasis during which little or no morphological change takes place. Species B does undergo minor structural modifications with time, but these modifications represent merely oscillations around the mean.

Redrawn by Phil Mettes, in Volpe, E. P., *Biology and Human Concerns,* 4th ed., 1993, Dubuque, IA: WCB. fig. 41.2, p. 443.

INSTANTANEOUS SPECIATION

The process leading to the formation of a new species generally extends over a great reach of time. As we have seen, the origin of new species involves a long period of geographical isolation and the long-term influences of natural selection. A sudden and rapid emergence of a new kind of organism is scarcely imaginable. Yet, a natural mechanism does exist whereby a new species can arise rather abruptly. The process is associated with the phenomenon of *polyploidy,* or the multiplication of the chromosome complement of an organism. Many of our most valuable cultivated crop plants, such as wheat, oats, cotton, tobacco, and sugar cane, trace their origin to this instantaneous type of evolution.

POLYPLOIDY IN NATURE

Speciation is the "smoking gun" of evolution. Critics of evolution frequently demand proof that new species arise by natural forces, and polyploidy provides the most straightforward examples. That said, some examples of polyploidy are far from straightforward and frequently belie our mammalian-based notion of speciation (table 11.1).

In nature, polyploid happens by two similar mechanisms, *autopolyploidy* and *allopolyploidy.* When a single species multiplies its own chromosome complement to form a new species, *autopolyploidy* has occurred. Cultivated examples of autopolyploidy are potatoes (autotetraploid-$4n$) and bananas (autotriploids-$3n$). Autopolyploidy is a much less common mode of polyploidy than is allopolyploidy. *Allopolyploidy* involves two or more species hybridizing and contributing chromosome complements to the new species. Bread wheat is an allohexaploid ($6n$), an organism having six sets of chromosomes derived from plants whose haploid number was 7. Bread wheat has three doubled sets of chromosomes. (AABBCC). The A set of chromosomes came from Einkorn wheat ($n=7$), the B was contributed by the wild grass *Aegilops* ($n=7$) and the C set came from goat grass ($n=7$) (see table 11.2 and fig. 11.1). Some polyploids have an even number (e.g., $4n$, $6n$, $8n$) of chromosomes and some have an odd number of chromosomes (e.g. $3n$, $5n$, $7n$). In general, organisms with an odd number of sets of chromosomes are sterile and reproduce asexually.

TABLE 11.1	Polyploidy in Plants and Animals

Plant	Ancestral Haploid Number	Chromosome Number	Ploidy	Ploidy Name
Banana	11	22,33	2n,3n	Diploid, Triploid
Apple	17	34,51	2n,3n	Diploid, Triploid
Peanut	10	40	4n	Tetraploid
Tobacco	12	48	4n	Tetraploid
Potato	12	48	4n	Tetraploid
Cotton	13	52	4n	Tetraploid
Wheat	7	42	6n	Hexaploid
Strawberries	7	14,28,42,56,70	2n,4n, 6n,8n,10n	Diploid, Tetraploid Hexaploid, Octaploid Decaploid
Chysanthemums	9	18,36,54,70,90	2n,4n,6n,8n,10n 6n,8n,10n	Diploid, Tetraploid Hexaploid, Octaploid Decaploid
Sugar cane	10	80	8n	Octaploid
Animal				
Mole Salamander (*Ambystoma*)	14	28,42,56,70	2n,3n,4n,5n	Diploid, Triploid, Tetraploid, Pentaploid
Gray tree frog	12	24,48	4n	Haploid, Tetraploid
Whiptail	23	46–52	2n	Diploid
Lizard		69–71	3n	Triploid
		92	4n	Tetraoids
Sturgeon	Varies	500+	8n	Octaploid
Viscacha rat	28	102	4n	Tetraploid

TABLE 11.2	Species of Wheat (*Triticum*)

14 Chromosomes	28 Chromosomes	42 Chromosomes
T. aegilopoides (Wild Einkorn)	*T. dicoccoides* (Wild Emmer)	*T. aestivum* (Bread Wheat)
T. monococcum (Cultivated Einkorn)	*T. dicoccum* (Cultivated Emmer)	*T. sphaerococcum* (Shot Wheat)
	T. durum (Macaroni Wheat)	*T. compactum* (Club Wheat)
	T. persicum (Persian Wheat)	*T. spelta* (Spelt)
	T. turgidum (Rivet Wheat)	*T. macha* (Macha Wheat)
	T. polonicum (Polish Wheat)	

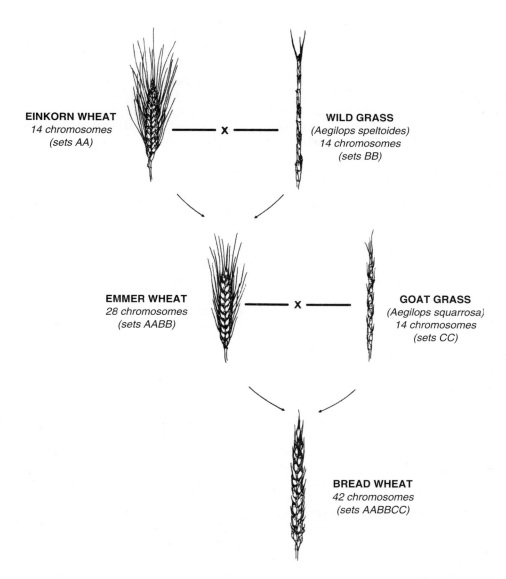

Figure 11.1 Evolution of wheat. Emmer wheat resulted from the hybridization of einkorn wheat with a wild grass, *Aegilops speltoides*. The common bread wheat is the product of hybridization of emmer wheat with a noxious species of goat grass, *Aegilops squarrosa*. The hybridizations were followed by chromosome doubling, a phenomenon discussed in the text.

Polyploidy is much more common in plants than in animals, accounting for up to 70 percent of all flowering plants (angiosperms) and up to 85 to 95 percent of all fern species. Polyploidy is also common in agricultural plants, some of which developed naturally and some of which have been manipulated by humans. *Brassica oleracea* is *paleopolyploid,* which means that historically it was a polyploidy that was "diplodized" ($2n{\rightarrow}4n{\rightarrow}2n$) by losing a set of duplicated chromosomes over evolutionary times. Varieties of *Brassica oleracea* have been developed as familiar foods from a diverse assortment of plant parts, including roots (turnips and swedes), leaves (cabbage and brussel sprouts), seeds (mustard seed and oilseed rape), stems (kohlrabi), and flowers (cauliflower and broccoli).

Speciation via polyploidy is less common in animals, but is known in some insects, amphibians, fish, mollusks, crustaceans, and reptiles. Some lizard species are all female sterile polyploids (usually triploids) that reproduce *parthenogenetically. Parthenogenesis* is the development of an egg without fertilization and is found in some members of other animal groups, including insects, snails, sharks, crayfish (and other crustaceans), birds, and mammals.

Whiptail lizards in the genus *Asphidoscelis* (formerly *Cnemidophorus*) form all-female triploid species that reproduce via parthenogenesis. Some parthenogenetic all-female triploid species of whiptail lizards are themselves hybrid species formed by the union of gametes from diploid parent species. During the mating season, fluctuation in the hormones estrogen and progesterone cause some female whiptails to behave like males and mount the female that is about to lay eggs. Studies have indicated that this behavior may be important in maximizing reproductive success.

Many fish species are polyploid, including salmon and sturgeons. Sturgeons are evolutionarily ancient fish and are known to be diploid ($2n$), tetraploid ($4n$), and octoploid ($8n$), with as many as 500 chromosomes. Some artificially produced salmon are the product of diploid ($2n$) and tetraploid ($4n$) crosses that produce sterile triploids ($3n$), which need to continually be restocked. There are also naturally occurring polyploid salmon.

Polyploidy is also well known in amphibians. The gray tree frogs *Hyla versicolor* and *Hyla chrysoscelis* of eastern North America are familiar voices of summer. Both gray tree frog species are phenotypically indistinguishable. Although their geographic ranges are largely distinct, they do overlap in parts of Wisconsin and Michigan. The eastern gray tree frog (*H. versicolor*) is a tetraploid ($4n$) and the Cope's gray tree frog (*H. chrysoscelis*) is a diploid ($2n$) species. Both tree frogs make a trill-like noise, but Cope's gray tree frog has a shorter, faster call.

An especially interesting example of amphibian polyploidy has evolved in the unisexual (all female) polyploid mole salamanders in the genus *Ambystoma*. Unisexual female may be diploid ($2n$), triploid ($3n$), tetraploid ($4n$), and even pentaploid ($5n$). In order for eggs of unisexual triploid females to develop, the females must make contact with the sperm of related diploid males, but in most instances the male genetic material is not incorporated into the offspring. Reproduction where another species sperm cells are required to activate the egg's cell division, but where the resulting offspring carry no genetic material from those sperm cells is called *gynogenesis*.

The tetraploid ($4n$) mammal, the red vizcacha rat, *Typanoctomy barrerae* (Rodentia, Octodontidae) with 102 chromosomes, is the first mammalian polyploid to be discovered. This *endemic* of the Monte Desert of west-central Argentina is not a true rat, but rather a member of the chinchilla family. Polyploidy occasionally occurs in other mammals, including humans, but the embryos die *in utero* or shortly after birth. Triploid ($3n=69$) and tetraploid ($4n=92$) chromosome complements are not uncommon among human spontaneous abortions (fig. 6.8).

WHEAT

The domestic wheats and their wild relatives have an intriguing evolutionary history. There are numerous species of wheat, all of which fall into three major categories on the basis of their chromosome numbers. The most ancient type is the small-grain einkorn wheat, containing 14 chromosomes in its body *(somatic)* cells. There are two species of einkorn wheat, one wild and the other cultivated, and both of them may be found growing in the hilly regions of southeastern Europe and southwestern Asia. Cultivated einkorn has slightly larger kernels than the wild form, but the yields of each are low and the grain is used principally for feeding cattle and horses.

Another assemblage of wheat, once widely grown, is the emmer series, of which there are at least six species. The chromosome number in the nuclei of somatic cells of emmer wheats is 28. These varieties, found in Europe and the United States, are used today principally as stock feed, although one of them, called durum wheat, is of commercial value in the production of macaroni and spaghetti.

The most recently evolved, and by far the most valuable agriculturally, are the bread wheats. The bread wheats have not been known to occur in the wild state; all are cultivated types. The bread wheats have 42 chromosomes. These wheats, high in protein content, comprise almost 90 percent of all the wheat harvested in the world today.

The various species of wheat thus fall into three major groups, with 14, 28, and 42 chromosomes, respectively. A list of the representatives of these three groups is given in table 11.2.

ORIGIN OF WHEAT SPECIES

Virtually all authorities are agreed on the sequence of evolutionary events depicted in figure 11.1. Einkorn wheat, possessing a chromosome number of 14, was doubtless one of the ancestral parents of the 28-chromosome emmer assemblage. A most remarkable, but generally accepted, thesis is that the other parent was not a wheat at all, but rather *Aegilops speltoides,* a wild grass with 14 chromosomes. This wild grass parent occurs as a common weed in the wheat fields of southwestern Asia. The cross of einkorn wheat and the wild grass would yield an F_1 hybrid that possesses 14 chromosomes, 7 from each parent. We may designate the 7 chromosomes from one parent species as set (or genome) *A,* and the 7 from the other parent species as set (or genome) *B.* Accordingly, the F_1 hybrids would have the *AB* genomes.

If the chromosome complement in the hybrid accidentally doubled, then the hybrid would contain 28 instead of 14 chromosomes and pass on the doubled set of chromosomes to its offspring. Such an event, strange as it may seem, accounts for the emergence of the 28-chromosome emmer wheat. This new species is characterized as having the *AABB* genomes.

In turn, the 28-chromosome emmer wheat was the ancestor of the 42-chromosome bread wheat. In the early 1900s, the British botanist John Percival hazarded the opinion that the bread wheat group arose by hybridization of a species of wheat of the emmer group (28 chromosomes) and goat grass, *Aegilops squarrosa,* a weed commonly found growing in the wheat fields in the Mediterranean area. Although this startling suggestion was initially viewed with skepticism, it is currently conceded that Percival was correct. *Aegilops squarrosa* possesses 14 chromosomes, and thus would transmit 7 of its chromosomes (set *C*) to the hybrid. The hybrid would contain 21 chromosomes (sets *ABC*), having received 14 (sets *AB*) from its emmer wheat parent. The subsequent duplication in the hybrid of each chromosome set provided by the parents would result in a 42-chromosome wheat species (*AABBCC*).

The initial F_1 hybrid between einkorn wheat and *Aegilops speltoides* (or between emmer wheat and *Aegilops squarrosa*) is sterile, but when the chromosome complement doubles, then a fully fertile species arises. Is it to be expected that the F_1 hybrid would be sterile? And what would account for the fertility of the hybrid when chromosome doubling occurred? This requires a deeper look into the phenomenon of polyploidy, to which we shall now turn.

MECHANISM OF SPECIATION BY POLYPLOIDY

Figure 11.2 illustrates the underlying basis of the fertility of a formerly sterile hybrid resulting from the doubling of its chromosome number. For ease

SPECIATION BY POLYPLOIDY

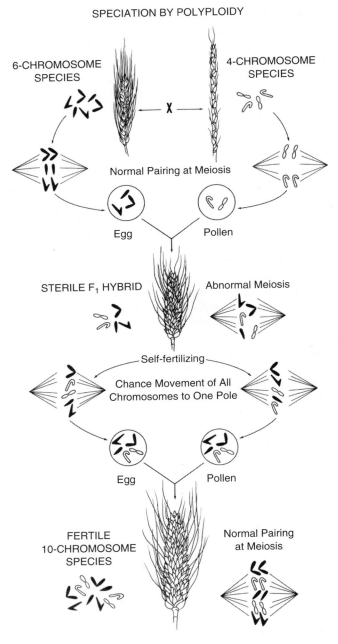

6-CHROMOSOME SPECIES

4-CHROMOSOME SPECIES

Normal Pairing at Meiosis

Egg

Pollen

STERILE F₁ HYBRID

Abnormal Meiosis

Self-fertilizing

Chance Movement of All Chromosomes to One Pole

Egg

Pollen

FERTILE 10-CHROMOSOME SPECIES

Normal Pairing at Meiosis

of presentation, the parental species are shown with a small number of chromosomes, 6 and 4, respectively. It should be noted that the chromosomes are present in pairs. The members of each pair are alike or homologous, but each pair is distinguishable from the other. Thus, the 6-chromosome parent possesses three different pairs of chromosomes; the 4-chromosome parent, two different pairs.

The gametes, egg cells and pollen cells (sperm), are derived by cell divisions of a special kind, called meiosis (see chapter 3). One of the essential features of the process of meiosis is that the members of each pair are attracted to each other and come to lie side by side in the nucleus. The meiotic cell divisions are intricate, but the pertinent outcome is the separation of like, or homologous, chromosomes, such that each gamete comes to possess only one member of each pair of chromosomes. Each gamete is said to contain a *haploid* complement of chromosomes, or half the number found in a somatic cell. The latter, in turn, is described as being *diploid* in chromosome number.

The sterility of the first-generation hybrid is now comprehensible. There are simply no homologous chromosomes in the F_1 hybrid. Each chromosome lacks a homologue to act as its pairing partner at meiosis. The process of meiosis in the hybrid is chaotic; the chromosomes move at random into the gametes. The eggs and pollen cells typically contain an odd assortment of chromosomes and are nonfunctional.

Occasionally, by sheer chance, a few gametes might be produced by the F_1

Figure 11.2 Sequence of events leading to a new, fertile species from two old species by hybridization and polyploidy. The F_1 hybrid plant derived from a cross of the two paternal species is sterile. The F_1 hybrid may occasionally produce viable gametes when all chromosomes fortuitously enter a gamete during the process of meiosis. The fusion of such gametes leads to a new form of plant, which contains two complete sets of chromosomes (one full set of each of the original parents).

hybrid that contain all the chromosomes (fig. 11.2). These gametes would be functional, and the fusion of such sex cells would give rise to a plant that contains twice the number of chromosomes that the first-generation hybrid possessed. The plant actually would contain two complete sets of chromosomes; that is, the full diploid complement of chromosomes of each original parent. Such a double diploid is termed a polyploid—more specifically, a *tetraploid.*

The tetraploid hybrid would resemble the first-generation hybrid, but the plant as a whole would tend to be larger and somewhat more robust as a consequence of the increased number of chromosomes. More importantly, the tetraploid hybrid would be fully fertile. The meiotic divisions would be normal, since each chromosome now has a regular pairing partner during meiosis (fig. 11.2). The tetraploid hybrid is a true breeding type; it can perpetuate itself indefinitely. It is, however, reproductively incompatible with its original parental species. If the tetraploid hybrid were to cross with its original parental species, the offspring would be sterile. Hence, the tetraploid hybrid is truly a new distinct species, reproductively isolated from its parental species. In but a few generations, we have witnessed essentially the fusion of two old species to form a single derived species.

EXPERIMENTAL VERIFICATION

To return to our wheat story, bread wheat (fig. 11.1) contains the chromosome sets of three diploid species—the *AA* of einkorn wheat, the *BB* of *Aegilops speltoides,* and the *CC* of *Aegilops squarrosa.* (The third set is typically referred to by botanists as the *D* genome. It is an accident of nomenclature that the third genome received the letter *D,* rather than *C.*) Technically, then, the bread wheat contains six sets of chromosomes; it is a *hexaploid.* The relationships of the three major groups of wheat can be shown as follows:

Einkorn = 14 = *AA* = diploid
Emmer = 28 = *AABB* = tetraploid
Bread = 42 = *AABBCC* = hexaploid

Experimental proof was lacking at the time John Percival proposed that the bread wheats originated from hybridization between the emmer wheat and goat grass, followed by chromosome doubling in the hybrid. Verification awaited an effective method of artificially inducing diploid cells to become polyploid. The search for an efficient chemical inducing agent culminated in the discovery in the late 1930s of *colchicine,* a substance obtained from the roots of the autumn crocus plant. Treatments of diploid plant cells with colchicine result in a high percentage of polyploid nuclei in the treated plant cells. Colchicine acts on the spindle apparatus of a dividing cell and prevents a cell from dividing into two daughter halves. The treated undivided cell contains two sets of daughter chromosomes, which ordinarily would have separated from each other had cell division not been impeded. The cell thus comes to possess twice the usual number of chromosomes.

The experimental production of polyploid cells through the application of colchicine paved the way for studies on wheat by E. S. McFadden and E. Sears of the United States Department of Agriculture. These investigators successfully hybridized a tetraploid species of emmer wheat with the diploid goat grass, *Aegilops squarrosa.* The chromosome number in the hybrids was doubled by treatment with colchicine. The synthetic hexaploid hybrids were similar in characteristics to natural hexaploid species of bread wheat, and they produced functional gametes. At almost the same time, Hitoshi Kihara of Japan obtained a comparable hexaploid wheat species, which spontaneously and naturally had become converted from a sterile hybrid to a fertile hybrid. Kihara's work reinforced the notion that doubling of chromosomes can occur accidentally.

To complete the proof, McFadden and Sears crossed their artificially synthesized hexaploid wheat species with one of the naturally occurring bread wheat *Triticum spelta.* Fully fertile hybrids resulted, removing any doubt that goat grass, a noxious weed, is indeed a parental ancestor of the bread wheats.

12

ADAPTIVE RADIATION

The capacity of a population of organisms to increase its numbers is largely governed by the availability of resources—food, shelter, and space. The available supply of resources in a given environment is limited, whereas the organism's innate ability to multiply is unlimited. A particular environment will soon prove to be inadequate for the number of individuals present. It might thus be expected that some individuals would explore new environments where competition for resources is low. The tendency of individuals to exploit new opportunities is a factor of major significance in the emergence of several new species from an ancestral stock. The successful colonization of previously unoccupied habitats can lead to a rich multitude of diverse species, each better fitted to survive and reproduce under the new conditions than in the ancestral habitat. The spreading of populations into different environments accompanied by divergent adaptive changes of the emigrant populations is called *adaptive radiation*.

GALÁPAGOS ISLANDS

One of the biologically strangest, yet most fascinating, areas of the world is an isolated cluster of islands of volcanic origin in the eastern Pacific, the Galápagos Islands. These islands, which Darwin visited for five weeks in 1835, lie on the equator 600 miles west of Ecuador (fig. 12.1). The islands are composed wholly of volcanic rock; they were never connected with the mainland of South America. The rugged shoreline cliffs are of gray lava and the coastal lowlands are parched, covered with cacti and thorn brushes. In the humid uplands, tall trees flourish in rich black soil.

Giant land-dwelling tortoises still inhabit these islands (see figs. 2.4 and 12.6). After many years of being needlessly slain by pirates and whalers, these remarkable animals now live protected in a sanctuary created in 1959. Still prevalent on the islands are the world's only marine iguana (fig. 12.2) and its inland variety, the land iguana (fig. 12.3). These two species of prehistoric-looking lizards are ancient arrivals from the mainland. The marine forms occur in colonies on the lava shores and swim offshore to feed on seaweed. The land iguana lives on leaves and cactus plants. Cactus fills most of the water needs of the land iguana.

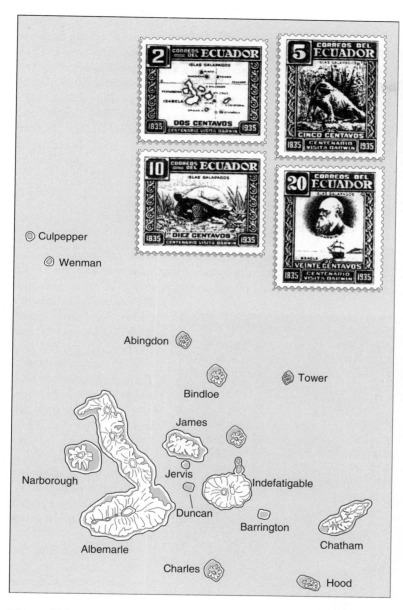

cormorants that cannot fly, found only on Narborough Island, and flamingos, which breed on James Island. Of particular interest are the small dark finches. These dusky birds exhibit remarkable variations in the structure of the beak and in feeding habits. The finches afford an outstanding example of adaptive radiation. It was the marked diversity within this small group of birds that gave impetus to Darwin's evolutionary views. Darwin had correctly surmised that the diverse finches were modified descendants of the early, rather homogeneous, colonists of the Galápagos.

The Galápagos finches, now appropriately called "Darwin's finches," have been studied since Darwin's visit to the islands. In *The Voyage of the Beagle*, Darwin astutely drew attention to the variation in the beaks of the finches that he (and Captain Fitzroy) had collected while visiting the islands. Darwin wrote:

These birds are the most singular of any in the archipelago…. It is very remarkable that a nearly perfect gradation of structure in this one group can be traced in the form of the beak, from one exceeding in dimensions that of the largest gros-beak, to another differing but little from that of a warbler.

Figure 12.1 Galápagos Islands ("Enchanted Isles") in the Pacific Ocean, 600 miles west of Ecuador. Darwin had explored this cluster of isolated islands and found a strange animal life, a "little world within itself." The four stamps shown were issued by Ecuador to commemorate the centenary of Darwin's visit in 1935.

At least 85 different kinds of birds have been recorded on the islands. These include rare

Since Darwin, many scientists have visited the Galápagos to study the unique assemblage creatures

Figure 12.2 Marine iguana. Endemic to the Galápagos archipelago, the marine iguana is the only modern lizard capable of living in and eating from the sea. After eating marine algae, their primary food source, marine iguanas expel excess salt via uniquely adapted nasal glands.

endemic to these enchanted islands. Among finch researchers, perhaps the foremost among them was Oxford's David Lack, who started his work in 1938. More recently, the British-born Princeton University team of Rosemary and Peter Grant conducted an unprecedented long-term study of Darwin's finches, which began in 1973. The Grants' epic research studies were chronicled in the celebrated 1995 Pulitzer Prize–winning book *The Beak of the Finch: A Story of Evolution in Our Time* by Jonathan Weiner.

DARWIN'S FINCHES

The present-day assemblage of Darwin's finches descended from small sparrow-like birds that once inhabited the mainland of South America. The ancestors of Darwin's finches were early migrants to the Galápagos Islands and probably the first land birds to reach the islands. These early colonists have given rise to 14 distinct species, each well adapted to a specific niche (fig. 12.4). Thirteen of these species occur in the Galápagos; one is found in the small isolated Cocos Island, northeast of the Galápagos. Not all 13 species are found on each island.

The most striking differences among the species are in the sizes and shapes of the beak, which are correlated with marked differences in feeding habits. Six of the species are ground finches, with heavy beaks specialized for crushing seeds. Some of the ground finches live mainly on a diet of seeds found

Figure 12.3 Land iguana on one of the Galápagos Islands. Despite their horrendous appearance, these bizarre inland lizards are mild, torpid, and vegetarian. They feed on leaves and cactus plants.
(Courtesy of American Museum of Natural History.)

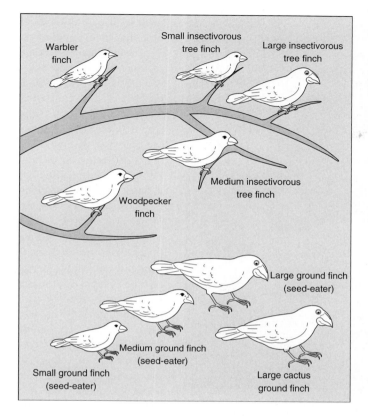

Warbler finch

Small insectivorous tree finch

Large insectivorous tree finch

Medium insectivorous tree finch

Woodpecker finch

Large ground finch (seed-eater)

Medium ground finch (seed-eater)

Small ground finch (seed-eater)

Large cactus ground finch

Figure 12.4   Representatives of Darwin's finches. There are 14 species of Darwin's finches, all confined to the Galápagos with the exception of one species that inhabits Cocos Island. Closest to the ancestral stock are the six species of ground finches, primarily seed-eaters. The others evolved into eight species of tree finches, the majority of which feed on insects.

on the ground; others feed primarily on the flowers of prickly pear cacti. The cactus eaters possess decurved, flower-probing beaks. Their beaks are thicker than those of typical flower-eating birds.

All the other species are tree finches, the majority of which feed on insects in the moist forests. One of the most remarkable of these tree dwellers is the woodpecker finch. It possesses a stout, straight beak, but lacks the long tongue characteristic of the true woodpecker. Like a woodpecker, it bores into wood in search of insect larvae, but then it uses a cactus spine or twig to

probe out its insect prey from the excavated crevice (fig. 12.5). Equally extraordinary is the warbler finch, which resembles in form and habit the true warbler. Its slender, warbler-like beak is adapted for picking small insects off bushes. Occasionally, like a warbler, it can capture an insect in flight.

FACTORS IN DIVERSIFICATION

No such great diversity of finches can be found on the South American mainland. In the absence of vacant habitats on the continent, the occasion was lacking for the mainland birds to exploit new situations. However, given the unoccupied habitats on the Galápagos Islands, the opportunity presented itself for the invading birds to evolve in new directions. In the absence of competition, the colonists occupied several ecological habitats, the dry lowlands as well as the humid uplands. The finches adopted modes of life that ordinarily would not have been opened to them. If true warblers and true woodpeckers had already occupied the islands, it is doubtful that the finches could have evolved into warbler-like and woodpecker-like forms. Thus, *a prime factor promoting adaptive radiation is the absence of competition.*

The emigration of the ancestral finches from the South American mainland was assuredly not conscious or self-directed. The dispersal of birds from the original home was a chance event. Birds (along with other animals and plants) are thought to have been carried westward by the Humboldt Current aboard seaweed mats or driftwood rafts from the continent. The survivors of the ordeal of raft voyages initiated an evolution of their own on the Galápagos Islands.

Figure 12.5 **Tool-using woodpecker finch** *(Camarhychus pallidus)* prodding for grub with a carefully chosen cactus spine of proper size and shape. (© W/Miguel Castro/Photo Researchers, Inc.)

The original flock of birds that fortuitously arrived at the islands was but a small sample of the parental population, containing at best a limited portion of the parental gene pool. It is likely that only a small amount of genetic variation was initially available for selection to work on. What evolutionary changes occurred at the outset were due mainly to random survival *(genetic drift)*. However, the phenomenon of drift would become less important as the population increased in size. Selection unquestionably became the main evolutionary agent, molding the individual populations into new shapes by the preservation of new favorable mutant or recombination types. More than one island was colonized, and the complete separation of the islands from each other promoted the genetic differentiation of each new local population.

The Galápagos Islands are the tips of enormous volcanoes, most of which rise from 7,000 to 10,000 feet above the floor of the sea. It is of especial interest that five distinct varieties of tortoises occupy regions corresponding to the five principal volcanic tops of the large Island of Albermarle (fig. 12.6). The five different forms of tortoises evolved independently of each other at a time when Albermarle Island was so deeply submerged that only its five major craters were above water. The subsequent lowering of the sea level in the geologic past had the effect of uniting the five volcanoes into a single land mass. Even to this day the five volcanoes are separated from each other by deep valleys. There is no intermingling of the five varieties of tortoises. This intriguing natural geologic circumstance serves to document the importance of long-term isolation for the genetic differentiation of the separated populations.

Today, all species of Galápagos tortoise are threatened with extinction due to historic overcollecting and, more recently, introduced predators such as feral pigs, dogs, and rats, which predate turtle eggs. In *The Voyage of the Beagle*, Darwin's descriptions foreshadow what has become a dire conservation situation when he noted that "*the old Bucaniers found this tortoise in greater numbers even than the present.*" Tortoises, like all of their kin, are long-lived species with a low reproductive capacity. Efforts are ongoing to forestall further

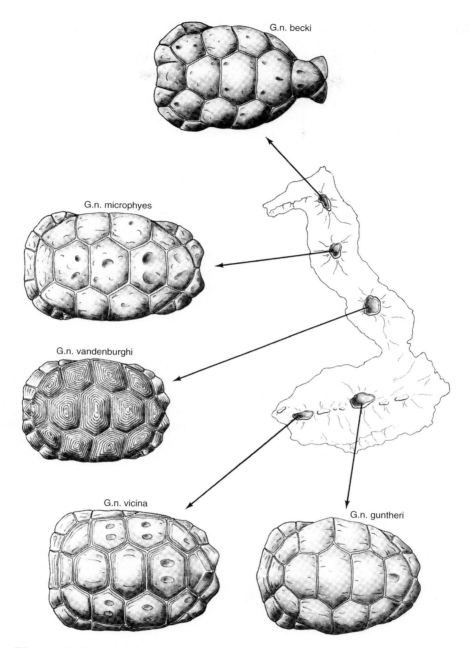

Figure 12.6 Distribution of the five subspecies of the giant land tortoise *(Geochelone nigra)* on Albemarle Island. Each of the five subspecies inhabits one of the five major volcanoes on Albemarle. Each subspecies evolved independently on the separate volcanic tops of the island. The Galápagos tortoises are currently in the midst of a taxonomic revision and each subspecies may soon be recognized as a separate species.

depopulation and to attempt to otherwise conserve this unique assemblage of magnificent creatures.

Students of evolution had debated whether such unique life forms as the marine and land iguanas could have evolved in the relatively short period of geologic time represented by the present-day Galápagos Islands, less than or equal to three million years old. The contentious question seems to have been resolved by the impressive finding that the chain of Galápagos Islands once included older islands that are now below the ocean's surface. The long-vanished islands are geologically nine million years old and are 370 miles closer to the South American mainland than the present-day islands. Accordingly, some of the inimitable animals, such as the two varieties of iguanas, would have had a much greater interval of time to evolve from a common mainland ancestor than was previously thought. Presumably, many of the life forms on the older, submerged islands found their way to the newer, present-day islands in the chain.

COMPETITIVE EXCLUSION

When two populations of different species are obliged (under experimental conditions) to use a common nutrient, the two will compete with each other for the common resource. Under competition because of identical needs, only one of the two species populations will survive; the other will be eliminated or excluded by competition. This is the principle of *competitive exclusion*. It is also known as *Gause's principle,* after the Russian biologist G. F. Gause, who first demonstrated experimentally that under controlled laboratory conditions, one of two competing species will perish.

In 1934, Gause studied the interactions under carefully controlled culture conditions of two protozoan species, *Paramecium caudatum* and *Paramecium aurelia.* When each species was grown separately in a standard medium in a test tube containing a fixed amount of bacterial food, each species flourished independently. When the two species were placed together in the same culture vessel, however, the growth of *P. caudatum*

gradually diminished until the population became eliminated. In enforced competition for the same limited food supply, *P. aurelia* was the more successful species.

Gause enunciated his principle primarily on the basis of observations of "bottle populations" in the laboratory. The competition experiments reveal that two species populations cannot exist together if they are competing for precisely the same limited resource. Alternatively, if two species in nature were to occupy the same habitat, the expectation is that each would have different ecological requirements, even though the degree of difference is slight. Ecologists have demonstrated the validity of this view. In virtually every natural situation carefully examined, two co-inhabiting species have been found to differ in some requirement. The heterogeneous resources of the environment in a given locality are typically partitioned among the co-inhabiting species to minimize direct competition and enable the two (or more) species to coexist.

Most of the Galápagos Islands are occupied by more than one species of finch. On islands where several species of finch exist together, we find that each species is adapted to a different ecological niche. The three common species of ground finch—small (*Geospiza fuliginosa*), medium (*Geospiza fortis*), and large (*Geospiza magnirostris*)—occur together in the coastal lowlands of several islands. Each species, however, is specialized in feeding on a seed of a certain size. The small-beaked *Geospiza fulginosa,* for example, feeds on small grass seeds, whereas the large-beaked *Geospiza magnirostris* eats large, hard fruits. Different species, with different food requirements, can thus exist together in an environment with varied food resources.

COEXISTENCE

Gause's principle may be stated in the following form: No two species with identical requirements can continue to exist together. However, it is exceedingly unlikely that two species in nature would have *exactly* the same requirements for food and habitat. The sum total of environmental requirements for a species to thrive and reproduce has been termed "the ecological niche" of that species population. The term *niche,* as ecologists use it, is more than simply the physical space that the species population occupies. It is essentially the way of life peculiar to a given species: its structural adaptations, physiological responses, and behavior within its habitat. Experience has shown that the likelihood of two species having identical niches is almost nil.

Direct evidence of the process of competition between species in nature is difficult to obtain. Observable competitive interaction is a relatively fleeting stage in the relation of two species populations. What the ecologist observes is the end result of competitive contact, when the actual or potential competitors have become differentially specialized to exploit different components of a local environment and accordingly live side by side. The outcome, then, of incipient competitive interaction is the *avoidance* of competition through differential specialization, or in the terminology of ecologists, through *niche diversification.*

Indirect arguments have been used to support the view that two closely related species populations come to exploit different ecological niches in the same locality after beginning a competitive interaction. We may envision a situation in which two closely related species, with almost similar ecological requirements, expand their geographic ranges and meet in a common habitat. It may be presumed that each of the two species populations has appreciable genetic variability and that the resources in the common habitat are varied. The two species populations would initially compete for suitable ecological niches in the new common habitat. However, we can expect that the members of the two species will ultimately become so different in structure and behavior that each species will become specialized to use different components of the environment. In other words, if genetic differences in morphological and behavioral characteristics tend at first to reduce competition between the two species, then subsequently natural selection

will act to augment the differences between the two competing species. It is especially noteworthy that the differences between the two species become more pronounced as a consequence of selection *reducing* competition rather than *intensifying* competition.

A striking case of niche diversification has been described by David Lack for certain species of *Geospiza* in the Galápagos. Where two species occur together on an island, there is a conspicuous difference in the size of the beaks and food habits. Where either species exists alone on other islands, the beak is adapted to exploit more than one food resource. Thus, on Tower Island (fig. 12.7), the cactus-feeding *Geospiza conirostris* and the large seed-eating ground finch *Geospiza magnirostris* live side by side. The former species occurs also on Hood Island, but the latter species is absent, presumably having failed to invade or reach the island. In the absence of competition, *Geospiza conirostris* on Hood Island has evolved a larger beak that is adapted to feeding on both cactus and seeds. The sharp separation in beak size of the two species on Tower Island is understandable if we assume that competition initially fostered the differentiation of the beaks to permit each of the two species to adapt to a limited or restricted range of the available food resources. Each species is now genetically specialized in food habits, and competition between the two is now avoided.

There are many examples to illustrate how two or more species avoid competition. Among warblers that inhabit the spruce forests of Maine, each species confines its feeding to a particular region of the spruce tree. The myrtle warbler, for example, preys on insects at the base of the spruce tree, whereas the Blackburnian warbler prefers those insects on the exterior leaves of the top of the tree. They can exist together because they use different resources of the same tree.

We may conclude by stating, paradoxically, that competition between two species populations achieves the avoidance or reduction of further competition, and not an intensification. In natural populations, coexistence of two species, rather

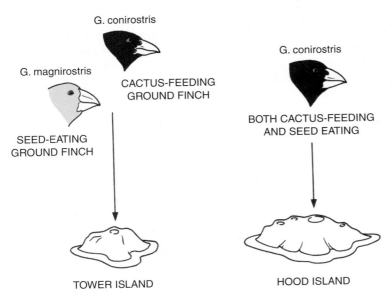

Figure 12.7 Niche diversification. In the absence of competition, the ground finch *Geospiza conirostris* on Hood Island has evolved a large beak that enables it to exploit a variety of food items (seed and cacti). On Tower Island, the same ground finch *Geospiza conirostris* has evolved a specialized beak adapted solely for cactus feeding, thus reducing or avoiding competition for food resources with the seed-eating ground finch *Geospiza magnirostris.* The latter species does not exist on Hood Island.

than competitive exclusion, is the general rule. The end result is that each species is part of a highly organized community in which each plays a constructive, or stabilizing, role. Nevertheless, there is one species—*Homo sapiens*—that seems unable to live in harmony with other species. Human overpopulation is directly responsible for the wholesale destruction of natural habitats and species, global pollution, and other worldwide impacts that many experts believe pose the real danger of destroying our own species. When severe overpopulation occurs, organisms tend to exceed the available natural resources, or *carrying capacity,* which can lead to a population collapse. In all corners of the Earth,

the number of habitats and species that are either already extinct or are threatened with extinction grows daily and at a rate not seen before in the history of our planet. In chapter 16 we will continue this discussion of the history and evolutionary implications of humans impacts on the global environment.

CLASSIFICATION

The original ancestral stock of finches on the Galápagos diverged along several different paths. The pattern of divergence is reflected in the biologists' scheme of classification of organisms. All the finches are related to one another, but the various species of ground finches obviously are more related by descent to one another than to the members of the tree-finch assemblage. As a measure of evolutionary affinities, the ground finches are grouped together in one genus *(Geospiza)* and the tree finches are clustered in another genus *(Camarhynchus).* The different lineages of finches are portrayed in figure 12.8. It should be clear that our classification scheme is nothing more than an expression of evolutionary relationships between groups of organisms.

In chapter 13 we shall see how adaptive radiation on a much larger scale than that which occurred in the finches led to the origin of radically new assemblages of organisms, distinguishable by the taxonomist as Orders and Classes.

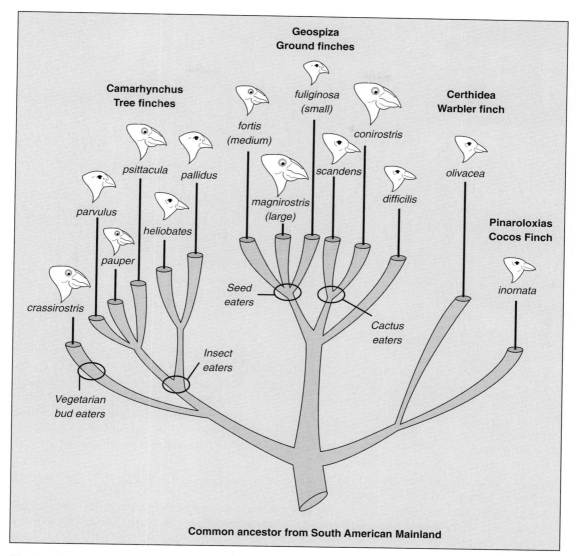

Figure 12.8 Evolutionary tree of Darwin's finches, graphically expressing what is known or surmised as to the degree of relationship or kinship between the different species of finches. Darwin's finches evolved from a common stock and are represented today by 14 species. The species are assembled in four genera, of which two (Certhidea and Pinaroloxias) contain only a single species each. The species associated together in a genus have significant common attributes, judged to denote evolutionary affinities.

(Based on the findings of David Lack.)

13 MAJOR ADAPTIVE RADIATIONS

The diversity of Darwin's finches had its beginning when migrants from the mainland successfully invaded the variety of vacant habitats on the Galápagos Islands. Adaptive radiation, such as manifested by Darwin's finches, has been imitated repeatedly by different forms of life. Organisms throughout the ages have seized new opportunities open to them by the absence of competitors and have diverged in the new environments. The habitats available to Darwin's finches were certainly few in comparison with the enormous range of ecological habitats in the world. The larger the region and the more diverse the environmental conditions, the greater the variety of life.

Approximately 400 million years ago, during a period of history that geologists call the *Devonian,* (see fig. 13.1 and table 16.1) the vast areas of land were monotonously barren of animal life. Save for rare creatures like scorpions and millipedes, animal life of those distant years was confined to the water. The seas were crowded with invertebrate animals of varied kinds. The fresh and salt waters contained a highly diversified and abundant assemblage of cartilaginous and bony fishes. The vacant terrestrial regions were not to remain unoccupied

for long. From one of the many groups of fishes inhabiting the pools and swamps in the Devonian period emerged the first land vertebrate. The initial modest step onto land started the vertebrates on their conquest of all available terrestrial habitats. The origin and diversification of the backboned land dwellers is one of the most striking examples of adaptive radiation on a large scale.

TRANSITION TO LAND

Prominent among the numerous Devonian aquatic forms were the lobe-finned fishes, the Crossopterygians, who possessed the ability to gulp air when they rose to the surface. These ancient air-breathing fishes represent the stock from which the first land vertebrates, the amphibians, are derived (figs. 13.1 and 13.2). The transition of these fish to land required a series of adaptations, including the ability to obtain oxygen from the air, the ability to support themselves on land with legs, and the ability to avoid desiccation.

Until recently, scientists believed that the air-breathing lobe-finned fishes (Crossopterygians) inched out of dwindling Devonian puddles to larger

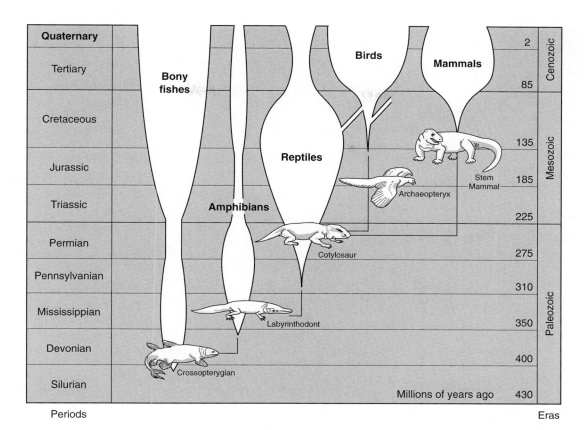

Figure 13.1 Evolution of land vertebrates in the geologic past. From air breathing, lobe-finned fishes (crossopterygians) emerged the first four-footed land inhabitants, the amphibians. Primitive amphibians (labyrinthodonts) gave rise to the reptiles, the first vertebrates to become firmly established on land. The birds and mammals owe their origin to an early reptilian stock (cotylosaurs). An important biological principle reveals itself: each new vertebrate group did not arise from highly developed or advanced members of the ancestral group but rather from early primitive forms near the base of the ancestral stock. The thickness of the various branches provides a rough measure of the comparative abundance of the five vertebrate groups during geologic history. The Devonian period is often called the *Age of Fishes;* the Mississippian and Pennsylvanian periods (frequently lumped together as the Carboniferous period) are referred to as the *Age of Amphibians;* the Mesozoic era is the grand *Age of Reptiles;* and the Cenozoic era is the *Age of Mammals.*

water bodies to escape dehydration, thus forming the stock from which the first land vertebrates, the amphibians, were derived. In 2006, paleontologists lead by Neil Shubin of the University of Chicago and Ted Daeschler of the Philadelphia Academy of Natural Sciences made discoveries of new fossils that have modified those previous deductions. *Tiktaakik* (Inuit for "big freshwater fish") (figs. 13.2 and 13.3) is a transitional fossil between the lobe-finned fish and the terrestrial four-limbed *tetrapods,* such as amphibians, and the other land vertebrates, the reptiles, dinosaurs, mammals, and birds. In the Late Devonian, roughly 380 million years ago, the land masses that are now the northern Canadian alpine tundra straddled the equator and were subtropical, akin to

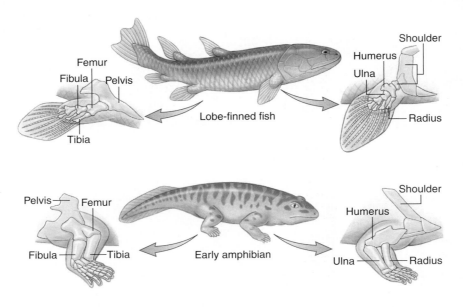

Figure 13.2 Fins to Limbs. Lobe-finned fish limb has the homologous limb bone structure to that found in the early tetrapod amphibian.

modern equatorial rainforests. The lush ecology of the tropical streams was ideally suited for *Tiktaakik* to transition to terrestrial haunts.

Tiktaalik (figs. 13.3. 13.4 and 13.6) and its kin possessed a mosaic of characteristics adapted for shallow water locomotion, which would prove to be important under the new conditions of life on land. Crocodilian-like, *Tiktaalik's* transitional characteristics included a flat head with eyes at the top of the skull (making it look more crocodile-like than fishlike), a functional neck (a feature absent in fish), ribs to support its body and aid in respiration, and limbs already equipped with wrist and elbow bones necessary to move and hold itself up. Figures 13.2 and 13.3 shows the evolutionary transitions leading toward the development of a vertebrate limb capable of walking on land. The challenge of walking on the land went through a series of refinements from

the lobe-finned fish to the transitional *Tiktaalik* to the early amphibian.

Paradoxically, the first adaptations for movements on land appear to have been associated with movement in shallow water rather than an attempt to abandon aquatic existence. Evolutionists speak of such potentially adaptive characters as *preadaptations*. It should be understood that such pre-adapted characters were not favorably selected with a view to their possible use in some future mode of life. There is no foresight or design in the selection process. A structure does not evolve in advance of impending events. Nor do mutational changes occur in anticipation of some new environmental condition. A trait is selected for only when it imparts an advantage to the organism in its immediate environment.

Accordingly, limb bones in *Tiktaalik* evolved, not with conscious reference toward a

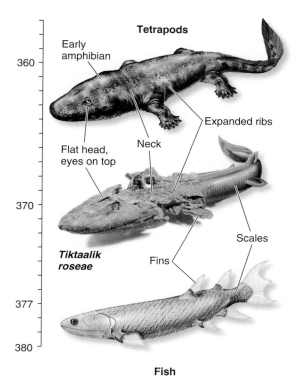

Figure 13.3   Transition to land. During a 20-million-year span in the Late Devonian, vertebrate fossils exhibit increasing adaptations to land. Illustration depicts tetrapod ancestors, a Devonian fish, the transitional fossil *Tiktaalik* and an early amphibian. Many scientists believe that *Tiktaalik* was capable of both crawling in shallow water and occasional terrestrial excursions onto land. Early amphibians were quite fishlike but possessed many more terrestrial adaptations.

possible future land life, but only because such structures were important, if not essential, to its survival in its shallow water habitat. In essence, the selection process does not foresee environments yet to come. The term *pre-adaptation* simply signifies that a structure evolves under one set of environmental circumstances and later becomes modified for a totally different function.

Like *Archaeopteryx* (figs. 13.1, 13.12), which is intermediate between a reptile and a bird,

Tiktaalik is a transitional form between fish such as 380 million year old *Panderichthys* and early tetrapods, such as *Ichthyostega,* known from 365 million years ago (figs. 13.4, 13.15 and 13.6). As such, the discovery of *Tiktaalik* is what might best be termed a "missing link" (or transitional fossil) in the chain of events leading to the first terrestrial tetrapod. *Tiktaalik,* with its mixture of fish and tetrapod characteristics, led one of its discoverers to characterize it as a "fishapod." *Tiktaalik's* features, adapted for movement in shallow waters, would prove necessary for life on land.

Before the close of the Devonian period, the transition from fish to amphibian was complete. The early land-living amphibians were slim-bodied with fishlike tails but possessed limbs capable of locomotion on land, such as the fish-like amphibians *Ichthyostega* (fig. 13.5). These four-footed amphibian-like creatures flourished in the humid swamps of Mississippian and Pennsylvanian times but never did become completely adapted to existence on land.

The most ancient amphibian-like creatures were the labyinthodonts, such as *Acanthostega* and *Ichthyostega* (fig. 13.6), which spent most of their lives in water. These intermediates between fish and more modern amphibians featured lung, limbs, and an amphibian-like build. Their modern descendants (salamanders, newts, frogs, and toads) must return to water to deposit their eggs. *Amphibia,* which means "double life," refers to the fact that the amphibians have one foot on land and another still in the water. Thus, the amphibians were the first vertebrates to colonize land but were, and still are, only partially adapted for terrestrial life.

CONQUEST OF LAND

From the amphibians emerged the reptiles, the true terrestrial forms. The appearance of a shell-covered egg, which can be laid on land, freed the

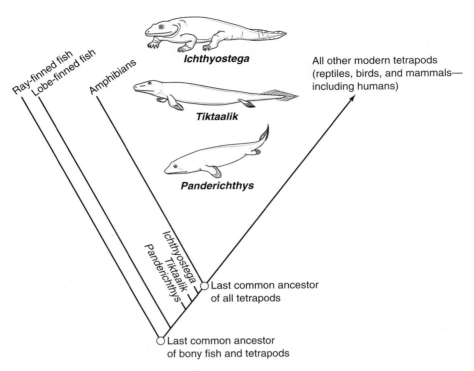

Figure 13.4 The fossil sequence of *Ichthyostega, Tiktaalik* and *Panderichythys* may represent the evolutionary lineage that led to modern tetrapods.

Figure 13.5 *Ichthyostega* (from the Greek meaning fish-roof) was a basal amphibian that retained many fishlike characteristics such as fin rays on its tail, which assisted its locomotion when in the water. *Ichthyostega* boasted effective limbs (with seven digits on its hind limbs) capable of walking on land, ears adapted for detecting airborne sounds, and an improved sense of smell on its elongated snout.

reptile from dependence on water. The elimination of a water-dwelling stage was a significant evolutionary advance. The first primitive reptile most likely arose during Carboniferous times. The ancestral reptile probably possessed the body proportions and well-developed limbs of the more-advanced terrestrial forms of amphibians, such as *Seymouria* (fig. 13.7). *Seymouria* is reptile-like, but its skeletal features suggest terrestrial habits. This advanced amphibian may have been a descendant of an earlier amphibian group in the lower Carboniferous period, a group that was ancestral to the reptiles.

The terrestrial egg-laying habit evolved very early in reptilian evolution. A key feature of reptiles (and higher vertebrates) is the *amniotic egg* (fig. 13.8). The egg of the reptile (or bird) contains a large amount of nourishing

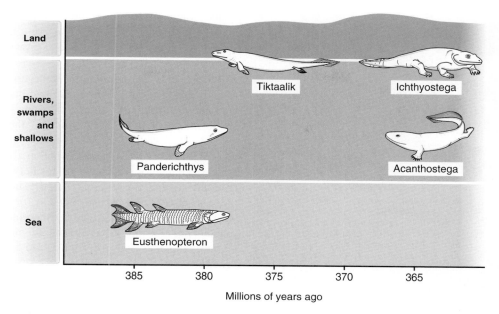

Figure 13.6 **Transition from Late Devonian lobe-fined fish to early tetrapod.** The Late Devonian was a period of rapid transition. Open-ocean dwelling lobe-finned fish, such as *Eusthenopteron,* had modified fins that were pre-adaptations to the tetrapod limb. *Panderichtys* was a fish with additional transitional features. Transitional fossil *Tikaaalik's* limblike fins made it better able to venture onto land, and amphibian-like *Ichthyostega* had lungs and limbs. *Acanthostega* had eight webbed digit appendages but lacked wrists, making it poorly adapted to movement on land.

yolk. Moreover, development of the embryo takes place entirely within a thick shell. These circumstances call for a special provision whereby food derived from the yolk and oxygen obtained from an external source can be made accessible to all parts of the developing embryo. The embryo itself constructs a complex system of membranes, known as *extraembryonic membranes,* which serve for protection, nutrition, and respiration. These are the *amnion, chorion, yolk sac,* and *allantois.*

When fully developed, the *amnion* is a thin membrane loosely enclosing the embryo. It does not fit the embryo snugly. The space between the amnion and the embryo, the *amniotic cavity,* is filled with a watery fluid, the *amniotic fluid.* The fluid is a cushion that protects the embryo from mechanical impacts and, at the same time, allows it freedom of movement. The chemical composition of the amniotic fluid resembles the chemical makeup of blood. Thus, during its development the embryo is bathed by a fluid that is compatible in its chemical nature with the embryo's blood. An interesting fact is that for the embryos of birds and reptiles, as well as for the human fetus, the amniotic fluid resembles seawater in chemical composition, in terms of elements (ions), such as sodium, potassium, calcium, and magnesium. After approximately 400 million years of evolution on land, vertebrates still carry essentially the chemical composition of seawater in their fluids—amniotic fluid, blood, and other body fluids. This has been interpreted as indicating that life originated in the sea and that the balance of salts in various body fluids did not change very much in subsequent evolution. The amnion has been picturesquely characterized as a sort of private aquarium in which the embryos of land-living

Figure 13.7 *Seymouria,* an advanced amphibian with reptilian body proportions (short trunk and well-developed limbs). *Seymouria* is known only from Permian rocks (275 million years ago), long after the reptiles had appeared. It may represent a relict of an earlier amphibian group that was ancestral to the reptiles.
(Photo by L. Botin and R. Logan Department of Library Services, American Museum of Natural History.)

vertebrates recapitulate the water-living mode of existence of their remote ancestors. The presence of an amnion *(amniotes)* or absence of an amnion *(anamniotes),* is a major dividing line between vertebrates. Fish and amphibians, with their aquatic development, lack an amnion; the terrestrially developing birds, reptiles, and mammals all possess this membrane.

As figure 13.8 shows, the enormous mass of yolk has been enclosed by an internal circular membrane, the *yolk sac.* This membrane, which has formed by growing over the yolk, is attached to the embryonic body by a narrow stalk. The yolk sac is highly vascular, its many blood vessels communicating with the blood channels of the embryo proper. The blood circulating through the vessels of the yolk sac carries dissolved yolk materials to all parts of the embryo, thus making the yolk available for chemical activities and growth in all regions.

From the hind region of the embryonic body, a sac bulges out on the underside and pushes its way between the yolk sac and chorion. This sac is the *allantois* (fig. 13.8). Much of the allantoic blood vessels lies close to the inner surface of the shell. The porous shell permits the ready exchange of respiratory gases between the external air and the internal blood. The *allantoic sac* also serves as a receptacle for urinary wastes. Waste fluids excreted by the embryonic kidneys pass into the cavity of the allantois. The allantois thus has both respiratory and excretory functions.

Finally, there is an outermost investing membrane, the *chorion,* which is abundantly supplied with blood vessels. The embryo depends, in part, on this vascularized enveloping membrane for carrying on gaseous exchange with the outer air through the porous shell.

The amnion, chorion, allantois, and yolk sac in the human embryo are similar in essence to these extraembryonic membranes in the reptiles (and birds). Curiously, the yolk sac does develop in the human embryo in spite of the absence of any appreciable amount of yolk. The human yolk sac, however, remains small and largely functionless, providing early blood circulation to the developing mammalian embryo. There is also no elaborate development of the allantois in the human embryo; the allantoic sac never becomes more than a rudimentary tube of minute size that contributes to the formation of the umbilicus. Nevertheless, the very appearance of the yolk sac and allantois in the human embryo is one of the strongest pieces of evidence documenting the evolutionary relationships among the widely different kinds of vertebrates. To the student of evolution this means that the mammals, including human beings, are descended from animals that reproduced by means of externally laid eggs rich in yolk.

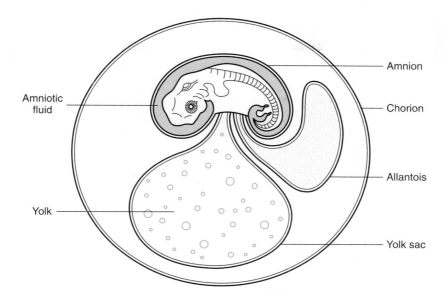

Figure 13.8 The amniotic egg. The developing embryo is enclosed by the membranous *amnion* and cushioned by amniotic fluid. The large reserve supply of food (yolk) is contained within the *yolk sac*. The sacklike *allantois* serves as a receptacle for the embryo's waste products. The outermost, vascularized enveloping membrane is the *chorion*.
(From a restoration by Charles R. Knight. Courtesy of American Museum of Natural History.)

ADAPTIVE RADIATION OF REPTILES

With the evolution of the amniotic egg, the reptiles exploited the wide expanses of land areas. The ancestral reptilian stock initiated one of the most spectacular adaptive radiations in the history of life on Earth. The reptiles endured as dominant land animals of the Earth for well over 100 million years. The Mesozoic era, during which the reptiles thrived, is often referred to as the "Age of Reptiles."

Figure 13.9 reveals the variety of reptiles that blossomed from the basal stock of *amniota*. As described in the previous section, the *amniota* are those vertebrates with an amnion as part of their extraembryonic membranes and include the birds, reptiles, and mammals. The *anamniota* are those vertebrates without an amnion and include the amphibians and fishes. The Class Reptilia is typically divided into three subclasses based on skull characteristics (fig. 13.10): the *Anapsiada,* the *Synapsida* and the *Diaspida.* The Anapsida have solid skulls except for the eye orbit and include the extinct stem reptiles, the extinct aquatic freshwater reptiles, and perhaps the modern turtles. Exactly where exactly the turtles fit into the evolutionary scheme remains murky, with conflicting data from DNA analysis and paleontological sources. The Synapsids diverged from the Anapsids and have a single hole in their skulls behind their eye sockets. Synapsids are the sail-backed pelycosaurs and the mammal-like therapsids, which eventually evolved into mammals. The Diaspsids are the largest group and also contain a pair of holes in their skull posterior to the eye and consist of the archeosaurs, the dinosaurs (Saurischia and Ornithischia) (figs. 13.9 and 13.11), the Mesozoic marine reptiles such as the plesiosaurs, the mollusk-eating placodonts, the fishlike ichthyosaurs, the flying reptiles (pterosaurs), the

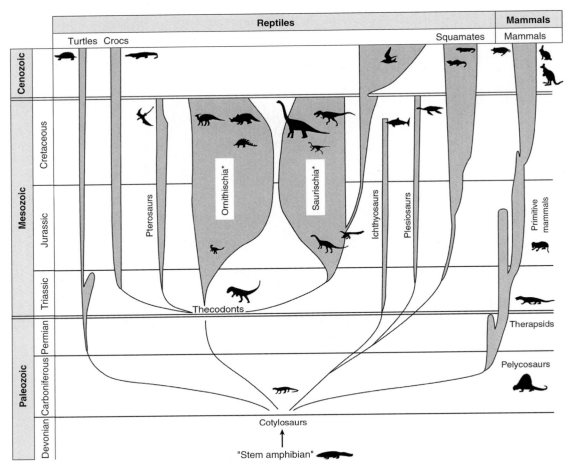

Figure 13.9   The ancestors of the stem amphibian consist of the modern land mammals, the turtles, crocodilians, lizard and birds as well as their extinct ancestors such as the dinosaurs which all went extinct around 65 million years ago. These vertebrates diverged early in the fossil record and encompass the living anapsids, synapsids and diaspids shown below in figure 13.10.

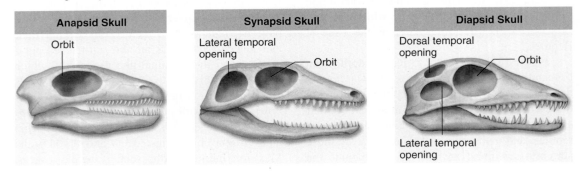

Figure 13.10   Reptile skulls. The classification of reptiles is based on the number and type of holes on the side of their skulls behind the orbit of the eye. Anapsids have no holes, Synapsids have one hole and Diapsids have two holes. See text for details.

sphenodonts, the lizards and snakes (Squamata), and the crocodilians. The remaining tetrapods are the amphibians.

The dinosaurs (fig. 13.11) were by far the most awe-inspiring and famous of the early reptiles. They reigned over the land until the close of the Mesozoic era, when they became extinct, roughly 65 million years ago, as part of a massive extinction pulse that most scientists believe was caused by a meteor or comet striking the Earth. The dinosaurs were remarkably diverse; they varied in size, bodily form, and habits. There are two Orders of dinosaurs, the lizard-hipped *Saurischians* and the bird-hipped *Ornthischians*. Some of the dinosaurs were carnivorous, such as the huge *Tyrannosaurus*, whereas others were

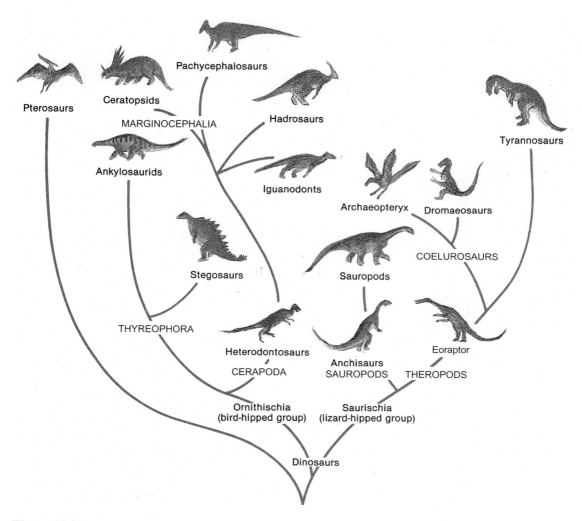

Figure 13.11 Simplified dinosaur family tree. The major dinosaur groups are classified based on hip structure. The bird-hipped dinosaurs are the Ornithischians and the lizard-hipped dinosaurs Saurischians. The flying pterosaurs do not fit either of these major classes.

vegetarians, such as the feeble-toothed but ponderous *Apatosaurus*. The prodigious body of *Apatosaurus* weighed 30 tons and measured nearly 70 feet in length. Not all dinosaurs were immense; some were no bigger than chickens. Some dinosaurs strode on two feet; others had reverted to four. The exceedingly long necks of certain dinosaurs were adaptations for feeding on the foliage of tall coniferous trees.

The dinosaurs descended from the *archosaurs* about 230 million years ago during the Middle to Late Triassic. Early dinosaur fossils of the genus *Eoraptor* (fig. 13.11) were meter-long bipedal predators that ran upright on their hind legs. *Eoraptor's*

Avian features:
1. Feathered wing and tail

Reptilian features:
1. Long bony tail
2. Claws on wing
3. Toothed beak

Figure 13.12 Transition reptile. *Archaeopteryx* (Greek for "ancient feather") is now a series of transitional fossils that are intermediate between dinosaurs and birds. Its reptilian features include a long, bony tail, claws on its wings, and a toothed beak. Its avian or birdlike features are its feathered wings and tail. The first specimen was found in 1861, just two years after Darwin published *The Origin of Species*, making it a key piece of evidence in the support of evolution.

forelimbs were about half as large as its legs, and there were five claws on each hand, which it presumably used to hold prey. Recent paleontological discoveries suggest that many of the first dinosaurs were also small predatory bipeds. The archosaurs also gave rise to the bizarre reptiles that took to the air, the pterosaurs (fig. 13.11). These "dragons of the air" possessed highly expansive wings and disproportionately short bodies. The winged pterosaurs succumbed before the end of the Cretaceous when the dinosaurs went extinct.

Another independent branch of the archosaurs led to eminently more successful flyers, the birds. The link between the reptiles and the birds is seen in the celebrated transitional fossil *Archaeopteryx*, a Jurassic form that was essentially an airborne lizard (figs. 13.11 and 13.12). This feathered creature possessed a slender, lizard-like tail and a scaly head equipped with reptilian teeth. Nearly all authors agree that birds descended from reptiles, which were already warm-blooded (*endothermic*) rather than, as often assumed, cold-blooded (*ectothermic*). Under this view, the endothermic birds are merely aerial extensions of a terrestrial endothermic stock.

Certain reptiles returned to water. The streamlined, dolphin-like *ichthyosaurs* and the long-necked, short-bodied plesiosaurs were marine, fish-eating reptiles. The ichthyosaurs were proficient swimmers; their limbs were finlike and their tails were forked.

The *plesiosaurs* were efficient predators, capable of swinging their heads 40 feet from side to side and seizing fish in their long, sharp teeth. These aquatic reptiles breathed by means of lungs; they did not redevelop the gills of their very distant fish ancestors. Indeed, it is axiomatic that a structure once lost in the long course of evolution cannot be regained. This is the doctrine of irreversibility of evolution, or Dollo's law, after Louis Dollo, the eminent Belgian paleontologist to whom the principle is ascribed. The reversal of the long evolutionary path from lungs to gills would demand the implausible precise retrieval of an untold number of steps. Among the early reptiles present before Mesozoic days were the *pelycosaurs,* such as *Dimetrodon,* notable for their peculiar sail-like extensions of the back (fig. 13.13). The function of the gaudy sail is unknown, but it should not be thought that this structural feature was merely ornamental or useless. As we have emphasized, traits of organisms were selected for their adaptive utility. It may be that the pelycosauran sail was a functional device to achieve some degree of heat regulation. Be that as it may, the pelycosaurs gave rise to an important group of reptiles, the *therapsids.* These mammal-like forms bridged the structural gap between the reptiles and the mammals.

The extant reptiles inhabit six of the seven continents and include four Orders. The Order Testudines includes the turtles, tortoises, and terrapins. The Squamata embraces the lizards, snakes, and wormlike amphisbaenids. The Crocodilia comprise the crocodiles, alligators, gavials, and caimans. Finally, the Sphenodontia consist of just two living species of living fossils, the tuataras, which are endemic to New Zealand (fig. 13.9).

EXTINCTION AND REPLACEMENT

The history of the reptiles attests to the fact that the ultimate fate of most groups of organism is extinction. The reptilian dynasty collapsed before the close of the Mesozoic era. Of the vast array of Mesozoic reptiles, relatively few survived to modern times; the ones that have include the lizards, snakes, crocodiles, and turtles. The famed land dinosaurs, the great marine plesiosaurs and ichthyosaurs, and the flying pterosaurs all became extinct. Some groups of reptiles, such as the ichthyosaurs, pterosaurs, and placodonts, went extinct before the two major groups of true dinosaurs (figs. 13.9, 13.11). The cause of the decline and death of the true dinosaurs (Ornithischians and Saurischians) 65 million years ago has been linked to a large extraterrestrial impact (comet or meteor) in the vicinity of the Yucatan Peninsula. Whatever radical environmental changes this impact created, dinosaurs and many other reptiles were unable to adapt to the altered environmental conditions.

As one group of organisms recedes and dies out completely, another group spreads and evolves (fig. 13.1). The decline of the reptiles provided evolutionary opportunities for the birds and the mammals. The vacancies in their habitats could then be occupied by warm-blooded vertebrates. Small and inconspicuous during the Mesozoic era, the mammals arose to unquestionable dominance in the years immediately following the extinction of the dinosaurs. The extent to which mammals contributed to the decline of the dinosaur

Figure 13.13 Dimetrodon, one of the carnivorous pelycosaurs that flourished during Permian times. The gaudy "sail" may have served as a heat-regulating device. (From a restoration by Charles R. Knight, courtesy of American Museum of Natural History.)

assemblage is unanswerable. Our knowledge of evolutionary history suggests that the ascendancy of the mammals was the result, rather than the cause, of the fall of the great reptiles. In this regard, the adaptive radiation of mammals that led to the diversification and expansion of the primates and to the eventual evolution of humans might not have occurred were it not for the extinction of the dinosaurs 65 million years ago.

In mammals, a major evolutionary advance was the development of the placenta, an intimate apposition of maternal and fetal tissues. The placenta provides optimum conditions for the growth of the fetus—protection, a continuous supply of oxygen and food, an efficient system for removing waste products, and the transmission of antibodies. Of paramount significance, mammals have added a higher order of parental care of the young. Many mammal newborns are relatively helpless and obtain nourishment in the form of milk through special mammary glands. The most primitive mammals are the *monotremes* (duckbill platypus and spiny anteater), which lay reptile-like yolky eggs. The monotremes nourish their young with milk secreted by modified sweat glands. The *marsupials,* which carry their developing early infants in a pouch, are also mammals and include the North American opossum. The third major mammalian group is the *placental mammals,* which includes humans.

The mammals diversified into marine forms (for example, the whale, dolphin, seal, and walrus), *fossorial* forms living underground (such as the mole), flying and gliding animals (such as the bat

and the flying squirrel), and *cursorial* types well-adapted for running (for example, the horse). The mammalian appendages are highly differentiated for the different modes of life. An important lesson may be drawn from the variety of specialized appendages. Superficially, there is scant resemblance between the human arm, the flipper of a whale, and the wing of a bat. And yet a close comparison of the skeletal elements (fig. 13.14)

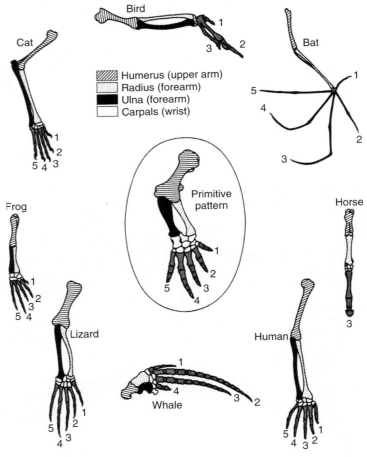

Figure 13.14 Varied forelimbs of vertebrates, all of which are built on the same structural plan. The best explanation for the fundamentally similar framework of bones is that humans and all other vertebrates share a common ancestry. Homologous bones are shaded appropriately. The number of phalanges in each digit is indicated by a numeral, beginning with the first digit (thumb).

shows that the structural designs, bone for bone, are basically the same. The differences are mainly in the relative lengths of the component bones. In the forelimb of the bat, for instance, the *metacarpals* and *phalanges* (except for the thumb) are greatly elongated. Although highly modified, the bones of the bat's wing are not fundamentally different from those of the forelimbs of other mammals. The conclusion is inescapable; the limb bones of the human, the bat, and the whale are modifications of a common ancestral pattern. The facts admit no other logical interpretation. Indeed, as seen in figure 13.14, the forelimbs of all tetrapod vertebrates exhibit a unity of anatomical pattern intelligible only on the basis of common inheritance. The corresponding limb bones of tetrapod vertebrates are said to be *homologous* because they are structurally identical with those in the common ancestor. *Homology* implies evolutionary divergence from a common ancestor (descent with modification). In contrast, the wing of a bird and the wing of a butterfly are *analogous;* both are used for flight, but they are built on an entirely different structural plan and are of independent origin *(convergence)*. Many other examples of divergence from a common ancestor and adaptations to similar ecology leading to convergence can be found throughout the domains of life (fig 13.15).

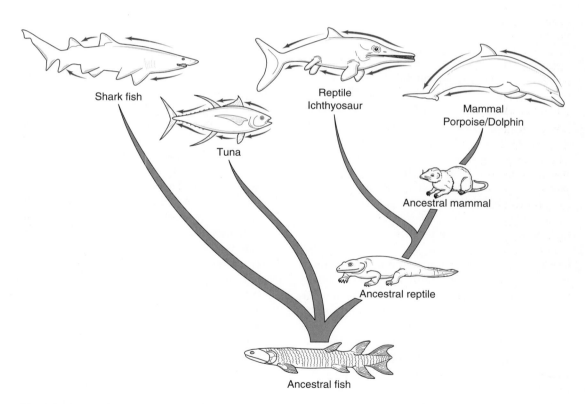

Figure 13.15 Convergent evolution among fast-swimming predators. Body shapes adapted to fast movement in the water have evolved numerous times. Sharks and tuna are fish, the ichthyosaur is an extinct reptile, and the dolphin is a marine mammal.

14

ORIGIN OF LIFE

T he polemics surrounding Darwin's *The Origin of Species* was matched by another monumental controversy of that day, a controversy regarding the unbelievable proposition that germs (bacteria) cause disease. The father of germ theory was Louis Pasteur, the chemist who was led into the study of germs through problems associated with the fermentation of wine. In the early nineteenth century, the notion of *spontaneous generation* was still flourishing— that is, most people believed that living things could originate from lifeless matter. In 1862 Pasteur provided proof that new life can come only from pre-existing life. Following years of bitter wrangling with skeptics, Pasteur finally laid to rest the age-old belief that life could appear out of nowhere.

Since Pasteur's time, countless generations of students have been taught not to believe in spontaneous generation. However, Pasteur's experiments revealed only that life cannot arise spontaneously under conditions that exist on Earth today. Conditions on the primeval Earth billions of years ago were assuredly different from present conditions, and the first form of life, or self-duplicating particle, likely arose spontaneously from chemical, inanimate substances.

We are unable to witness the unique molecular events that led to the first form of life ages ago. The scientist's reconstruction of life's origin is largely circumstantial, but persuasive.

In an 1871 letter to botanist and friend Joseph Hooker, Darwin conjectured that life may have evolved in a "warm little pond"; as we shall see, he may not have been far off the mark. For some evolutionists, the question of the origin of life is unimportant, as it is clear that all life on Earth today shares a common ancestor. Therefore, the process of descent with modification is the critical issue for students of evolution, not how or when the first life came to be. Be that as it may, exploring the question of the origin of life is a question that arouses intense curiosity and provides us with valuable insights into contemporary evolution as well as the history of planet Earth.

Astronomers agree that the Big Bang, the explosion that caused the formation of the universe, occurred about 13.7 billion years ago, that our solar system was formed about 4.6 billion years ago, and that our planet came into existence at roughly 4.5 billion years ago. Scientists are able to date the oldest rocks to 3.8 billion years old (from Canada) and the first minerals date to nearly 4.3 billion years ago (from Australia). Early Earth was so inhospitable to any form of life that even if life had evolved, it would have immediately been exterminated. The first evidence of life comes from fossils that are roughly 3.6 billion years old (fig. 14.1, table 16.1). Presupposing that life did evolve on Earth, the first questions are: what is life, when did it evolve, and how?

WHAT IS LIFE

The line between the living and nonliving is sometimes blurred by organisms such as viruses, which rely on host organisms to exist. Nonetheless, the conventional view is that contemporary life is made up of:

- organic matter (especially carbon) that is capable of reproduction (e.g., mitosis and/or meiosis), and
- proteins that are made up of the 20 essential amino acids, and
- genetic information that encodes proteins carried in the form of a universal DNA-RNA genetic code.

It is implied in these attributes that all living organisms are capable of accumulating biomass (growth and development), can obtain and use energy, can reproduce, and are likewise subject to mutation and selection (can respond to the environment). Reproduction and natural selection require the accumulation of biomass, energy (in the form of chemical reactions), and some way to store and transmit heritable variation. Life is also cellular (see Prokaryotic and Eukaryotic Cells and table 14.1). Presupposing that life did evolve on Earth, we can explore the question of what were the conditions on primitive Earth.

PRIMITIVE EARTH

The view that life emerged through a long and gradual process of chemical evolution was first

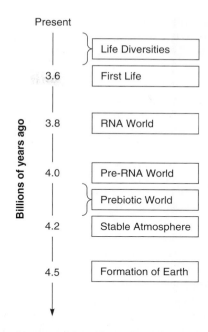

Figure 14.1 **Approximate timeline** of events in the early history of life (in billions of years before the present).

convincingly set forth by the Russian biochemist Alexander I. Oparin, in 1924, in an enthralling booklet entitled *The Origin of Life*. The transformation of lifeless chemicals into living matter extended over some one billion years. Such a transformation, as Oparin points out, is no longer possible today. If by pure chance a living particle approaching that of the first form of life should now appear, it would be rapidly decomposed by oxygen of the air or quickly destroyed by the countless microorganisms presently populating the Earth.

Our Earth is reliably estimated to be 4.6 billion years old. It was formerly thought that the Earth originated as a fiery mass that was torn away from the sun. Astronomers now generally acknowledge that the Earth (like other planets in the solar system) condensed out of a swirling cloud of gas surrounding the primitive sun. The atmosphere of the pristine Earth was quite unlike our present atmosphere. Oxygen in the free gaseous state was virtually absent; it was bound in water and in metallic oxides on surface rocks and particles. Accordingly, any complex organic compound that arose during this early time would not be subject to degradation by free oxygen. Moreover, there was no layer of ozone to absorb the stark ultraviolet rays from the sun, which would be lethal to modern animal life.

The early gas cloud was especially rich in hydrogen. The hydrogen (H_2) of the primordial Earth chemically united with carbon to form methane (CH_4), with nitrogen to form ammonia (NH_3), and with oxygen to form water vapor (H_2O). Thus, the early atmosphere had a strongly reducing (nonoxygenic) character, containing primarily hydrogen, methane, ammonia, and water. The atmospheric water vapor condensed into drops and fell as rain; the rains eroded the rocks and washed minerals (such as chlorides and phosphates) into the seas. The stage was set for the combination of the varied chemical elements. Chemicals from the atmosphere mixed and reacted with those in the waters to form a wealth of *hydrocarbons* (that is, compounds of hydrogen and carbon). Water, hydrocarbons, and ammonia are the raw materials of amino acids, which, in turn, are the building blocks for the larger protein molecules. Thus, in the primitive seas, amino acids accumulated in considerable quantities and became linked together to form proteins.

Complex carbon compounds, such as proteins, are termed organic because they are made by living organisms. Our present-day green plants use the energy of sunlight to synthesize organic compounds from simple molecules. What, then, was the energy source in the primitive Earth, and how was synthesis of organic compounds effected in the absence of living things? It is generally held that ultraviolet rays from the sun, electrical discharges, such as lightning, and intense dry heat from volcanic activity furnished the energy to join the simple carbon compounds and nitrogenous substances into amino acids. Is there a valid basis for such a widely accepted view?

EXPERIMENTAL SYNTHESIS OF ORGANIC COMPOUNDS

In the early days of chemistry, it was believed that organic compounds could be produced only by living organisms. But in 1828, Friedrich Wöhler succeeded in manufacturing the organic compound *urea* under artificial conditions in the laboratory. Since Wöhler's discovery, a large variety of organic chemicals (amino acids, monosaccharides, purines, and vitamins) formerly produced only in organisms have been artificially synthesized.

In 1953, Stanley Miller, then at the University of Chicago and a student of Nobel Laureate Harold Urey, synthesized organic compounds under conditions resembling the primitive atmosphere of the Earth. He passed electrical sparks through a mixture of hydrogen, water, ammonia, and methane (fig. 14.2). The electrical discharges duplicated the effects of violent electrical storms on primitive Earth. In the laboratory, the four simple inorganic molecules interacted, after a mere week, to form several kinds of amino acids, among them alanine, glycine, aspartic acid, and glutamic acid. Hydrogen cyanide and cyanoacetylene were also identified among the products, and those compounds can serve as intermediates in the formation of nitrogenous bases (purines and pyrimidines) of nucleic acids. Miller's instructive experiment has been successfully repeated by a number of investigators; amino acids and nucleotides can also be generated by irradiating a similar mixture of gases with ultraviolet light.

It is very likely that proteins were randomly produced abiotically on the primitive Earth. However, life itself apparently did not begin with a self-sustaining network of proteins. Polypeptide chains of amino acids are not self-complementary and, accordingly, cannot serve as templates for their own replication. On the other hand, nucleic acids are capable of self-replication. As discussed in the next section entitled "Life's Beginnings", ribonucleic acid (RNA) probably was present before the emergence of protein synthesis. Moreover, RNA evidently preceded DNA in the evolution of life.

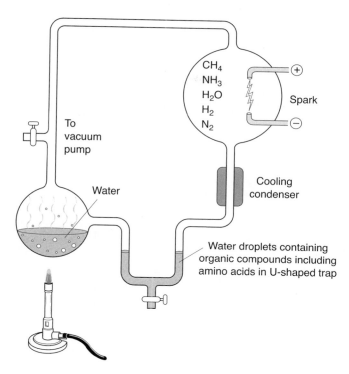

Figure 14.2 **Stanley Miller's apparatus simulating conditions of the primordial Earth.** The upper chamber contains a mixture of gases (methane, ammonia, and hydrogen gas plus water vapor). An electrical discharge (simulating lightning) is passed through the vaporized mixture. The mixture is cooled by a water condenser, and any complex molecules formed in the atmosphere chamber are dissolved in the water droplets. The organic molecules accumulate in the U-shaped trap, where samples are withdrawn for analysis.

LIFE'S BEGINNINGS

It is clear that organic compounds can be formed without the intervention of living organisms. Thus, it appears likely that somewhere in the primitive Earth's sea a rich mixture of organic molecules spontaneously accumulated. In the absence of living organisms, the organic compounds would have been stable and would have persisted for countless years. The sea became a sort of dilute organic soup in which the molecules collided and associated to form larger molecules (*polymers*) of increasing levels of complexity. Proteins capable of catalysis, or enzymatic activity, had to evolve, and nucleic

acid molecules capable of self-replication must also have developed.

The living cell is an orderly system of chemical reactions (directed by enzymes) that has the ability to reproduce. It seems reasonable that the machinery for self-replication (that is, self-duplicating nucleic acids) evolved before the development of the metabolic (enzymatic) machinery of the cell. Whereas amino acids are the building blocks of proteins, nucleotides are the basic units of nucleic acids. The linear *polynucleotides* occur in nature in the form of deoxyribonucleic acid (DNA) or ribonucleic acid (RNA). Polynucleotides have the unique property of specifying the sequence of nucleotides in a new molecule by acting as templates for the assembly of the nucleotides. Because preferential binding occurs between pairs of nucleotides (cytosine coupled with guanine and adenine coupled with uracil), new daughter polynucleotide molecules can arise in which the nucleotide sequences are complementary to the parent molecules. A second round of copying, in which the complementary strand serves as a template, restores the original sequence. Thus, the original random sequence is multiplied many times.

It is reasonable to presume that self-replicating *polynucleotides* slowly became established in the primordial Earth 3.5 to 3.8 billion years ago. In the absence of proteinaceous enzymes in the prebiotic environment, less efficient catalysts in the form of minerals or metal ions would be sufficient to promote the reactivity of the nucleotide precursors. More importantly, however, RNA itself can act as a catalyst. The linear RNA molecule can fold up to form complex surfaces that catalyze specific reactions. Accordingly, RNA molecules have enzymatic properties in addition to their well-known properties as templates. It is highly probable that RNA guided the primordial synthesis of protein (fig. 14.3). The nucleotides

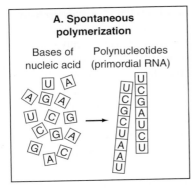

A. Spontaneous polymerization

Bases of nucleic acid

Polynucleotides (primordial RNA)

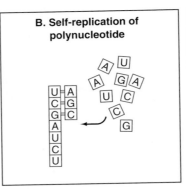

B. Self-replication of polynucleotide

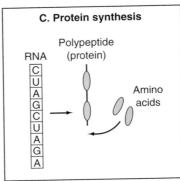

C. Protein synthesis

RNA   Polypeptide (protein)

Amino acids

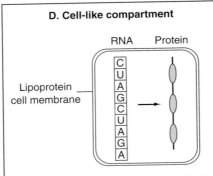

D. Cell-like compartment

RNA   Protein

Lipoprotein cell membrane

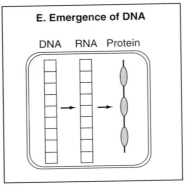

E. Emergence of DNA

DNA   RNA   Protein

Figure 14.3 Postulated stages in the origin of living cells. *(A)* Polynucleotides (primordial RNA) emerge from the spontaneous association of nitrogenous bases (adenine, cytosine, guanine, and uracil). *(B)* RNA has enzymatic properties and catalyzes its own replication. *(C)* Information encoded in the base sequences of RNA specifies the amino acid sequence in a polypeptide chain (protein). *(D)* A spontaneously associated lipoprotein serves as a cell membrane enclosing RNA and its protein product. *(E)* DNA replaces RNA as the repository of hereditary information.

of RNA (not DNA) were probably the first carriers of genetic information. RNA established a primitive genetic code for ordering amino acids into proteins. It warrants emphasis that the genetic code—the translation of nucleotide sequences into amino acid sequences—became established at a very early stage of organic evolution. The universality of the genetic code in present-day organisms attests to the early origin of the code.

We may presume that errors continually occurred in the copying process, and that the sequence of nucleotides in the original polynucleotide molecule became altered on numerous occasions. Large numbers of polynucleotide variations were undoubtedly maladapted. Just as organisms

today compete for available resources, molecules with different nucleotide sequences competed for the available nucleotide precursors in promoting copies of themselves. Any new mutant sequence with a replication rate higher than the antecedent sequence would be the "fittest" and would prevail. Natural selection (differential reproduction) thus operates on populations of molecules as it does on populations of organisms.

A selectively advantageous RNA molecule would be one that directs the synthesis of a protein that accelerates the replication of that particular RNA. However, a protein specified by a particular variant of RNA could not foster the reproduction of that kind of RNA unless it was restrained in the immediate vicinity of the RNA. There may have been a time in Earth's early history before proteins when information storage and enzyme activity (including self-replication) were accomplished exclusively by RNA. Such a prebiotic world—this hypothetical stage in the origin of life—is called "the RNA World." The term *RNA World* was coined by Nobelist Walter Gilbert in 1986 and expounded upon by another fellow Nobelist, Francis Crick. The basis of this concept arose from discoveries by Thomas Cech and Stuart Altman, who shared the Nobel Prize for Chemistry in 1989, that RNA could function as a biochemical catalyst, a property that previously was thought to be restricted to proteins. Cech and Altman named these RNA enzymes *ribozymes.*

Since the RNA World was first conceived, the list of ribozyme abilities has grown, and that new information has propelled thought on the origin of earthly life. Today we know that RNA catalysis is much more fundamental to biology than Cech, Altman, or Gilbert first conceived, including controlling the expression of genes. We also now know that RNA can also make DNA via a process called *reverse transcription,* which is common among the RNA viruses, such as the human immunodeficiency virus (HIV). In this instance, RNA serves as the template for making DNA, and an enzyme, *reverse transcriptase,* accomplishes the transformation of single-stranded RNA to double-stranded DNA (fig 4.10).

Under particular circumstances based on an RNA World scenario, RNA enzymes could catalyze the synthesis of new RNA molecules. Mutation and natural selection would exert constant selective pressure to harness favorable mutations that improved or enhanced the capabilities of RNA. This may have been particularly important in the self-replication of polynucleotides and in early protein synthesis (fig. 14.3).

The free diffusion of proteins could be forestalled if some form of a compartment evolved to enclose or circumscribe the specific protein made by a particular RNA (fig. 14.3). All present-day cells have a limiting (plasma) membrane composed principally of phospholipids. It is not implausible that the first cell was formed when polarized films of phospholipids formed soaplike bubbles enclosing aggregations of other complex macromolecules. Specifically, a membranous structure (plasma membrane) might have arisen that encircled a self-replicating aggregation of RNA and protein molecules. Once bounded by a limiting membrane, a given RNA molecule could be assured of propagating its own special protein molecule.

DNA is more chemically stable than RNA. In the hypothetical RNA World, at some later stage in the evolutionary process DNA took over the function of data storage, leaving RNA to continue to function in enzyme production and later to serve as a component of modern protein synthesis that utilizes all of the RNA polynucleotides. If RNA carried on the dual functions of storage and replication of genetic information and had the ability to serve as a molecular catalyst to accomplish primitive metabolism and the biochemical reactions required, then it may have been pivotal in the establishment of the first life on Earth. Today RNA molecules serve essentially their primal function: directing protein synthesis. Thus, in modern cells, genetic information is stored in DNA, transcribed into RNA, and translated into protein.

The RNA World is not without its unanswered questions, but the concept remains an exciting line of contemporary research into both basic cell function and the origin of life on Earth. Ultimately, an endless permutation of nucleic acids (RNA and DNA) and proteins has fostered an enormous richness of life.

We may reasonably assume that the first living systems drew upon the wealth of organic materials in the sea. Organisms that are nutritionally dependent on their environment for ready-made organic substances are called *heterotrophs* (Greek, *hetero,* "other," and *trophos,* "one that feeds"). The primitive one-celled heterotroph probably had little more than a few genes, a few proteins, and a limiting cell membrane. The heterotrophs multiplied rapidly in an environment with a copious supply of dissolved organic substances. However, the ancient heterotrophs could survive only as long as the existing store of organic molecules lasted. Eventually living systems evolved the ability to synthesize their own organic requirements from simple inorganic substances. Over the course of time, *autotrophs* (*auto,* meaning "self") arose, which were able to manufacture organic nutrients from simpler molecules. Under the deep-sea vent hypothesis for the origin of life discussed in a later section entitled "Deep Sea Vents", the order would have been reversed, with chemoautotrophs evolving before ancient heterotrophs.

AUTOTROPHIC EXISTENCE

The first simple autotroph arose in an *anaerobic* world, one in which little, if any, free oxygen was available. The primitive autotrophs obtained their energy from the relatively inefficient process of fermentation (the breakdown of organic compounds in the absence of oxygen). Thus, the early fermentative autotrophs were much like our present-day anaerobic bacteria and yeast. The metabolic processes of the anaerobic autotrophs resulted in the liberation of large amounts of carbon dioxide into the atmosphere. Once this occurred, the way was paved for the evolution of organisms that could use carbon dioxide as the sole source of carbon in synthesizing organic compounds and could use sunlight as the sole source of energy. Such organisms would be the photosynthetic cells.

Early photosynthetic cells probably split hydrogen-containing compounds, such as hydrogen sulfide. In other words, hydrogen sulfide was cleaved into hydrogen and sulfur. (This is still done today by sulfur bacteria.) The hydrogen was used by the cell to synthesize organic compounds, and the sulfur was released as a waste product (as evidenced by the Earth's great sulfur deposits). In time, the process of photosynthesis was refined so that water served as the source of hydrogen. The result was the release of oxygen as a waste product. At this stage, free oxygen became established for the first time in the atmosphere.

The first organisms to use water as the hydrogen source in photosynthesis were the blue-green algae. Since blue-green algae were active photosynthesizers, atmospheric oxygen accumulated in increasing amounts. Oxygen would be toxic to anaerobic bacteria. A massive extinction of anaerobic bacteria was likely coincidental with the evolution of photosynthesis. Many primitive anaerobic bacteria, incapable of adapting to free oxygen, remained in portions of the environment that were anaerobic, such as sulfur springs and oxygen-free muds. However, new kinds of bacteria arose that were capable of utilizing the free oxygen. Today, there are bacterial types that are anaerobic as well as *aerobic.*

The Earth's atmosphere gradually changed from a reducing, or hydrogen-rich, atmosphere to an oxidizing, or oxygen-rich, atmosphere. Geologic evidence indicates that free oxygen began to accumulate in the atmosphere about 2 billion years ago. The rising levels of atmospheric oxygen set the stage for the appearance of one-celled eukaryotic organisms, which arose at least 1 billion years ago (fig. 14.4). Then, within the comparatively short span of the last 600 million years, the one-celled eukaryote evolved in various directions to give rise to a wealth of multicellular life forms inhabiting the Earth.

PROKARYOTIC AND EUKARYOTIC CELLS

Some of the simplest cells in nature include bacteria and cyanobacteria ("blue-green algae"), and archaea, which have been classified as *prokaryotes.* The Greek expression for nucleus is *karyon;* hence, prokaryote means "before a nucleus." Prokaryotic cells (table 14.1) have no nuclear membrane by which the hereditary materials (DNA) are set apart from the cytoplasm and they lack specialized cytoplasmic bodies (organelles), such as mitochondria and choloroplasts. In a bacterial cell, the DNA forms a simple loop that is attached to the inside of the cell's membrane. In the absence of mitochondria, the enzymatic machinery in bacteria for converting energy into usable form

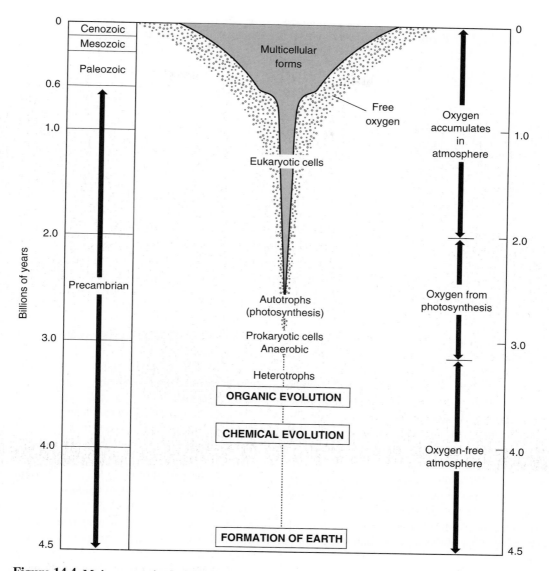

Figure 14.4 Major events in the history of life. Chemical evolution preceded biological evolution. The first self-duplicating life forms were heterotrophs. Oxygen began to accumulate in the atmosphere with the appearance of photosynthetic autotrophs. Rising levels of atmospheric oxygen were associated with the emergence of eukaryotes. The last 600 million years have witnessed a great diversity of life in an oxygen-rich atmosphere.

is found in the plasma membrane. The replication of a bacterial cell occurs rapidly by the simple division of the cell into two. One bacterial cell can divide every 20 minutes and can leave 4 billion descendants within 11 hours.

In contrast to the prokaryotes, the cells of the morphologically more complex plants and animals, or *eukaryotes* (table 14.1), have a distinct nuclear membrane that encloses discrete DNA-containing chromosomes. The eukaryotes include

TABLE 14.1	Prokaryotes vs. Eukaryotes	
Property	**Prokaryotes**	**Eukaryotes**
Cell size	Small (0.2–2.0 um)	Large (10–100 um)
Nucleus	No nuclear membrane	True nucleus present
Nucleoli	Absent	Present
Membrane-bound cytoplasmic organelles	Absent	Present
Cell division	Fission	Mitosis
Chromosomes	Small, circular, without histone proteins	Linear chromosomes complexed with histone proteins
Cell division	Binary fission	Mitosis
Sexual reproduction	No meiosis	Involves meiosis
Ribosomes	Small	Large and small
Mitochondria	Absent	Present
Flagella (when present)	Lack microtubules	Contain microtubules

not only multicellular animals and plants but also unicellular organisms, such as protozoa, fungi, and many algae. Eukaryotic cells also have an elaborate system of membrane-bound cytoplasmic organelles such as mitochondria, chloroplasts, Golgi apparatus and the endoplasmic reticulum. The enzymes that are responsible for releasing large amounts of energy are packed inside the mitochondria. In the eukaryotic plant cell, a prominent organelle is the chloroplast, the photosynthesizing agent that converts light energy into chemical energy. One of the most intriguing speculations is that mitochondria arose not by a gradual evolutionary process but abruptly in an unusually striking manner.

ORGANELLES AND EVOLUTION

Mitochondria have a number of interesting properties that suggest they were once free-living, or independent, bacteria-like organisms. Mitochondria have small amounts of their own DNA distinct from nuclear DNA. Mitochondrial DNA exists as a loop-shaped molecule like the DNA of bacteria. In the early 1970s, Lynn Margulis, then at Boston University, enunciated the provocative hypothesis that mitochondria may have been derived from primitive aerobic bacteria that were engulfed by predatory organisms, probably fermentative bacteria, destined to become eukaryotic (fig.14.5). The predatory hosts became dependent on their enslaved mitochondria, and the later, in turn, became dependent on their hosts. Thus, the association was of mutual advantage to the predatory cell and the engulfed prey. Such a close association, or partnership, is called *symbiosis*. In essence, the mitochondria (formerly oxygen-respiring bacteria) established permanent residence within the hosts. The predatory hosts became the first eukaryotic cells.

Eukaryotic plant cells contain both mitochondria and chloroplasts. Chloroplasts, like mitochondria, have their own unique DNA and the associated protein-synthesizing machinery. We can now speculate that those predatory cells that engulfed but did not digest the aerobic bacterial cells became the eukaryotic animal cells. Those predators that captured both the photosynthetic blue-green algae cells and the aerobic bacteria evolved into eukaryotic plant cells (fig. 14.5).

Accordingly, the engulfed prey became permanent symbiotic residents—either as mitochondria or chloroplasts—within the predatory cell. The startling idea that present-day mitochondria and chloroplasts were descendants of ancestral aerobic bacteria and blue-green algae, respectively, has gained widespread acceptance in the scientific community through the years.

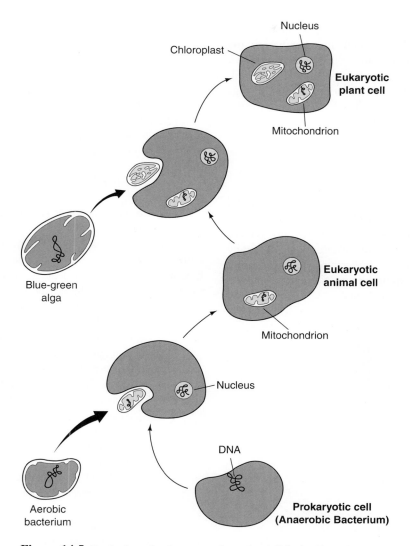

Figure 14.5 Evolution of eukaryotes through symbiosis. The eukaryotic animal cell may have arisen through a symbiotic relationship between a prokaryotic cell and an aerobic bacterium. The eukaryotic plant cell may have originated by a comparable symbiotic relationship betwen a blue-green alga and a eukaryotic animal cell.

Although the relatively abrupt origin of mitochondria and choloroplasts in eukaryotes is now generally accepted, it is likely that the evolution of the nucleus and endoplasmic reticulum proceeded through a series of gradual changes. Continual modification of the surface membrane of a prokaryotic cell may have been the means by which the nucleus and endoplasmic reticulum of a eukaryotic cell originated. In other words, the nucleus and endoplasmic reticulum may have evolved by the invagination, or drawing inward, of the surface membrane of a primitive cell. Figure 14.6 shows the postulated mechanism for the origin of the nucleus and endoplasmic reticulum from the cell surface membrane.

CLASSIFICATION OF ORGANISMS

Aristotle (384–322 B.C.), the great philosopher and naturalist, aptly declared that Nature is marvelous in each and all her ways. Although Aristotle did not view different kinds of organisms as being related by descent, he arranged all living things in an ascending ladder with humans at the top. A formal scheme of classification was developed by the Swedish naturalist Karl von Linné (1707–1778), whose name generally appears in latinized form, Carolus Linnaeus (chapter 10). Life was divided by Linnaeus into two grand kingdoms, *Animals* and *Vegetables,* broadly defined as follows:

ANIMALS adorn the exterior parts of the Earth, respire, and generate eggs; are impelled to action by hunger, congeneric affections, and pain; and by preying on other animals and vegetables, restrain within proper proportion the numbers of both.

They are bodies organized, and have life, sensation, and the power of locomotion.

VEGETABLES clothe the surface with verdure, imbibe nourishment through bibulous roots, breathe by quivering leaves, celebrate their nuptials in a genial metamorphosis, and continue their kind by the dispersion of seed within prescribed limits.

They are bodies organized, and have life and not sensation.

Linnaeus was convinced that all species of animals and plants were fixed, unchanging entities, and that "there are just as many species as there were created in the beginning." Modern taxonomists have departed completely from Linnaeus' stand. Present-day taxonomy is based on the proposition of evolutionary change. The modern classification scheme is an expression of the evolutionary relationships among groups of organisms. All systems of classification are imperfect to the extent that our knowledge is imperfect. The classification scheme is continually being modified as new information on the evolutionary relations of organisms comes to light.

Darwin used a tree metaphor for the organization and divergence of life from a common ancestor. The only figure Darwin presented in *The Origin of Species* (our fig. 14.7) embodies his concept of descent with

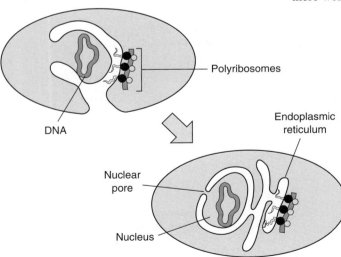

Figure 14.6 Hypothetical scheme of the origin of the nucleus and endoplasmic reticulum of a eukaryotic cell by the invagination, or drawing inward, of the surface membrane.

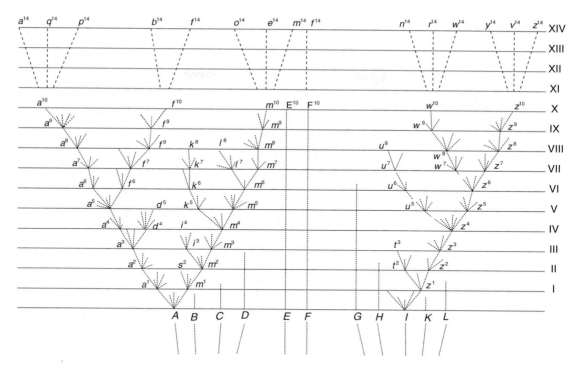

Figure 14.7 **Tree diagram from *The Origin of species*,** the only figure in Darwin's monumental monograph.
From Origin of Species by Charles Darwin

modification. Darwin introduces his tree concept as follows:

> *We may assume that the modified descendant of any one species will succeed so much the better as they become more diversified in structure and are thus enabled to encroach on places occupied by other beings. Now let us see how this principle of benefit derived from divergence of character, combined with the principles of natural selection and of extinction, tend to act.*

Today Darwin's insight seems matter of fact because the branching tree concept is now a commonplace icon of evolution (e.g., figs. 14.8 and 14.9). However, in Darwin's day the branching tree showing divergence and common descent was transformative and revolutionary.

In 1969, R. H. Whittaker suggested a radical departure from the traditional two-kingdom system. He grouped organisms into five separate kingdoms (fig. 14.8). Whittaker's system takes into consideration the fundamental differences between prokaryotic and eukaryotic levels of organization and stresses the principal modes of nutrition—photosynthetic, absorptive, and ingestive. The five kingdoms are the Monera (unicellular, prokaryotic organisms), Prostista (unicellular, eukaryotic organisms), Plantae (multicellular higher algae and green plants), Fungi (multinucleate plant-like organisms lacking photosynthetic pigments), and Animalia (multicellular animals). Whittaker's scheme was favorably received and adopted by most taxonomists for many years. We have become accustomed to dividing all life forms into two major categories based on cell types, the prokaryotes (bacteria and blue-green algae) and the eukaryotes.

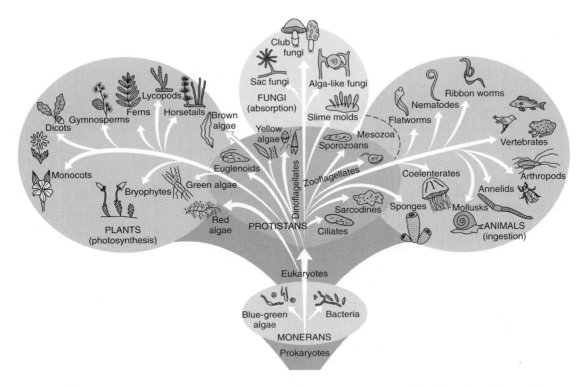

Figure 14.8 **The five-kingdom classification.** Primitive unicellular prokaryotes (kingdom *Monera*) gave rise to more complex single-celled eukaryotes (kindom *Protista*). The divergence of the Protistans led to the three kingdoms of *Plantae, Fungi,* and *Animalia.*

(Original design contributed by Stuart S. Bamforth, Tulane University.)

In 1997, Carl R. Woese at the University of Illinois brought to light an unusual group of one-celled organisms so different from other forms of life, including bacteria, that he placed them in a separate domain of life. These unusual microbes, which flourish in harsh conditions, were initially called *archaebacteria* but are now referred to as *archaea.*

Woese proposed a novel universal phylogenetic tree (fig. 14.9). Instead of five major kingdoms, there are three broad domains: Bacteria, Archaea, and Eukarya. In Woese's scheme, the Archaea and the Bacteria appear first and diverge from their common ancestor relatively soon in the history of life. Later, the Eukarya branches off from the Archaea. The one-celled organisms seem to dominate the tree of life. The multicellular

organisms (Eukarya) appear as a mere twig in a great microbial world!

Woese used the tools of molecular biology in reassigning life into three broad domains. He relied heavily on the nucleotide sequences found in ribosomal RNA, which are highly conserved in all organisms and would aptly serve to chronicle the past evolutionary history. Woese subsequently undertook the sequencing of the complete genome of one of the archaea, *Methanococcus jannaschii.* His work, reported in 1996, showed that *M. jannaschii* has a host of genes unlike those of any organism known. Specifically, nearly 56 percent of the gene sequences are completely different from any so far described from either bacteria or eukaryotes. Of the 44 percent of the genome that does

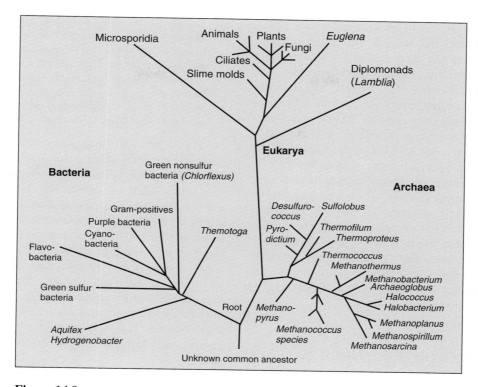

Figure 14.9 The three broad domains of Carl Woese's classification scheme: Bacteria, Archaea, and Eukarya. The predominant forms of life appear to be one-celled.
(Modified from *Science Magazine*, vol 276. No. 5313, 2 May 1997, p. 701.)

match other life forms, some are recognizable as bacteria-like while others, particularly those controlling transcription and translation, resemble eukaryotic genes.

DEEP-SEA VENTS

One view of early Earth arose from our exploration of the deep oceans. In 1977 *Alvin*, a manned submersible vessel capable of surviving the extreme pressure of the deep sea, discovered a thriving and diverse ecosystem surrounding cracks in the Earth surface along mid-ocean ridges called *hydrothermal vents*. The water in these vents reaches temperatures exceeding 300 degrees Celsius and is enriched by hydrogen sulfide (H_2S) and metal ions. Where this unique water meets colder water,

a thermal gradient is created. At these thermal gradients are a series of ecosystems unlike any other ecosystem on Earth. These communities do not rely on the Sun or photosynthesis for their source of energy but instead capture the energy of the heat of the Earth. Organisms capable of producing food via chemical synthesis are called *chemoautotrophs*. These organisms form the primary production component of the deep-sea hydrothermal vent food chain. Around deep-sea vents, various bivalves such as clams and mussel and assorted tube worms feed upon the food produced from the action of the chemoautotrophs, thus forming the consumer level of the deep-sea vent food chain. The discovery of *extremophile* bacteria (Archaea) capable of living in the completely dark, extremely hot, high-pressure hydrothermal vents has lead to conjecture that the first life on Earth may have evolved in or around the gradient of temperatures between these deep sea vents and the colder ocean waters.

The Archaea are classified into several major types, depending on the conditions they live under. The *thermophiles* thrive at very hot deep sea vents and hot springs and around other geothermal sources in excess of 100 degrees Centigrade. The *halophiles* flourish in extremely saline environments, and the *acidophiles* tolerate acids. The *methanogens* (methane generators) prosper under anaerobic conditions, such as landfill sites and the rumens of cows, where they gain energy by oxidizing hydrogen. Methanogens are plentiful wherever organic matter is present in anaerobic conditions. Even though today's Archaea live under environmental conditions that resemble those that may have existed on early Earth, it is presumptuous to assume that the Archaea were the first microorganisms.

EXTRATERRESTRIAL ORIGINS

There is a minority perspective among competent scientists that life on Earth was seeded or infected from extraterrestrial sources. The theory of *panspermia* posits that meteors or comets carried the precursors of life, or life itself, to Earth. Untold numbers of meteorites and comets collided with early Earth, some of which may have carried the essentials of life. Other impacts of extraterrestrial origin were likely profoundly damaging and may have caused the sterilization of the Earth's surface.

An elaboration on the panspermia concept, envisioned by DNA co-discoverer Francis Crick, is called *directed panspermia*. Directed panspermia speculates that life was seeded to Earth on purpose from other places in the universe. According to Crick:

> It now seems unlikely that extraterrestrial living organisms could have reached the Earth either as spores driven by the radiation pressure from another star or as living organisms imbedded in a meteorite. As an alternative to these nineteenth-century mechanisms, we have considered Directed Panspermia, the theory that organisms were deliberately transmitted to the Earth by intelligent beings on another planet. We conclude that it is possible that life reached the Earth in this way, but that the scientific evidence is inadequate at the present time to say anything about the probability. We draw attention to the kinds of evidence that might throw additional light on the topic.

The Murchison meteorite, which fell in Australia in 1969, contains organic substances including dozens of earthly amino acids and some amino acids not known from Earth. Of the nearly 100 amino acids thus far identified from the Murchison meteorite, only about 20 are also found on Earth. The remainder have never before been found on our planet. In addition to amino acids, sugars, also necessary for life, have been recovered from the Murchison meteorite. These discoveries have honed the debate and led scientists to consider the possibility that Earth's life originated somewhere else in the universe and may have been transported to Earth via extraterrestrial comets or meteors.

Astronomers estimate that the universe contains 100 billion galaxies; our galaxy—the Milky Way—has about 400 billion stars, putting an estimate of the total number of stars in the universe at 1 followed by 22 zeros. Many argue that it is inconceivable that there is not life somewhere else.

Since the 1960s, the Search for Extraterrestrial Intelligence (SETI) has kept a watchful ear to the universe, trying to detect any communication from beyond our planet. Radio SETI utilizes radio telescopes to try to detect narrow-bandwidth radio signals from space. These signals are not known to occur naturally, so detecting them would provide evidence of life beyond our planet. Since 1999, *SETI@home,* based at the University of California, Berkeley, has enlisted volunteers to add the power of their home computers to the search, thereby expanding the capabilities of SETI. Thousands of people in hundreds of countries have contributed to what is now the fastest supercomputer on Earth, scanning the heavens for possible transmissions from outer space.

Life may well exist elsewhere within our own solar system. A fundamental goal of the scientific search for life on other planets has been to identify extraterrestrial water. In 1995 the NASA spacecraft *Galileo* photographed Jupiter's four moons, including *Europa,* which appeared to be covered in ice. The presence of icebergs suggests a liquid ocean beneath the frozen ice surface. Efforts to identify lunar ice have thus far been unsuccessful.

Mars has also been explored for signs of life. In 2000 NASA's *Mars Global Surveyor* discovered evidence that within the last million years liquid had flowed on Mars, creating geologic formations such as gullies. Beginning in 2003, NASA's search for Martian water continued with its Mars Exploration Rover (MER) Mission, utilizing twin golf-cart-size robotic rovers (*Spirit* and *Opportunity*) to explore the Martian surface and geology and search for liquid water that may contain life.

To date, there is no hard evidence to support an extraterrestrial origin to life on Earth, but many reputable scientists continue to explore this line of inquiry. More importantly, although panspermia addresses the question of the origin of life on Earth, it does not address the fundamental question of the origin of life somewhere else! The search for the origins of life from outside the planet is a matter of conjecture and is not likely to result in any consensus without some new information or discoveries.

The question of whether life originated on Earth or was transported to Earth from elsewhere remains an open question that may never be known with certainty. How the first life came to exist is an interesting theoretical question, but to a large degree, the origin of life is inconsequential to the study of evolution. Evolutionary biology is the study of what happened after life appeared on this planet and not about how that life came to be. The organic molecules that came to be assembled and associated with a cell membrane and a common genetic code have prospered. On Earth today, all life originates from pre-existing life and requires the dynamic interaction between nucleic acids (DNA and/or RNA) and proteins for its continued existence and perpetuation. However life arose, it came to exist on planet Earth and has flourished over billions of years into many millions of species both alive and extinct.

15

MOLECULAR EVOLUTION

O ne of the dramatic breakthroughs in the 1950s was an understanding of the interrelations of protein structure and the genetic information carried in deoxyribonucleic acid (DNA). There emerged the fundamental tenet that the gene (a linear array of nitrogenous bases in the DNA) codes for the precise sequence of amino acids that compose a protein. There was also the realization that gene mutation often involves no more than an alteration of a single base in the DNA sequence that specifies the amino acid. Such a single base alteration can result in the substitution in the protein molecule of one amino acid for another. This interferes with the activity of the protein. This is vividly exemplified by the aberrant hemoglobin of sickle-cell anemia, in which a highly localized change in one of the bases of the DNA molecule leads to the substitution of only one amino acid residue in the hemoglobin molecule.

Over evolutionary time, DNA sequences diverge and proteins change in composition. The amino acid sequences of proteins have gradually been modified with time, yielding the arrays that are found today in extant species. Comparisons of the amino acid compositions in present-day organisms enable us to infer the molecular events that occurred in the past. The more distant in the past that an ancestral stock diverged into two present-day species, the more changes will be evident in the amino acid sequences of the proteins of the contemporary species. Viewed another way, the number of amino acid modifications in the lines of descent can be used as a measure of time since the divergence of the two species from a common ancestor. In essence, protein molecules incorporate a record of their evolutionary history that can be just as informative as the fossil record.

NEUTRAL THEORY OF MOLECULAR EVOLUTION

Electrophoretic methods of detecting differences in the amino acid sequences of proteins in organisms have disclosed a wealth of amino acid substitutions in the course of evolution. DNA sequencing and the analysis of protein structure have also been informative in our understanding of past evolutionary affinities. Ardent advocates of Darwinian selection, popularly called *selectionists,* maintain that a mutated form of a gene responsible for an altered amino acid must pass the stringent test of natural selection. Selectionists assert that a mutant allele is favored only when an amino acid substitution confers an advantage to the organism. In the late 1960s, the Japanese population geneticist Motto Kimura championed the view that the majority of amino acid substitutions are not likely to provide either an evolutionary advantage or disadvantage and are preserved by sheer chance. In essence, Kimura's postulate is that most amino acid replacements are not fostered by natural selection but result from *neutral* mutations that are fixed by *random drift.* The frequency of a new neutral mutant allele fluctuates over time, increasing or decreasing fortuitously. Many of these alleles will be lost purely by chance in a few generations. But an occasional neutral mutant will spread through the population to reach fixation, or a frequency of 100 percent.

Neutralists do not discount a role for natural selection at the molecular level. The role of selection is to protect a molecule from *deleterious* mutations. Indeed, most newly arising mutations are deleterious and are weeded out by natural selection. Natural selection also, in its positive role, permits the incorporation of *beneficial* mutant alleles. Beneficial mutant alleles are, of course, evolutionarily very important, but they are rare. Accordingly, the majority of variant alleles witnessed today at the molecular level are likely to have been selectively neutral mutations that have been fixed by random drift.

In response to critics, Kimura and his colleague, Tomoko Ohta, a gifted population geneticist in her own right, continually sharpened the neutral thesis of molecular evolution. Kimura and Ohta theorized that new neutral genes insinuate themselves at a steady, or constant, rate in a given lineage. Over periods of millions of years, new forms of genes randomly establish themselves at a steady pace in the population and may themselves be gradually replaced. This turnover of alleles, governed purely by chance, is responsible for the coexistence of the multitude of amino acid alternatives in a protein.

With this theoretical background, we shall turn to the experimental studies on varied proteins and examine the data that provide support for the rather unorthodox doctrine that molecular evolution is not primarily the result of the accumulation of advantageous mutations, but rather the steady accumulation of selectively neutral mutations that do not impair protein function.

CYTOCHROME *c*

Cytochrome *c* is an ancient, evolutionarily conservative molecule that serves as an essential enzyme in respiration. It is a relatively small protein of a chain length slightly over 100 amino acid residues, which apparently exists in all eukaryotic organisms. No other protein has been so fully analyzed for so many different organisms. This ubiquitous protein has been extracted, purified, and analyzed in numerous eukaryotic species. It has 104 amino acids in vertebrates and a few more in certain more primitive species. Cytochrome *c* performs the same vital function in all organisms.

Approximately a third of the amino acid residues of cytochrome *c* have not varied at all throughout time. In particular, the same amino acids have been found in 35 positions in all organisms tested, from molds to humans. Evidently, certain amino acids at specific positions are irreplaceable or *invariant.* Any substitutions at these invariant sites are likely to interfere with the chemical integrity or folding properties of the protein molecule. Mutational changes at these sites are lethal and are rejected by natural selection. No such evolutionary constraints have been placed on the acceptance of chance alterations at other sites. Amino acid substitutions are tolerated elsewhere since they preserve the essential chemical properties of the molecule. It is these tolerable amino acid substitutions that are promulgated by Kimura's selectively neutral mutations.

The number of amino acid replacements in cytochrome *c* of several species are compared in table 15.1. First, it is evident that the cytochromes

TABLE 15.1	Evolution of Cytochrome *c**

Comparison with Human Cytochrome *c*

Organisms	Number of Variant Amino Acid Residues
Chimpanzee	0
Rhesus monkey	1
Rabbit	9
Cow	10
Kangaroo	10
Duck	11
Pigeon	12
Rattlesnake	14
Bullfrog	18
Tuna	20
Fruit fly	24
Moth	26
Wheat germ	37
Mold *(Neurospora)*	40
Baker's yeast	42

*Compiled from studies by R. E. Dickerson.

of closely related vertebrates differ in only a few residues, or not at all. There is no difference at all in the composition of the amino acid residues between humans and the chimpanzee. Cytochrome *c* of the rhesus monkey differs from the human only at position 66 of the molecule, where threonine is present instead of isoleucine. Secondly, the greater the phylogenetic differences, the greater the likelihood that the cytochrome *c*ompositions differ.

MOLECULAR CLOCK

Amino acid substitutions accumulate at fairly steady rates over long periods of evolutionary time. The uniform rate is understandable because it is governed primarily by the rate of mutation without the restraint of natural selection. The clocklike rate of molecular evolution was first detected by the Nobel laureate Linus Pauling and Emile Zuckerkandl at the California Institute of Technology, in their now classical analysis of the hemoglobin

molecule. The differences in amino acid sequences between two living species can be used to estimate the period of time that has elapsed since the two species diverged from a common ancestor. The number of allelic substitutions generated per time serves as a *molecular clock*. Figure 15.1 shows the constant rate at which the cytochrome *c* gene has evolved.

The molecular clock does not tick at the same rate for all protein molecules. Indeed, the rates of amino acid substitutions are appreciably different in several protein molecules that have been extensively studied. Cytochrome *c* has evolved very slowly in comparison to the fibrinopeptide *A*, in which the rate of amino acid substitution has been exceedingly rapid. It is likely that evolutionary change at the molecular level is slow where there are strong functional constraints and faster where changes are least likely to disrupt protein function.

Those who adhere to the neutralist position point with satisfaction to the analysis of the fibrinopeptides. These are two 20-residue fragments (A and B) that are cleaved from the protein fibrinogen when it is activated to form fibrin in a blood clot. Among several mammalian species, there is an extraordinarily high level of amino acid substitutions in fibrinopeptide A. This peptide apparently functions equally well with numerous different amino acid sequences. Indeed, virtually any amino acid change appears to have been acceptable. It has been argued persuasively that the mutations responsible for the many different amino acids must be selectively neutral. Since the fibrinopeptides have little known function except to permit the activation of fibrinogen by being discarded, they evidently can tolerate any amino acid change.

There are a few proteins in which nearly the entire amino acid sequence is critical for protein function. In such cases, it would be difficult to deny the strong role of natural selection. Certain histones that bind to DNA in the nucleus have been highly conserved in their amino acid sequence. For example, histone IV from such divergent

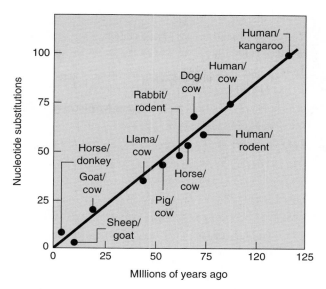

Figure 15.1 The number of differences ("substitutions") in nucleotide sequences in the cytochrome *c* genes of pairs of organisms is plotted against the time evolutionists estimate had elapsed since the pairs of organisms diverged. The resulting straight line signifies a constant rate of nucleotide substitutions, which is interpreted as indicating that cytochrome *c* is evolving at a steady, clocklike pace in all lineages.

organisms as the pea plant and the calf differ in only 2 of 102 amino acid residues. Nearly all amino acid substitutions would impair the activity of histone IV, and any substitutions, save a few, that may have appeared in the evolutionary past were most likely eliminated by natural selection. Since histone IV plays a vital role in the expression of the hereditary information encoded in DNA, it is hardly surprising that this protein would be closely specified and scrutinized.

If proteins evolve at a constant rate, then the data on amino acid substitutions can be used to construct—or reconstruct—phylogenetic relationships among organisms. As a generality, most molecular phylogenetic trees are in accord with the tempo of evolution based on fossil findings. There are a few notable exceptions. Indeed, in one instance, the molecular data have dramatically

modified our understanding of taxonomic affinities and evolutionary time scales. In the mid-1960s, Vincent M. Sarich, working in the laboratory of Allan C. Wilson at the University of California at Berkeley, compared the albumin molecules of humans and African apes. He calculated that the divergence time between humans and African apes was only a scant five million years. At the time of Sarich's pronouncement, the prevailing view among anthropologists was that humans and apes diverged at least 15 million years ago. This view was based on a fossil finding, *Ramapithecus* (now classified as *Sivapithecus*) (see chapter 17), who was purported to be one of the earlier members of the human family. Ironically, the human status of *Ramapithecus* was undermined by new fossil finds. *Ramapithecus* has been dethroned as a forebear of hominids and is presently considered an ancestor of the orangutan. This paved the way for the acceptance of Sarich's thesis that the road to humankind began 5 million years ago, or at least less than 10 million years ago. Equally important, as seen in figure 15.2, we have come to the realization that the African apes (chimpanzee and gorilla) are more closely related to humans than they are to Asian apes (orangutan).

GENE DUPLICATION

The process by which genes have evolved through duplication, followed by mutation and selection, is depicted in figure 15.3. When a particular locus duplicates itself, two genes with identical base sequences coexist in the same individual. Inasmuch as only one locus is necessary to produce the polypeptide chain, an additional locus might be regarded as superfluous. On the contrary, the duplicated locus might prove valuable in protecting the individual against the hazards of mutational changes. Ordinarily a mutation that interferes

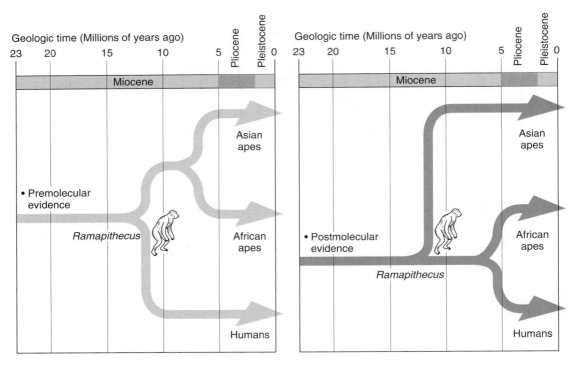

Figure 15.2 On the basis of anatomical features and fossil findings *(the premolecular evidence; left illustration),* the three great apes (chimpanzee, gorilla, and orangutan) were judged to be more closely related to each other than to humans. Genetic analyses *(postmolecular evidence; right illustration)* indicate that the African apes (chimpanzee and gorilla) are more closely related to humans than to the Asian apes (orangutan).

adversely with the functioning of a locus would be detrimental or lethal. However, an otherwise lethal mutation could be accepted if it occurred in the duplicated locus, since the original gene is unimpaired and would continue to synthesize the essential polypeptide. With the continual occurrence of mutations, the duplicated locus would produce a polypeptide whose sequence of amino acids is different from the original polypeptide. This new polypeptide, although still related to the original one, would be sufficiently altered to have a different activity and specificity. In this manner, two polypeptide chains, each with its own gene, would ultimately evolve from a single polypeptide and a single gene.

A particular gene may be duplicated more than once, and large families of related proteins may eventually evolve from a common progenitor. Indeed, it is generally accepted that the varied hemoglobin genes, alpha (α), beta (β), gamma (γ), and delta (δ), arose by duplication from a single common ancestral gene. Each duplication was followed by the accumulation of mutations in the duplicated locus. The evolution of different molecular species of a protein thus involved the interplay of gene duplication and mutation. Duplications that originated early in the evolutionary history were frequently separated into different chromosomes by translocations, whereas recently duplicated genes have remained linked on the same chromosome. The α and β genes of hemoglobin presently reside in separate chromosomes, whereas the more recently evolved δ gene has remained closely adjacent to the β gene.

Selective advantage of gene duplication

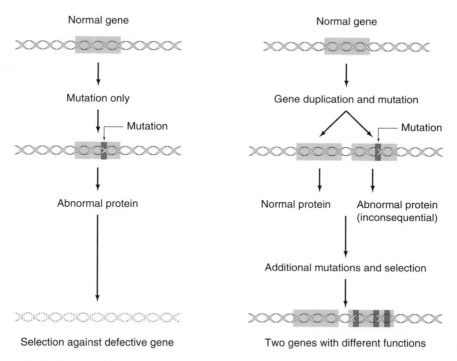

Figure 15.3 Duplication of a gene followed by random mutations and natural selection can lead to two genes with different functions. In the absence of gene duplication, a mutation would be lethal if the mutant gene produces an aberrant protein *(left)*. With a duplicated gene, the same mutation that would otherwise be lethal is not selectively disadvantageous, since the original gene continues to produce the normal protein *(right)*. With time, several mutations can accumulate in the duplicated gene.

EVOLUTIONARY HISTORY OF HEMOGLOBIN

We have thus far suggested that the genes involved in the syntheses of the polypeptide chains of hemoglobin were derived through duplications of a common ancestral locus, followed by divergent mutations. The original hemoglobin molecule probably resembled the chemically related myoglobin molecule. Myoglobin does not function as a respiratory protein; it is used mainly for the storage of oxygen in metabolically active tissues, especially within muscles. The storage of oxygen is very important in diving animals, which

have exceptionally high concentrations of myoglobin in their muscles. Mammalian myoglobin is a single polypeptide chain of 153 amino acid residues.

In the distant past, it is likely that myoglobin in tissue cells and hemoglobin in blood cells were identical. Myoglobin evolved into a storage protein with a high oxygen affinity, and hemoglobin became progressively adapted to fulfill its primary role of oxygen transport in different species. Based on the degree of similarity (or homology) of the amino acid residues between any two polypeptide chains, an evolutionary tree can be constructed (fig. 15.4). Each branching

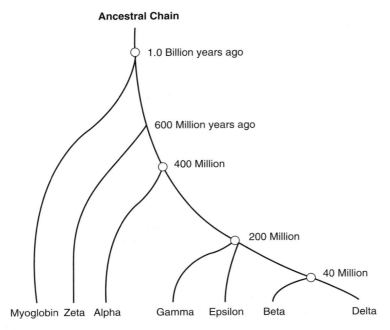

Figure 15.4 Molecular evolution of hemoglobin chains. The small circles at the foci of branching indicate where the ancestral genes were duplicated to give rise to new globin chains.

During mammalian evolution, some 200 million years ago, the β locus duplicated once again to give rise to the γ (gamma) gene that codes for fetal hemoglobin with its additional oxygen-binding capacity. The most recent duplication resulting in the δ (delta) gene occurred only 40 million years ago. The delta chain has been found only in higher primates. The divergence of the δ gene occurred prior to the separation of the phylogenetic lineage leading to the Old World monkey assemblage and the line represented by Old World monkeys, great apes, and humans. Curiously, present-day Old World monkeys do not synthesize delta chains, at least in detectable amounts. Apparently, the δ gene became silenced recently in Old World monkeys.

point in the tree indicates where an ancestral locus was duplicated, giving rise each time to a new gene line.

Linus Pauling calculated that one amino acid substitution has occurred every 7 million years. The numerous differences in the amino acid sequences between myoglobin and hemoglobin indicate that the two molecules had a common origin at least 600 million years ago, well before the first appearance of the vertebrates. A one-chain hemoglobin still persists in primitive fish, the *cyclostomes*. While the cyclostomes (lamprey and hagfishes) have a single hemoglobin chain, the carp (bony fish) has two chains. Accordingly, the differentiation of the alpha and beta chains of hemoglobin occurred about 400 million years ago, when the first bony fishes appeared. At that time, the alpha and beta chains became aggregated to produce the efficient four-chain hemoglobin molecule characteristic of most vertebrates.

PSEUDOGENES

Genetic studies have established that the α gene and the β gene reside in different chromosomes. Indeed, a cluster of α-like genes has evolved from a single ancestral α gene by a series of duplications in one chromosome, and a comparable family of β genes has emerged in another chromosome. As shown in figure 15.5, the linked group of α-like genes on human chromosome 16 contains two active embryonic zeta (ζ) and two active alpha (α) genes. The β-like cluster on human chromosome 11 comprises five active genes—namely, one epsilon (ε), two gamma (γ), one delta (δ), and one beta (β).

Both families of hemoglobin genes contain loci called *pseudogenes* (depicted as *psi,* ψ) that cannot encode a functional polypeptide chain. Although each pseudogene shares many base sequences with its corresponding normal gene, the

Human hemoglobin genes

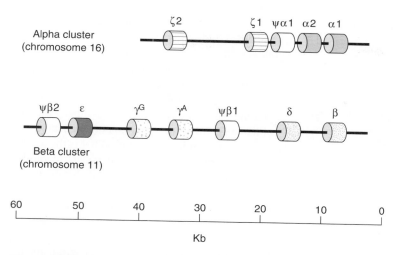

Figure 15.5 **The two families of human hemoglobin genes** form clusters on different chromosomes. The *alpha* cluster of linked genes occupies a length of about 30 Kilobases (30,000 base pairs). The *beta* cluster of genes is spread over a region of about 50 Kb (50,000 base pairs).

presence of a so-called "frameshift" (see chapter 4) mutation has shifted the triplets of bases so that polypeptide synthesis is prematurely halted. Pseudogenes apparently are products of gene duplication that have accumulated debilitating base changes during sequence divergence. If the $\psi\alpha_1$ gene were to become functional in the future, then the alpha chain would be coded by three operational α genes.

The occurrence of two immediately adjacent genes, both of which are expressed, is a common feature of gene clusters. Thus, two α genes (α_1 and α_2) are expressed in the production of alpha polypeptides. Such duplication of virtually identical genes has proved to be providential to some patients afflicted with the blood disorder, *thalassemia,* characterized by a reduced rate of alpha chain synthesis as a consequence of a defective α gene. Some synthesis of alpha chains becomes possible if the impairment involves only one of the several α loci.

The genes in both hemoglobin linkage groups are separated by substantial tracts of intergenic DNA. In other words, only about 10 percent of the DNA

in these clusters actually code for hemoglobin polypeptides. The function of the remaining 90 percent of intergenic DNA between the genes is presently enigmatic. Some investigators have suggested that these extensive intergenic regions might have no function and actually represent "junk" DNA. If such DNA is meaningless, then the base sequences could, during evolution, change rapidly without consequences. In this vein, it is of interest that the arrangements of the entire β cluster of genes are indistinguishable in the gorilla, baboon, and humans. This indicates that the arrangement of the human cluster has been fully established 20 to 40 million years ago and has been faithfully preserved since. This conservation of the arrangement suggests that the base sequences throughout the β region have been substantially constrained by selection, at least during recent primate evolution. Such conservative evolution is incompatible with the notion that intergenic DNA is "junk." One of the pioneers of molecular biology, Sidney Brenner, has cogently reminded us of a valuable difference between "junk" and "garbage"—garbage is thrown away whereas junk is stored (in the garage) for some unforeseen future use.

HOMEOTIC GENES

The name of Edward B. Lewis of the California Institute of Technology and 1995 Nobel Prize Winner immediately brings to mind his pioneering studies more than 60 years ago on genetic changes in the fruit fly, changes known as *homeotic* mutations. These mutations are responsible for such dramatic developmental effects as the transformation of parts of the body into structures appropriate to other positions. Legs may form in a position normally bearing a wing, or a second set of wings may emerge in addition to the usual single pair of wings

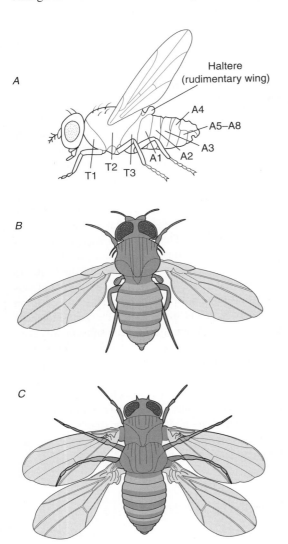

Figure 15.6 *(A)* A drawing of a normal fly. The third thoracic segment *(T3)* has a rudimentary wing, known as a *haltere*. *(B)* A normal fly with a single set of wings. *(C)* Homeotic mutations transform segment *T3* into a structure similar to *T2*. A second, fully developed set of wings emerges from this transformed segment. (Based on studies of E. B. Lewis, 1978, *Nature* 276:565.)

(fig. 15.6). Some evolutionists viewed such drastic morphological changes as major mutations, or *macromutations*. In the 1940s, the prominent geneticist Richard Goldschmidt contended that homeotic mutants illustrate a possible mode of origin of novel types of body structure and hence could be responsible for the rapid emergence of a new species in the distant past. Goldschmidt's views are considered untenable today, but Lewis's superb findings remain a landmark in developmental genetics.

The inevitable question is: What is the normal role of a homeotic gene? The recent revolution in molecular biology has revealed that homeotic genes do have assignments that are the antithesis of the execution of bizarre disruptions of structures. Homeotic genes, in their normal state, are distinctly "master" genes that orchestrate the orderly unfolding of the genetic program for the development of the embryo. Perhaps unthinkable, these master developmental genes are shared in animals ranging from the fruit fly to humans, which suggests that present-day life forms utilize in their development an ancient basic genetic network. Stated another way, the complex genetic network for development apparently has arisen only once during evolution.

In the early development of organisms as diverse as the fruit fly, mouse, and ourselves, one particular set of homeotic genes is responsible for determining the anterior to posterior (head-to-tail) axis of the embryo. In the embryo of the fruit fly, as seen in figure 15.7, eight homeotic genes are arranged in a tandem fashion and are expressed in an anterior to posterior direction that mirrors their spatial order in the chromosome. If the most anterior gene of the complex, the *labial* gene (abbreviated "lab" in fig. 15.7) were to mutate, then the development of head structures would be dramatically altered. Moreover, failure of expression of the *labial* gene by virtue of a mutation could lead to an overexpression of adjacent genes in the complex, resulting in strange imperfections in more posterior parts of the body. Certain homeotic genes normally act by suppressing the development of structures. One of the homeotic genes

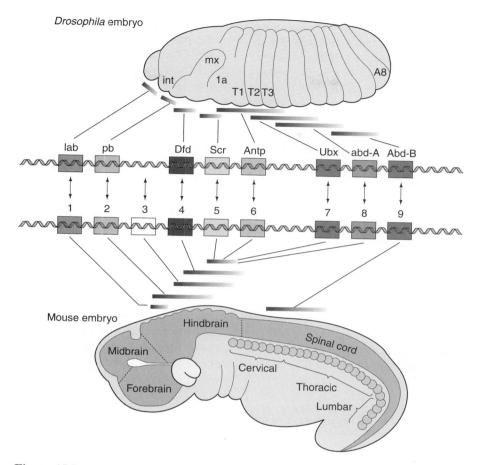

Figure 15.7 Similar spatial organization of homeotic genes in the fruit fly and mouse embryos. Segmental expression of the cluster of *Hox* genes in the mouse embryo is most pronounced in the developing central nervous system (nerve cord) and developing bony spinal column.
(From John Gerhart and Marc Kirschner, *Cells, Embryos, and Evolution,* Blackwell Science, 1997, fig. 7.12.)

expressed in the posterior thorax inhibits the development of a posterior (second) wing. Consequently, *halteres* (balancing organs resembling a reduced wing) are produced at that site rather than a second pair of wings.

The mechanism by which a homeotic gene exerts its effects has been actively pursued. It is now known that the protein product of a homeotic gene acts as a *transcription factor.* A transcription factor allows the cell to differentially turn on the expression of a particular target gene. It does so by binding to specific DNA sequences adjacent to the target gene it regulates. In essence, homeotic genes are master switches that bring into play, and coordinate the action of, other developmental genes. Each homeotic gene serves to engender a cascade of developmental effects.

Genes homologous to the homeotic genes of the fruit fly have been demonstrated in the mouse and humans. The normal homeotic genes of the fruit fly are collectively known as the homeotic complex, or HOM-C. The homologous genes in vertebrates are said to be *Hox* genes. Irrespective of nomenclature, the important message is that all HOM-C and *Hox* genes share a common ancestry. There exists in nature sets of structurally similar genes, originally discovered in the fruit fly, that are the prime regulators of embryonic development in life forms.

During the 1980s, homeotic genes were isolated and cloned in several species. A specific nucleotide sequence, called the *homeobox* sequence, is common to the coding region of all HOM-C and *Hox* genes. This sequence, 180 nucleotides in length, can be used as a genetic probe. The homeobox probe cross-hybridizes with the genomic DNA of many organisms, including yeast and humans. The homeobox sequences of the fruit fly show 90 percent homology with the homeobox sequences discovered in human genes. The evolutionarily conserved homeobox sequence apparently is a general feature of many developmental regulatory genes.

The homeobox sequence encodes a protein segment with 60 amino acid residues. The protein segment is known as the *homeodomain,* the part of the protein that binds strongly to recognition sites in other genes. Approximately 30 percent of the amino acids in the homeodomain are basic—arginine and lysine. This is a hallmark of DNA-binding regions of proteins. As seen in table 15.2, the homeodomains of various species are so highly conserved that they differ only in tolerable amino acid substitutions—that is, the replacement of one amino acid by another of similar chemical properties.

OCULAR MALFORMATIONS AND EVOLUTION

Increasingly, evolutionary developmental biology (also called "evo-devo") is providing insights into evolutionary questions. One example comes from another family of developmental control genes, called *Pax* genes that encode transcription factors. The vertebrate Pax genes are key regulators of the development of various organs of the body, notably the eye. One particular gene member of the family,

---

TABLE 15.2 | **Extensive Homology in Amino Acid Sequences of Five Homeodomains**

	1																			20
Mouse (*MO-010*)	Ser	Lys	Arg	Gly	Arg	Thr	Ala	Tyr	Thr	Arg	Pro	Gln	Leu	Val	Glu	Leu	Glu	Lys	Glu	Phe
Frog (*MM3*)	Arg	Lys	Arg	Gly	Arg	Gln	Thr	Tyr	Thr	Arg	Tyr	Gln	Thr	Leu	Glu	Leu	Glu	Lys	Glu	Phe
Fruit fly (*Antennapedia*)	Arg	Lys	Arg	Gly	Arg	Gln	Thr	Tyr	Thr	Arg	Tyr	Gln	Thr	Leu	Glu	Leu	Glu	Lys	Glu	Phe
Fruit fly (*Fushi tarazu*)	Ser	Lys	Arg	Thr	Arg	Gln	Thr	Tyr	Thr	Arg	Tyr	Gln	Thr	Leu	Glu	Leu	Glu	Lys	Glu	Phe
Fruit fly (*Ultrabithorax*)	Arg	Arg	Arg	Gly	Arg	Gln	Thr	Tyr	Thr	Arg	Tyr	Gln	Thr	Leu	Glu	Leu	Glu	Lys	Glu	Phe

	21																			40
Mouse (*MO-10*)	His	Phe	Asn	Arg	Tyr	Leu	Met	Arg	Pro	Arg	Arg	Val	Glu	Met	Ala	Asn	Leu	Leu	Asn	Leu
Frog (*MM3*)	His	Phe	Asn	Arg	Tyr	Leu	Thr	Arg	Arg	Arg	Arg	Ile	Glu	Ile	Ala	His	Val	Leu	Cys	Leu
Fruit fly (*Antennapedia*)	His	Phe	Asn	Arg	Tyr	Leu	Thr	Arg	Arg	Arg	Arg	Ile	Glu	Ile	Ala	His	Ala	Leu	Cys	Leu
Fruit fly (*Fushi tarazu*)	His	Phe	Asn	Arg	Tyr	Ile	Thr	Arg	Arg	Arg	Arg	Ile	Asp	Ile	Ala	Asn	Ala	Leu	Ser	Leu
Fruit fly (*Ultrabithorax*)	His	Thr	Asn	His	Tyr	Leu	Thr	Arg	Arg	Arg	Arg	Ile	Glu	Met	Ala	Tyr	Ala	Leu	Cys	Leu

	41																			60
Mouse (*MO-10*)	Thr	Glu	Arg	Gln	Ile	Lys	Ile	Trp	Phe	Gln	Asn	Arg	Arg	Met	Lys	Tyr	Lys	Lys	Asp	Gln
Frog (*MM3*)	Thr	Glu	Arg	Gln	Ile	Lys	Ile	Trp	Phe	Gln	Asn	Arg	Arg	Met	Lys	Trp	Lys	Lys	Glu	Asn
Fruit fly (*Antennapedia*)	Thr	Glu	Arg	Gln	Ile	Lys	Ile	Trp	Phe	Gln	Asn	Arg	Arg	Met	Lys	Trp	Lys	Lys	Asp	Arg
Fruit fly (*Fushi tarazu*)	Ser	Glu	Arg	Gln	Ile	Lys	Ile	Trp	Phe	Gln	Asn	Arg	Arg	Met	Lys	Ser	Lys	Lys	Asp	Arg
Fruit fly (*Ultrabithorax*)	Thr	Glu	Arg	Gln	Ile	Lys	Ile	Trp	Phe	Gln	Asn	Arg	Arg	Met	Lys	Leu	Lys	Lys	Glu	Ile

From W. J. Gehring. 1985. *Sci. Am.* 253/4:159.

Pax-6, has earned notoriety for causing malformations of the eye in mice and humans. Mutations in the Pax-6 gene are responsible in the mouse for diminutive eyes in the heterozygous state and the complete lack of eyes in the homozygous condition. The Pax-6 homologue in the human has been shown to be mutated in patients with congenital aniridia (lack of iris).

The Pax-6 gene is characterized by a highly conserved homeobox-like sequence, a span of 130 amino acids coded by 390 nucleotides. The homologue of Pax-6 has been isolated in the fruit fly, which shares a 94 percent identity with the amino acid sequences of mice and humans. Not unexpectedly, the Pax-6 homologue maps to the eyeless (eye) locus on the fourth chromosome of *Drosophila.*

Notwithstanding the great anatomical differences between the eyes of insects and vertebrates, the mouse Pax-6 gene induces ectopic eye development when transplanted into fruit fly embryos. It would seem that the fundamental role of the Pax-6 gene in eye development has been conserved throughout the animal kingdom. After hundreds of millions of years of evolutionary independence of the fruit fly and mouse, the astounding similarity between the developmental genes of an insect and those of a mammal seems possible *only* if these major transcriptional genes arose *only once* in a distant common ancestor.

The time-honored view is that the compound eye of insects and the camera-like eye of vertebrates evolved independently. Historically, evolutionary biologists have interpreted the origin of eyes in well-known lineages—squid, insects, and vertebrates—as illuminating examples of *convergence* (chapter 13), or the evolution of similar traits in genetically unrelated species, chiefly as a result of similar environmental selective pressures. At the University of Basel in Switzerland, a team of developmental biologists led by Walter J. Gehring proposed that the appearance of eyes in varied groups reflects *shared ancestry* (homologies of developmental genetic networks) and not *independent evolution* (convergence). This revisionistic paradigm has been embraced by some outstanding evolutionists but has been challenged by other prominent evolutionists who contend that photoreceptive organs have evolved independently on innumerable occasions, perhaps 60 or more times. Not surprisingly, there are some who think that some components of the eye evolved via homology while other components are examples of convergence. For example, it may be that the basic developmental pattern for eye development comes from *Pax* genes (homology) but that different groups of organisms have independently used this generalized genetic program to form vastly different types of eyes in dozens of different ways (convergence). Under this view, there are components eyes that evolved from shared common ancestry or homology and other components of eyes that evolved via convergence. If true, this coalescence of apparently conflicting ideas ranks as one of the more exquisite examples of the unity of knowledge arising from evolutionary theory.

MULTILEGGED FROGS REVISITED

We have seen that embryonic development is governed by a repertoire of transcription factors coded by master genes. To add an additional facet to the story, there are signals from the outside of developing cells that can modulate the activity of the transcription factors. These signals take the form of *signaling macromolecules* that enter cells and profoundly influence the differentiation of cells, tissues, and organs. For example, normal limb development in vertebrates depends upon a threshold concentration of endogenous retinoic acid (an analog of vitamin A) reaching the developing limb bud at the precise time that *Hox* genes are being transcribed.

In 1993, Malcolm Maden of King's College in London reported the striking homeotic transformation of tails into hindlimbs of frog tadpoles treated with retinoic acid. At a certain stage of development, the tails of tadpoles of a European species of frogs *(Rana temporaria)* were amputated and the experimental animals were placed for several days in varied concentrations of retinoic acid. At a particular dose and treatment time, the tail actually

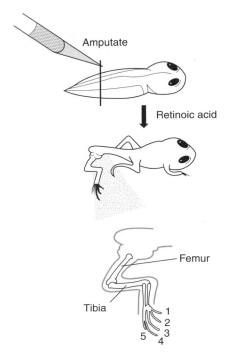

Figure 15.8 The tip is amputated from a tadpole's tail and the stump is exposed to retinoic acid. The tail stump regenerates a hindlimb rather than a tail. The anatomy of the hindlimb is normal; the pictorial enlargement of the limb shows the normal five digits. (Based on studies by M. Maden.)

regenerated a normal hindlimb (fig. 15.8). Control animals—with tails amputated but reared in the absence of retinoic acid—regenerated normal tails.

Retinoic acid may be viewed as a developmental teratogen in the sense that elevated levels can alter, or redirect, the developmental pathways. In our introductory first chapter, we evoked the usual environmental suspects—pesticides, for example—to account for the multilegged anomaly in frogs collected in nature. Now we invite the reader to consider the possibility that the causative factor in nature may have been the sensitivity of *Hox* genes to inappropriate levels of retinoic acid (or some other *Hox*-sensitive substance). Parenthetically, it may be noted that retinoic acid is occasionally used as a medication to treat skin disorders in humans, disorders such as acne. Pregnant women have been

cautioned to avoid taking retinoic acid in pill form. When absorbed in large quantities by a developing fetus, retinoic acid is highly teratogenic, causing both limb and cardiac malformations.

MITOCHONDRIAL EVE

In chapter 14 we stated that mitochondria probably began their existence as free-living, aerobic bacteria that had been engulfed by a primitive eukaryote. Mitochondria have a number of unique features in modern eukaryotic cells. For the present discussion, we shall cite two prominent features. First, the mutation rate of mitochondrial DNA is high, primarily because mitochondria have no effective DNA repair mechanism. The high mutation rate may also reflect the large concentration of mutagens (such as superoxide radicals) resulting from the intense metabolic functions performed by mitochondria. Second, mitochondrial DNA is inherited exclusively from the mother, and thus complications due to allelic segregation (as with nuclear genes) do not exist. The unfertilized egg contains some 200,000 maternal mitochondria, whereas a sperm cell contains a relatively small number of mitochondria, localized in the middle piece. The sperm's middle piece disintegrates during the act of fertilization, so that the father cannot contribute to the ultimate mitochondrial genotype of the offspring.

Maternal inheritance coupled with the high mutation rate of mitochondrial DNA makes it possible to distinguish among closely related individuals and, in fact, to trace human origins. All present-day humans may have shared a common ancestral mother (our *"mitochondrial Eve"*) or more likely, a population of women from whom we all descended. One can attempt to reconstruct human female phylogeny by detecting and quantifying nucleotide mutational changes by means of restriction enzymes. There are 37 genes, encompassing 16,569 pair bases, in each circle of mitochondrial DNA.

By analyzing mitochondrial DNA variations in women around the world, researchers have been able to approximate where the most recent common female ancestor lived. The major result from these genetic analyses of women from around the world is that sub-Saharan African populations show the greatest genetic diversity, suggesting that they are the most recent common ancestor of human females. Populations other than those in Africa retain just a subset of this genetic diversity, as would be expected when a smaller group breaks away from a founding population, taking only a small sample of the full range of genetic variation.

The ancestral mitochondrial sequence apparently arose in sub-Saharan Africa near the modern-day countries of Kenya and Ethiopia (fig. 15.9). Stated another way, based on mitochondrial DNA types, the intercontinental migrations of women to their present-day geographic distribution is best explained by postulating that the source gene pool of all human mitochondrial genes is peoples who once lived in west-central Africa. This is in accord with the thesis, subscribed to by many evolutionists, that humans first emerged in Africa and then emigrated to other continents. The population of females that represents mitochondrial Eve is presumed to have lived in Africa some time before 100,000 years ago (estimates range from 290,000 to 140,000 B.C.E.). It is likely that there has been more than one migration out of Africa. This theme will be explored in chapter 17.

Data on mitochondrial DNA have enabled investigators to reconstruct the emigration to the Americas by northeast Asian migrants. Clearly, American aboriginal peoples did not evolve *de novo* on the American continents. The Americas were populated by migrant groups from Asia, and the first of these migrations probably occurred 15,000 to 40,000 years ago. The most ancient migrants are represented today by descendants who speak "Amerind" languages, which includes most Native Americans. The Amerinds entered the Americas in a series of migratory waves 20,000 to 40,000 years ago. The Navajo, Apache, and other members of a Native American group known collectively as the "Na-Dene" are latecomers. The ancestors of modern speakers of Na-Dene entered the continent in a later migration, a mere 5,000 to 10,000 years ago. South America appears to have been colonized relatively quickly after the Americas were first occupied, with estimates ranging from 12,000 to 20,000 years before the present. Although it is intriguing to utilize these approximates dates as reference points, it is important to realize that they are just estimates based on what is a relatively new area of study.

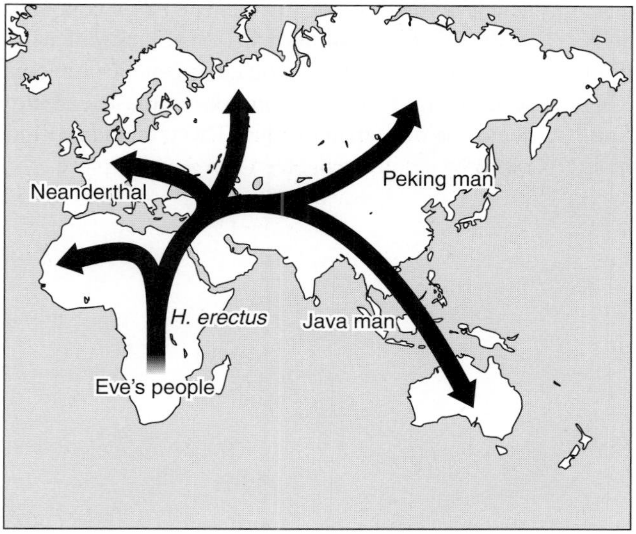

Figure 15.9 The migration of Eve's people, a group of modern humans who emerged in sub-Saharan Africa 150,000 years ago and then swept across Europe, India, Asia, and Australia. Presumably, the invading Eve's people replaced previously existing human groups, such as the Neanderthals in Europe, Peking man in Asia, and Java man in Australia and Indonesia (see chapter 17).

(Illustration from James Shreeve, "Argument Over a Woman," *Discover* [August 1996]:52–59.)

Y-CHROMOSOME ADAM

The Y chromosome is the male counterpart to mitochondrial DNA, and like mitochondrial DNA it has been used to trace the ancestry of the human species. The Y chromosome is unique to males, is inherited exclusively from father to son, and carries the genes that determine maleness. Like mitochondrial DNA, much of the Y chromosome does not recombine during meiosis. Furthermore, mutations to the Y chromosome tend to accumulate at a relatively constant rate. These characteristics have made comparative studies of mutations on the Y chromosome an excellent tool for tracing the human male ancestry.

In the same way as we have used the biblical metaphor of a "mitochondrial Eve," this imagery has been extended to the male's Y chromosome, resulting in the term "*Y chromosome Adam*." Put simply, the concept of a Y chromosome Adam is that there was a population of men whom we can identify as the most recent common ancestors of all modern human males. Like mitochondrial DNA, analysis of the mutations and diversity across modern-day human men's Y chromosome point to an origin in a sub-Saharan region of Africa.

Estimates project that all human males alive today are descended from a population of men who lived in Africa about 60,000 to 150,000 years ago. Although Y chromosome data reveal a slightly younger common ancestor than we find for our mitochondrial common ancestor, the apparent discrepancies are not very significant. Rather, these divergent lines of study both support the proposition that all people living today share common ancestors who lived roughly between 100,000 and 200,000 years ago in central or southern Africa. Likewise, both lines of inquiry provide data on when and where these common ancestors subsequently migrated. Thus there is a consensus of archeological and genetic data independently supporting the same places and same relative time frames for the common ancestry and subsequent migration of all people alive today.

It is important to emphasize that there was almost certainly not a single male "Adam" or a single female "Eve." Most likely, Mitochondrial Eve and Y chromosome Adam represent populations of our ancestors of perhaps 50 to 100 individuals; maybe even 1,000 individuals, but not two single human beings. Furthermore, although mitochondria and the Y chromosome allow for unique types of analyses, we must be mindful that DNA from mitochondria and the Y chromosome are only a small fraction of our genome, most of which is made up of the nuclear DNA contained in the X chromosome and the 22 pairs of autosomes. Contemporary molecular genetics research seeks to find better resolution to these complex questions of human origins. These themes will be explored again in chapter 17.

HISTORY OF LIFE

Our interpretation of the ever-changing procession of life is based in great measure on the fossils of animals and plants entombed in the Earth's rocks. An organism becomes preserved as a fossil when it is trapped in soft sediment that settles at the bottom of a lake or ocean. The deposits of mud and sand harden into the sedimentary, or stratified, rocks of the Earth's crust. Fossils, then, are the remains of past organisms preserved or imprinted in sedimentary rocks. Not all fossils are mere impressions of buried parts. Exceptionally, the whole organism may be preserved, as attested by the woolly mammoths embedded in the frozen gravels of Siberia. In the case of petrified tree trunks, the wood was infiltrated with mineral substances that crystallized and hardened. Incredibly, such mineral-impregnated wood can be sliced in very thin sections and studied microscopically for details of cellular structures.

The fossil remains of past life are incomplete and uneven. Innumerable organisms with soft tissue (jellyfishes and flatworms, for example) are not good candidates for preservation. In contrast, the woody parts of plants, the shells of mollusks, and the bones and teeth of vertebrates are relatively common as fossils. Complete intact organisms can become trapped and preserved in amber, the fossilized remains of tree resins. Similar processes occur with organic peat and natural asphalt. Other substrates and mechanisms by which ancient organisms can be preserved include coal fossils, petroleum, fossil carbonization, and recrystallization. In many places of the Earth, the sedimentary rocks have been subjected to such pressure and heat that their fossil holdings have been irretrievably lost. The Earth's rocks have also been exposed to the relentless forces of weathering and erosion. Despite the destructive work of geologic processes, the available fossil assemblage contains an extraordinary amount of information. The older strata of rock, those deposited first, bear only relatively simple kinds of life, whereas the newer, or younger, beds contain progressively more and more complex types of life. The increasing complexity of fossils in progressively more recent rock strata is one of the more profound evidences for evolutionary changes in organisms with time. Each species of organisms now living on the Earth has emerged from an earlier ancestral form.

GEOLOGIC AGES

The arrangement of geologic time is recorded in table 16.1. The largest division of time is the *eon* which is divided *eras* which embrace a number of subdivisions, or *periods,* and these periods are further refined into *epochs*. Fossils appear in abundance in the rocks of the Cambrian period, which dawned about 540 million years ago. The Cambrian strata bear numerous fossils of marine plants and animals; seaweeds, sponges, shellfishes, starfishes, and bizarre arthropods called trilobites. One of the more interesting Cambrian fossil sites is the *Burgess Shale,* located in British Columbia in Canada. Dated about 530 million years ago, the Burgess Shale includes an impressive assemblage of hard-bodied and soft-bodied creatures. The soft-bodied Burgess fossils owe their preservation to quick burial under anaerobic (non-oxygen) conditions.

The Cambrian was a period of great geologic upheaval of the Earth's surface. Mountains were being formed as continents collided. Earthquakes and ice ages added to the flux of environmental conditions. Oxygen levels in the oceans and the atmosphere were on the rise. Whether these earthly upheavals fostered the burst and diversification of life is unknown. Yet, it is likely that the Cambrian infusion of diversity was a function of the vast openness of ecological niches. The most explosive period of Cambrian life, estimated between 530 and 520 million years ago, has been aptly described as the *Cambrian explosion* or "biology's Big Bang."

Cambrian life did not come into existence abruptly. The evolutionary epic began at a far more distant time. At least 3 billion years of slow organic evolution preceded the diversity of organisms of the Cambrian period. The Precambrian seas swarmed with a great variety of microscopic forms of life. Studies in recent years have dispelled the notion that Precambrian rocks are devoid of fossils. Precambrian life was dominated by prokaryotic bacteria and blue-green algae (cyanobacteria). The oldest microfossil life has been uncovered from Warrawoona in Western Australia and dates back 3.6 billion years ago. The bacteria-like microfossils of Warrawoona are remarkably similar to present-day blue-green algae. In addition, the electron microscope has been used to examine samples of deep sedimentary rock from South Africa (the so-called Fig Tree sediments). The microscopic analysis uncovered impressions of rod-shaped bacteria that existed 3.2 billion years ago. Further, minute traces of two chemical substances, phytane and pristane, have been found in the Fig Tree rocks. These chemicals are the relatively stable breakdown products of chlorophyll. Thus, photosynthesis by algae, in which oxygen is released, may have begun about 3 billion years ago. Other ancient rocks—such as the 2-billion-year-old Gunflint rock formation in Ontario, Canada—contain an assemblage of fossil microorganisms that appear to be blue-green algae, possibly red algae, and even some fungi. Eukaryotic organization is clearly in evidence in deposits that are approximately one billion years old. The Bitter Springs formation of Australia contains eukaryotic fossils that represent higher algae and fungi. The transition from prokaryotes to eukaryotes apparently occurred during the late Precambrian era.

We may visualize the vast span of time by compressing the history of life into 60 minutes. In the 60-minute clock depicted in figure 16.1, each second represents one million years. The Precambrian era is conservatively viewed as originating at 12 o'clock, or 3.6 billion years ago. Primitive bacteria have left imprints in the sediments 3.2 billion years ago, or about 7 minutes after the start of the clock. The Precambrian era stretches over 5/6 of the Earth's entire history, during which time primitive life was confined to the sea.

Toward the close of the Precambrian 700 million years ago (or 11 minutes and 40 seconds before the hour), the fossil-bearing rocks show traces of soft-bodied invertebrates of simple organization. At the onset of the Paleozoic era 600 million years ago, the fossil record improves immeasurably. The Paleozoic era begins with only 10 minutes left in the hour and occupies a span of only 6 minutes and 5 seconds (or 365 million years). Nevertheless, within this short interval, we witness the invasion of land by plants, the emergence of great swamp trees, and the dominance of fishes and amphibians.

TABLE 16.1 | Geologic Time Scale

Eon	Era	Period	Epoch	Millions of Years Ago	Major Events and Characteristics
Phanerozoic	Cenozoic (age of mammals)	Quaternary	Holocene (Recent)	0.01 to present	Dominance of modern humans Much of Earth modified by humans
			Pleistocene	1.8–0.01	Extinction of many large mammals Most recent period of glaciation
		Tertiary	Pliocene	5–1.8	Cool, dry period Grasslands and grazers widespread Origin of ancestors of humans
			Miocene	23–5	Warm, moist period Grasslands and grazers common, Radiation of apes
			Oligocene	38–23	Cool period—tropics diminish Woodlands and grasslands expand, Abundant grazing mammals
			Eocene	54–38	All modern forms of flowering plants present All major groups of mammals present
			Paleocene	65–54	Warm period first placental mammals Many new kinds of birds and mammals,
	Mesozoic (age of reptiles)	Cretaceous		144–65	Meteorite impact causes mass extinction Dinosaurs dominant Many new kinds of flowering plants and insects
		Jurassic		208–144	Giant continent begins to split up Dinosaurs and other reptiles dominant First flowering plants and birds
		Triassic		245–208	Giant continent—warm and dry Explosion in reptile and cone-bearing plant diversity First dinosaurs and first mammals
	Paleozoic	Permian		286–245	90 % of species go extinct at end of Permian New giant continent forms Gymnosperms, insects, amphibians, and reptiles dominant
		Carboniferous		360–286	Gymnosperms and reptiles present by end Vast swamps of primitive plants—formed coal Amphibians and insects common
		Devonian		408–360	Glaciation the probable cause of extinction of many warm-water marine organisms Abundant fish, insects, forests, coral reefs, first amphibian
		Silurian		436–408	Melting of glaciers caused rise in sea level Numerous coral reefs First land animals (arthropods), jawed fish, and vascular plants
		Ordovician		505–436	Sea level drops causing major extinction of marine animals at end of Ordovician Primitive plants present—jawless fish common
		Cambrian		540–505	Major extinction of organisms at end of Cambrian All major phyla of animals present Continent splits into several parts and drift apart
Proterozoic	This period of time is also known as the Precambrian			2,500–540	First multicellular organisms about 1 billion years ago First eukaryotes about 1.8 billion years ago Increasing oxygen in atmosphere Single large continent about 1.1 billion years ago
Archaean				3,800–2,500	Continents form—oxygen in atmosphere Fossil prokaryotes—about 3.5 billion years ago
Hadean				4,500–3,800	Crust of Earth in process of solidifying Origin of the Earth

This chart shows the various geologic time designations, their time periods, and the major events and characteristics of each time period.

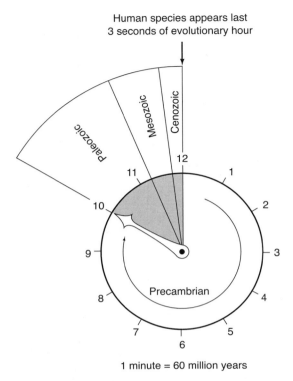

Human species appears last
3 seconds of evolutionary hour

1 minute = 60 million years

Figure 16.1 Model of evolutionary time, where one second equals one million years. The Precambrian is represented as originating at 12 o'clock, or 3.6 billion years ago. The human species appeared about 3 million years ago, or a mere terminal three seconds of the evolutionary hour.

The Mesozoic era, during which the reptiles reigned, commences with about 4 1/2 minutes left to the hour and lasts for 3 minutes (180 million years). Mammals and flowering plants dominate the landscape of the Cenozoic era during the last 1 minute and 15 seconds of the clock hour. Not until the closing seconds of the evolutionary hour does the human species appear on the scene. The whole span of recorded human history covers the last 3 seconds!

FOSSIL RECORD OF PLANTS

The general character of plant life from the dawn of the Cambrian period to recent times is depicted in figure 16.2. Only the broader aspects of past

evolution are portrayed. It is evident that the flora has changed in composition over geologic time. New types of plants have continually appeared. Some types thrived for a certain time and then disappeared. Others arose and flourished. Still others have survived until the present, only in much reduced numbers.

The Cambrian and Ordovician periods were characterized by a diverse group of algae in the warm oceans and inland seas. One of the most significant advances, which occurred during the Silurian period, was the transition from aquatic existence to life on land. The first plants that established themselves on land were diminutive herbaceous forms, the *psilophytes,* literally "naked plants," in allusion to their bare, leafless stems. The primitive woody stems sent up swollen aerial shoots to scatter their spores and thrust branches down into the Earth for water. The existence of psilophytes made possible the subsequent emergence of an infinite variety of tree-sized plants that flourished in the swamps of Carboniferous times. Two prominent trees that luxuriated in Carboniferous forests were the giant club moss, *Lepidodendron,* which soared to heights of 130 feet on slender trunks, and the coarse-leaved *Cordaites,* progenitors of the modern conifers. The first seed-bearing plants, *Pteridosperms,* arose in the humid environment of the vast swamps. These ancient groups of tall trees dwindled toward the close of the Paleozoic era and shortly became extinct. Their decomposed remains led to the formation of extensive coal beds throughout the world. Figure 16.3 shows the appearance of a Carboniferous swamp as reconstructed from fossils.

By far the greater number of Paleozoic species of plants failed to survive. The Paleozoic flora was largely replaced by the seed-forming gymnosperms, which became prominent in the early Mesozoic era. A diverse assemblage of cycads, ginkgoes (maidenhair trees), and conifers formed elaborate forests. During the closing years of the Mesozoic, the flowering plants *(angiosperms),* which began very modestly in the Jurassic period, underwent a phenomenal development and today constitute the dominant plants of the Earth. The

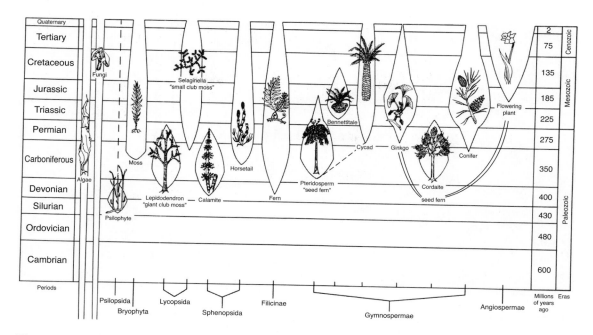

Figure 16.2 Historical record of plant life. Plant life during the Cambrian and Ordovician, the first two periods of the Paleozoic era, was confined to the water; seaweeds *(algae)* of immense size, often several hundred feet in length, dominated the seas. Land plants came into existence in Silurian time, in the form of strange little vascular plants, the psilophytes. In the Carboniferous period, imposing spore-bearing trees *(Lepidodendrids and Calamites)* and primitive naked-seeded plants *(Pteridosperms and Cordaites)* reached their peak of developement. The end of the Paleozoic era marked the extinction of the majority of the luxuriant trees of the Carboniferous coal swamps. The Mesozoic era was the *Age of Gymnosperms,* as evidenced by the abundance of cycads, ginkgoes, and conifers. Flowering plants *(Angiosperms)* rose to ascendancy toward the close of the Mesozoic era and established themselves as the dominant plant group on the Earth.

angiosperms have radiated into a variety of habitats, from sea level to mountain summits and from the humid tropics to the dry deserts. Associated with this diversity of habitat is a great variety of general form and manner of growth. Many angiosperms have reverted to an aquatic existence. The familiar duckweed (Family Lemnaoideae), which covers the surface of a pond, is a striking example.

FOSSIL RECORD OF ANIMALS

In the animal kingdom, we witness a comparable picture of endless change, with waves of extinction and replacement (fig.16.4). The deep Cambrian rocks contain the remains of varied invertebrate animals—sponges, jellyfishes, worms, shellfishes, starfishes, and crustaceans. One of the major developments of the Cambrian period was the advent of hard parts in the form of protective shells and plates. The shelled invertebrates were already so well developed that their differentiation must have taken place during the long period preceding the Cambrian. That this is actually the case has been attested by the discoveries of rich deposits of Precambrian fossils in rocks in the Ediacara Hills in southern Australia and on Victoria Island in the Canadian Arctic. These fossil-bearing rocks lie well below the oldest Cambrian strata and contain imprints of jellyfishes, tracks of various kinds of worms, and traces of soft corals. Since the rocks on Victoria Island are 700 million years old, these discoveries extend the fossil record of invertebrates 100 million years beyond the Cambrian.

Figure 16.3 Luxuriant forests of giant trees with dense undergrowth flourished in the Great Coal Age (Carboniferous period), between 350 million and 275 million years ago. The massive trunks at the left are the Lepidodendrids, an extinct group whose modern relatives include the small, undistinguished club mosses *(Selaginella)* and the ground pines *(Lycopodium)*. The tall, slender tree with whorled leaves at the right is a Calamite, represented today by a less prominent descendant, the horsetail *Equisetum*. The fernlike plants bearing seeds (at the left) are seed ferns *(Pteridosperms),* which resembled ferns but were actually the first true seed-bearing plants *(Gymnosperms).*

(Courtesy of Field Museum of Natural History, Chicago.)

Dominating the scene in early Paleozoic waters were the crablike trilobites and the large scorpion-like eurypterids (fig.16.5). Most trilobites were not more than 3 inches long; the so-called giant trilobite did not exceed 18 inches. The most powerful dynasts of the Silurian waters were the eurypterids (sea scorpions) which occasionally attained a length of 12 feet. Common in all Paleozoic periods were the nautiloids, related to the modern *Nautilus.* The clamlike brachiopods, or lampshells, constitute nearly one-third of all Cambrian fossils, but are relatively inconspicuous today. The odd graptolites, colonial animals whose carbonaceous remains resemble pencil marks, attained the peak of their development in the Ordovician period and then faded away. No land animals are known from Cambrian and Ordovician times. Seascapes of the early Paleozoic are shown in figure 16.5.

The bony fishes arose in the Devonian, the Age of Fishes. The lobe-finned fishes, or *crossopterygians,* had strong muscular fins by which they crawled, in times of drought, from pool to pool along arid streambeds. The end of the Paleozoic era witnessed a decline in amphibians and the initiation of reptilian diversification.

Many of the prominent Paleozoic marine invertebrate groups became extinct or declined sharply in numbers before the Mesozoic era. During the Mesozoic, shelled ammonoids thrived in the seas and insects became prominent. Some of the ancient insects were enormous, like *Meganeuron,* a form of dragonfly with a wingspread of 29 inches. Among vertebrates, the reptiles radiated swiftly to dominate the land. At the close of the Mesozoic, the once successful marine ammonoids perished and the reptilian dynasty

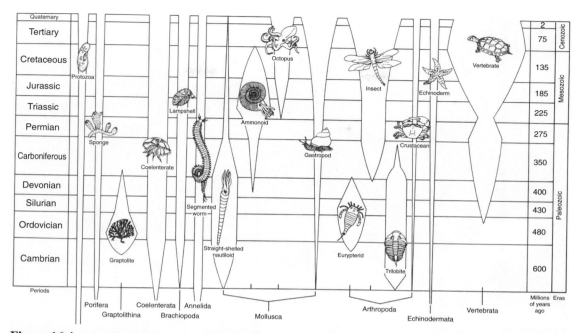

Figure 16.4 Historical record of animal life. Many of the invertebrate groups were already highly diversified and abundant in the Cambrian, the first period of the Paleozoic era, approximately 600 million years ago. The Paleozoic era is often called the *Age of Invertebrates,* with its multitude of nautiloids, eurypterids, and trilobites. Brachiopods with hinged valves were the commonest shellfish of the Paleozoic seas. In the Mesozoic era, air-breathing insects and vertebrates, notably the widely distributed reptiles, held the center of the stage. The Mesozoic seas were populated with large, shelled ammonoids, now extinct. Warm-blooded vertebrates (birds and mammals) became prominent in the Cenozoic era, and the human species arrived on the scene in the closing stages of this era.

collapsed, giving way to the birds and mammals. Insects have continued to thrive and have differentiated into a staggering variety of species. Well over a million different species of insects have been described, and conservative estimates place the total number of living species today at 10 to 14 million.

During the course of evolution, plant and animal groups interacted to each other's advantage. There is little doubt, for example, that the rise and spread of flowering plants fostered the diversification and dispersal of insects. As flowering plants became less and less dependent on wind for pollination, a great variety of insects emerged as specialists in transporting pollen. The colors and fragrances of flowers evolved as adaptations to attract insects. Flowering plants also exerted a major influence on the evolution of birds and mammals. Birds, which feed on seeds, fruits, and buds, evolved rapidly in intimate association with the flowering plants. During the Cenozoic era, the emergence of herbivorous mammals coincided with the widespread distribution of nutritious grasses over the plains. In turn, the herbivorous mammals furnished the setting for the evolution of carnivorous mammals. The interdependence between plants and animals continues to exist in nature today.

CONVERGENCE

The major pattern of the evolutionary panorama is divergence—the tendency of a population of organisms to become highly diversified when

Figure 16.5 Highly diversified assemblage of invertebrates of early Paleozoic seas. *Top:* Ordovician period. Large organism in foreground is a straight-shelled nautiloid. Other prominent forms are trilobites, massive corals, smaller nautiloids, and a snail. *Bottom:* Cambrian period. Conspicuous animals are the trilobite *(center foreground)*, eurypterid *(center background)*, and the jellyfish *(left)*. Other animals include brachiopods, annelid worms, sea cucumber, and varied shelled forms.

(*Top,* courtesy of Field Museum of Natural History, Chicago: *bottom,* courtesy of American Museum of Natural History, New York City.)

the population spreads over an area with varied habitats. The population can diverge into radically different lines, each modified for a specific ecological role. Today, essentially similar habitats may be found in widely separate parts of the world. It would thus not be surprising to find that two groups of organisms, unrelated by descent but living under similar environmental conditions in different geographic regions, can exhibit similarities in habits and general appearances. The tendency of one group of organisms to develop superficial resemblances to another group of different ancestry is called *convergence* (see chapter 13 and 15). Stated another way, convergence is the result of similar selective forces causing common adaptive features of genetically independent origin. Contrarily, similarities of features in two groups due to a shared common genetic ancestry constitute *homology.*

Convergence is not an uncommon phenomenon in nature (see figs. 13.5, 16.6). Many unrelated, or remotely related, organisms have converged in appearance as a consequence of exploitation of habitats with similar ecological makeup. The relationship of convergence to adaptive radiation should be evident; the former is the inevitable result of the countless series of adaptive radiations that have taken place in scattered parts of the globe. A striking example of convergence is afforded by the living marsupials of Australia (fig. 16.6).

The marsupials, or pouched mammals, are thought to have evolved in North America and expanded south, diverging from the placental mammals approximately 90 million years ago. In the late Cretaceous or early Tertiary periods (see table 16.1), marsupials began to migrate to Australia and New Zealand from North America via Antarctica. Later, when Australia separated from Antarctica and moved northward, it became a completely isolated island, allowing for the independent evolution of marsupials in Australia and New Zealand, free from competition from the more efficient placental mammals. In the absence of a land migration route between Asia and Australia, the latter land mass was inaccessible to practically all the placental mammals. On other continents the marsupials perished, save for the peculiar American opossum and their kin.

Imprisoned in Australia, the marsupials radiated into a variety of habitats and open ecological niches. Several live in the open plains and grasslands; some are tree dwellers; others are burrowers; and still others are gliders (fig. 16.6), Most kangaroos are terrestrial, but one variety, the monkey-like kangaroo, is arboreal. The slow-moving nocturnal "teddy bear," or koala, lives and feeds on *Eucalyptus,* the dominant tree in Australia. The bandicoot, with rabbit-like ears, has sturdy claws adapted for digging in the ground in search of insects. Marsupial moles live in desert burrows, and flying phalangers have webs of loose skin stretched between the forelimbs and hind limbs. The flying phalanger cannot actually fly but is adept at gliding. The impressive diversity of marsupials thus represents an admirable example of adaptive radiation.

The marsupials of Australia also vividly illustrate the phenomenon of convergence. They have filled the ecological niches normally occupied by placental mammals in other parts of the world. The marsupial "mouse," "mole," "anteater," "wolf," flying phalanger, and groundhog-like wombat strikingly resemble the true placental types—mouse, mole, anteater, wolf, flying squirrel, and groundhog, respectively—in general appearance and in their ways of life (fig. 16.6).

It is interesting to note that a marsupial "bat" has not evolved in Australia. The opportunity was apparently denied by the invasion of placental bats from Asia, one of the few placental forms that managed, probably as a result of dispersal by flight and winds, to reach Australia. With the coming of humans, the secure existence of the marsupials has been threatened. Prehistoric humans introduced dogs, which ran wild (the dingoes); later human settlers brought a number of European placental mammals, such as the rabbit, hare, fox, cat, and the Norway rat. Massive extinctions followed the arrival of these non-natives. One of

Placentals

Marsupials

Wolf
(Canis)

Tasmanian wolf
(Thylacinus)

Ocelot
(Felis)

Native cat
(Dasyurus)

Flying phalanger
(Petaurus)

Flying squirrel
(Glaucomys)

Ground hog
(Marmota)

Wombat
(Phascolomys)

Anteater
(Myrmecophaga)

Anteater
(Myrmecobius)

Mole
(Talpa)

Mole
(Notoryctes)

Mouse
(Mus)

Mouse
(Dasycercus)

Figure 16.6 Phenotypic convergence is a common theme in evolution and is exemplified by comparing the marsupial mammals of Australia to the placental mammals of the Americas.

the marsupials that succumbed to extinction in the twentieth-century is the marsupial "wolf," the largest known carnivorous marsupial of the modern era.

THE PLAINS-DWELLING MAMMALS

When the great majority of land mammals first arose, South America was just as much an island continent as Australia. An impressively rich assemblage of mammals evolved in isolation in South America during the 70 million years that the continent remained separated from the rest of the world. Such bizarre placental mammals emerged as the giant ground sloth, the enormous toothless anteater, and the armored armadillo-like glyptodont. The predatory wolves that evolved in South America were not true carnivores; rather, they resembled the predaceous marsupials of Australia.

The spread of grasses across the flatlands of the Americas in the Tertiary period was associated with the rapid expansion of herbivorous hoofed mammals (*ungulates*) in both continents. The crowd of grazing mammals that arose is so bewildering that it is difficult to call up a true picture of the immense variety of ungulate types. These archaic hoofed mammals are quite unfamiliar to most people; they have no vernacular names and taxonomists have had to create special suborders for many of them. Selected examples of these curious hoofed mammals were shown earlier in figure 2.3 (chapter 2). These South American forms bear only superficial resemblances to the hoofed mammals in other parts of the world. *Toxodon* superficially resembles a rhinoceros, *Thoatherium* has the appearance of a horse, *Pyrotherium* may be likened to an elephant, and *Macrauchenia* looks like a camel. The similarities in external appearances signify convergence and not relationship by descent.

At the time the Isthmus of Panama rose from the depths, the land bridge across the Bering Straits lay completely dry. Large mammoths streamed in increasing numbers across the Bering Strait from Asia into North America during the Pleistocene. The mammoths, particularly the 10-foot high woolly mammoth, dominated the American plains. North America was also the recipient of some of the South American mammals. Among the immigrants from south of the border were *Boreostracon,* a giant glyptodont, encased in armor and an offshoot of the armadillo line, and *Megatherium,* a great ground sloth that weighed more than a modern elephant (fig. 16.7).

The South American hoofed mammals, evolving independently of mammalian types in North America, thrived for many millions of years. Then the water gap between the Americas was closed. The Isthmus of Panama rose above the sea, establishing a land corridor linking the two continents. A horde of northern invaders infiltrated through this land bridge. Predatory carnivores, such as saber-toothed tigers and wolves, entered South America from the north, as did progressive herbivorous ungulates, among which were tapirs, peccaries, llamas, and deer. All of the native South American hoofed mammals, without exception, were driven into extinction. South America also became deprived of its indigenous variants of ground sloths and glyptodonts. It then later lost several of the immigrants, notably the mastodons and saber-toothed tigers.

EVOLUTIONARY STABILITY

Some types of organisms have not changed appreciably in untold millions of years. The opossum has survived almost unchanged since late Cretaceous, some 75 or more million years ago. The horseshoe crab, *Limulus,* is not very different from fossils uncovered some 500 million years ago. The maidenhair, or ginkgo, tree of the Chinese temple gardens differs little from its ancestors 200 million years back. The treasured ginkgo has probably existed on Earth longer than any other

Woolly Mammoth

Saber-toothed Tiger

Glyptodont

Ground Sloth

Figure 16.7 Prominent land mammals that occupied the flatlands of the Americas in the Tertiary period.

present tree. Darwin called the ginkgo "a living fossil."

Long-standing stability of organization seems antithetical to the concept of evolution. Nevertheless, the notion that an established species may be rather stable is not completely at variance with conventional evolutionary thought. Long periods of relatively little change may be the outcome of stabilizing selection (see chapter 6), in which the typical members of a population are selectively favored relative to the extremes. In other words, the "status quo" is maintained by selection after an ecological niche in a habitat has been successfully filled. Admittedly, to some

investigators, it seems implausible that natural selection could be so conservative as to continue to favor the same phenotype for many millions of years.

MASS EXTINCTION

The multitude of different kinds of present-day organisms is impressive and by some estimates far exceeds 10 million species. Yet modern, the biological diversity is only a fraction of the tremendous array of organisms of earlier periods. In the history of life, extinction has been the rule and not the exception, and it has punctuated Earth's history since life first arose. A paradox of evolution is that even though more than 99 percent of all organisms that ever lived are extinct, there are today more different kinds of organisms than at any time in the history of the Earth (fig.16.8). This is due to repeated cycles of extinction and replacement (see chapter 13). Some extinction events have been massive, while others have been smaller in scale. As we saw in chapter 13, a sequence of massive adaptive radiations typically follows mass extinction events. In this way, ecological niches vacated due to extinction are reoccupied as survivors adapt and diversify. Historically, each undulation of mass extinction and adaptive radiation happened over a period of a million to several millions of years.

In most cases, the reason why an individual species went extinct is unknown and likely unknowable, and the intensity of species extinction is highly variable across time. *Background extinctions* occurring at a relatively constant rate account for the majority of extinct species and occurred in the setting of natural selection and biological phenomenon. Only those species that can continue to adapt to changing environmental conditions avoid extinction.

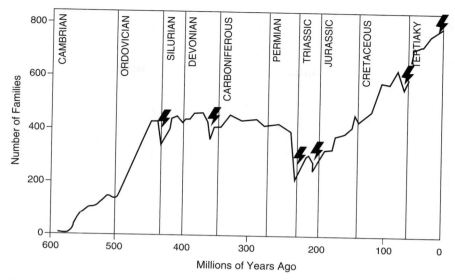

Figure 16.8 **The diversity of life** has increased over geological time despite the periodic declines caused by massive extinction events. Although an estimated 99 percent of all species have gone extinct, there is a higher level of biological diversity today than at any other time in the history of the Earth.

By contrast, mass extinctions occur over a comparatively short length of geological time and are due to abrupt catastrophic events such as volcanic eruptions, climatic changes, and changes in sea levels. Unlike other extinction events, mass extinctions occur randomly, without regard to the adaptive values of traits or of organisms. In general, less is known about background extinctions than is known about mass extinctions. As shown in figure 16.8, the rate of speciation has been higher than the rate of background extinctions over most of the last half billion years.

After the rise of animal diversity dubbed the Cambrian explosion (roughly 530 million years ago), there followed a series of mass extinction events that punctuated Earth's history. Especially significant is what are referred to as the five major mass extinction events, which occurred at or toward the end of the Ordovician, Devonian, Permian, Triassic, and Cretaceous periods. (fig. 16.8, table 16.2). Each of these five massive extinction events or

periods had profound consequences. Although many things can result in the extinction of an individual species, it takes a cataclysmic event such as global volcanism or an asteroid strike to alter the entire *biosphere*. In each instance, the consequences were greatest among species, followed by genera (groups of species), and then by families (groups of genera), and no phyla were completely eliminated. Some authors have argued that from the point of view of evolution, extinction may be a constructive phenomenon, eliminating organisms that are most susceptible to environmental stress.

HOLOCENE EXTINCTION

In addition to the above, a sixth, more controversial mass extinction event is the Holocene extinction event—the ongoing mass extinction of species occurring on planet Earth. Estimates vary on the number of species that have become extinct in the

TABLE 16.2 | **Six Major Mass Extinction Pulses Since the Cambrian Explosion**

Mass Extinction Event	When (million years before present)	Victims Casualties	Possible Cause(s)
Holocene	Recent	Megafauna, all taxa	Likely impact of human invasion of non-African continents Human population explosion and associates impacts—habitat destruction, invasive species, etc.
Cretaceous-Teritary	65 million	50 % of marine genera 76 % of marine species 18 % of land vertebrates inc. 100 % of dinosaurs.	Possible asteroid hitting Yucatan Siberian volcanos Global cooling
Late Triassic	199–214	25 % of all families 75 % of all species 50 % of marine genera 76 % of marine species End of mammal-like reptiles	Little known Possible comet or asteroid impacts Sea level decline; oceanic anoxia, (low dissolved oxygen) increased rainfall
Permian-Triassic	251	53 % of marine families 84 % of marine genera 70 % of all land species (plants, insects and vertebrates) 96 % of all species	Possible asteroid Global volcanism Dropping sea levels Oceanic anoxia (low dissolved oxygen)
Late Devonian	364	22 % of marine families 57 % of marine genera 83% of marine species	Unknown Possible asteroid or comet impact Oceanic anoxia (low dissolved oxygen) Possible glaciation followed by deadly temperature declines
Ordovician-Silurian	439	25 % of marine families 61 % of marine genera 85 % of marine species	Glaciation leading to decline in ocean levels Continental movements

past 40,000 years as humans spread out of Africa onto other continents (fig. 16.9). The causes of the most current pulse are many and complex, but all revolve around the exponential increase in the human population (fig. 16.10). The Holocene extinction is viewed as having three major waves or pulses.

The first wave is coupled with the spread of modern humans out of Africa and across the globe (see figs. 15.9 and 17.14). As humans spread out of Africa, wherever they landed they annihilated the large mammals and birds, who had no previous experience with our kind. Sometimes called the *Pleistocene overkill,* the megafauna, large mammals and flightless birds, vanished within 1,000 years of human colonization of each of the world's continents (fig. 16.9).

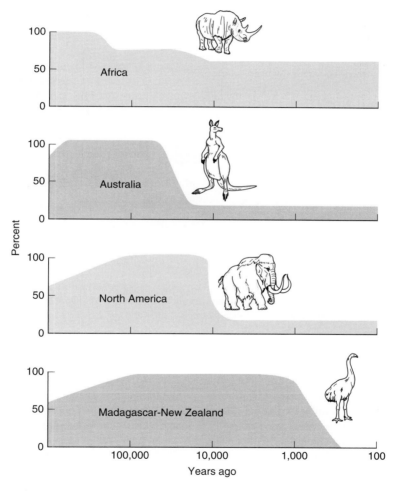

Figure 16.9 Pleistocene overkill. As humans migrated across the globe, the large mammals and flightless birds were the first to go extinct. In Africa, where humans and animal life evolved side by side, the extinction impacts was diminished.

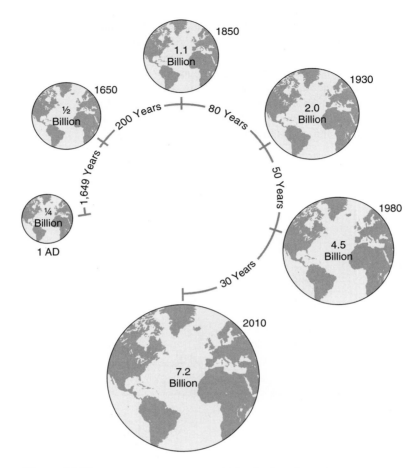

Figure 16.10 The global human population has been increasing at an exponential rate. According to population projections, the world population is likely to continue growing until at least 2050.

Animals not hunted to extinction were led to indirect extinction as a consequence of habitat and ecological disruptions. Similar scenarios played out on each continent as humans came to inhabit them. Because it was the homeland of our species and humans and the large mammals evolved together on the African plains, Africa was least affected (fig. 16.9).

The second Holocene extinction pulse is linked to the spread of Europeans, during which time habitat destruction quickened and the oceans, continents and islands all saw significant declines in species. In the oceans, the marine mammals and large fish were impacted; on continents, the remaining megafauna and many freshwater taxa were affected; on islands the casualties included birds, mammals, and tortoises.

The third most recent wave of extinctions in the Holocene began in the mid-twentieth century and continues to the present and is most directly linked to the exponential explosion of the human population (fig. 16.10) and associated widespread habitat loss and alteration. All taxa have been affected by this most recent mass extinction pulse.

Compared to other mass extinctions, which occurred over millions of years, the present mass extinction is occurring at an accelerated rate (tens of thousands of years). The rate of extinctions has been increasing, the closer to the present one looks, and is higher today than at any time in the past.

17

EMERGENCE OF THE HUMAN SPECIES

The human species is the outcome of the same natural forces that have shaped all animal life. As relatively recent arrivals on the Earth, humans represent one of the latest adaptive advances in the animal kingdom. We are clearly part of the fabric of nature and have ancestral ties with other animals. There is almost universal unanimity that our closest relatives are the apes. The line leading ultimately to humans diverged from the ape branch during Tertiary times. There are several candidates among Tertiary fossil apes that qualify as ancestors of the human species. One often hears stated, hesitatingly and perhaps in the form of an apology, that we are not really closely related to the apes but merely share a very distant ancestry with them. By indirection, the idea is conveyed that our remote generalized ancestor would not be at all apelike. This is sheer deception. There is simply no escaping the fact that our very early predecessors had many features in common with the apes.

PRIMATE RADIATION

The Order Primates, to which humans belong, underwent adaptive radiation in Cenozoic times,

approximately 85 million years ago. The primates are primarily tree dwellers; only humans are fully adapted for life on the ground. Many of the noteworthy characteristics of the primates evolved as specializations for an arboreal mode of life. Depth perception is important to a tree-living animal; the majority of the primates are unique in possessing binocular, or stereoscopic vision, wherein the visual fields of the two eyes overlap. The hands evolved as prehensile organs for grasping objects, an adaptation later useful to humans for manipulating tools. The use of the hands to bring objects to the nose (for smelling) and to the mouth (for tasting) was associated with a reduction of the snout, or muzzle, and a reduction of the olfactory area of the brain. Relative to other animals, primates generally have a poor sense of smell. Closely associated with enhanced visual acuity and increased dexterity of the hands was the marked expansion of the brain, particularly the visual and memory areas. Progressive enlargement of the brain culminated, in humans, in the development of higher mental faculties.

The primate stock arose and differentiated from small, chisel-toothed, insectivorous ancestors, perhaps as long as 85 million years ago. The Asiatic tree shrew, *Tupaia*, an agile tree climber that feeds on insects and fruits in tropical forests, survives as a model for the ancestor of the primates. Most authorities place the tree shrew in the mammalian Order Insectivora, while a few hold the view that the tree shrew is a primitive, generalized member of the primates. The important consideration is that the modern tree shrew bears likeness to the common ancestor of primates.

Primates are generally divided into two major groupings or suborders, the *Prosimii* (prosimians) and the *Anthropoidea* (anthropoids) (fig. 17.1). The prosimians are *arboreal* and largely nocturnal predators; they include such tropical forms as the lemurs, the lorises, and the tarsiers. As adaptations to an arboreal-nocturnal niche, these prosimians evolved large, forwardly placed eyes and strong, grasping hands and feet. The lemurs are largely confined to the island of Madagascar, the lorises are found principally in eastern Asia, and the tarsiers are limited to southeastern Asia. The tarsiers represent the most advanced group of prosimians and may be said to foreshadow the higher primate trends of the Anthropoidea.

The more advanced primates, the anthropoids, are composed of the New World monkeys, the Old World monkeys, the apes, and humans. All are able to sit in an upright position, and thus the hands are free to investigate and manipulate the environment. The monkeys are normally *quadrupedal,* running on all fours along branches of trees. Often referred to as the lesser apes, the gibbons and the siamangs habitually use their arms for hand-over-hand swinging in a motion known as *brachiation.* The great apes (orangutan, chimpanzee, and gorilla) can maintain prolonged semi-erect postures. When on the ground, the chimpanzee and gorilla are "knuckle-walkers," the weight being placed on the knuckles of the hands rather than on the extended fingers. The chimpanzee and the gorilla regularly travel quadrupedally on the ground. Humans alone are specialized for erect *bipedal* locomotion.

In 2009 a 47-million-year-old fossil, *Darwinius maisillae,* nicknamed Ida, was proclaimed

as a transitional fossil link between the prosimian and the anthropoid lineages and therefore could be ancestral to monkeys, apes, and humans (fig. 17.2). This controversial Eocene specimen was named to commemorate the 200th anniversary of Charles Darwin's birth and the ancestral Messel rain forest in Germany where the specimen was discovered. A juvenile specimen, Ida is a remarkably complete fossil with most of its skeleton, the outline of its soft body parts, and even the contents of the digestive track all intact. In the branching bush of the primate lineage, Ida would be near the base, although her precise location on the primate tree is contentious and awaits further study. Interestingly, Ida died at the fringe of a volcanic lake along with a host of other creatures that succumbed to the perils at the margins of a volcano. It is too early to tell whether Ida will join the ranks of Lucy (figs. 17.7 and 17.8) as an enduring icon of primate evolution.

Biochemical and genetic studies reveal that the two species of chimpanzees (the common chimpanzee, *Pan troglodytes,* and the bonobo—once called the pygmy—chimpanzee, *Pan paniscus*) and the two species of gorillas (western *Gorilla gorilla* and eastern *Gorilla grauer*) bear a special evolutionary relationship to humans. A wealth of molecular data (see chapter 15) reveals that the African great apes (chimpanzee and gorilla) are more closely related to humans than they are to Asian apes (the orangutans). Modern taxonomists firmly maintain that the molecular findings should be reflected in our classification scheme *by placing the African apes in the family of humans* (family *Hominidae*).

Higher primate taxonomy is fiercely debated, always in a state of flux and never agreed to by all taxonomists. By one proposed revision (fig. 17.1), all of the *hominoids* (living and extinct members of the superfamily *Hominoidea)* are in the two families, the Hominidae and the Hylobatidae. The family Hominidae is then divided into two subfamilies, the Homininae, which includes three tribes, the tribe

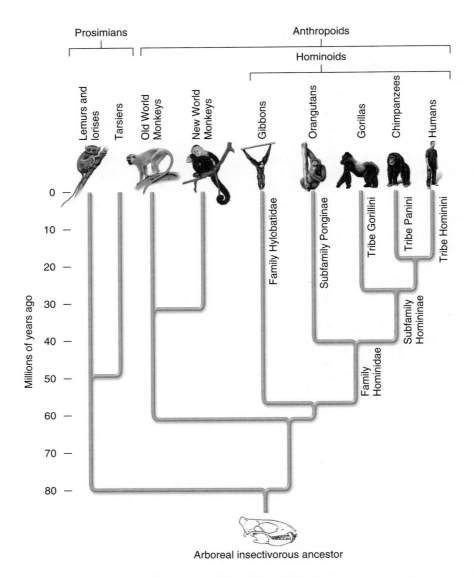

Figure 17.1   The adaptive radiation and classification of the living primates from a basal stock of small insect-eating placental mammals. Although the common ancestor of the modern primate dates from around 85 million years ago, the major adaptive radiation of the primates and other mammals coincides with the disappearance of the dinosaurs roughly 65 million years ago.

Gorillini (the gorillas in the genus *Gorilla*), the tribe Panini (the chimps in the genus *Pan*), and the tribe Hominini (the humans in the genus *Homo*). The other subfamily is the Ponginae (the orangutans in the genus *Pongo*). Gibbons (in the genus *Hylobates*) and siamangs (in the genus *Symphalangus*) form the second hominoid family, the Hylobatidae. Whether this reflects the true state of affairs awaits the next revision.

Figure 17.2 At nearly 95 percent complete, Ida is one of the most complete fossil primates ever discovered. Private collectors unearthed Ida in 1983 and split the fossil, selling it in two separate parts. Twenty-five years later, it was restored and recognized as a transitional basal primate.

© Mike Segar/Reuters/Corbis

ADAPTIVE RADIATION OF HUMANS

The morphological differences between humans and apes relate mainly to locomotory habits and brain growth. Humans have a fully upright posture and gait and an enlarged brain. The cranial capacity of a modern ape rarely exceeds 600 cubic centimeters, while the average human cranial capacity is 1,350 cubic centimeters. The mastery of varied environments by humans has been largely the result of superior intelligence gradually acquired throughout evolution.

It may be difficult to envision humans other than as they exist today. Nevertheless, it should not be imagined that fossil specimens of the early Pleistocene epoch possessed the attributes of contemporary forms. Present-day humans are certainly different from their predecessors, in much the same manner that any modern form of life is different from its forerunner. It is important to recognize that there have been different kinds of human ancestors. The evolutionary process of adaptive radiation produced a family of human ancestors in much the same way that other groups of organisms became highly diversified.

Our present-day species—*Homo sapiens*—is the sole surviving species of human. Until recently, we conceived of the human lineage as a tidy, straightforward succession of one species to the next, leading eventually to modern *Homo sapiens*. We scarcely permitted ourselves to evoke the proposition that two or more species could have *overlapped in place or time*. The saga of human evolution increasingly suggests that there were *coexisting species of human ancestors* throughout most of our evolutionary past.

FORERUNNERS OF THE GREAT APES

One of the best known of the early Miocene African hominoids is the fossils in the genus *Proconsul* (fig. 17.2). The name *Proconsul,* which literally means "before *(pro)* Consul" comes from a famous chimpanzee at the London Zoo named "Consul." *Proconsul* is believed to have been an arboreal quadruped, weighing 10 to 40 kg (roughly 20 to 80 lb), and like the modern apes it lacked a tail. The specimen shown in figure 17.3 is that of a small juvenile that weighed less than 20 kg (44 lb). Proconsul was a forest dweller, and it became extinct in the early Miocene.

The most famous of the late Miocene apes has been classified in the genus *Dryopithecus,* the oak-ape, so called because of the presence of oak leaves in the fossil beds. Known in more informal terms as the dryopithecines, these primitive oak-apes were a relatively successful group of 30 or more species that flourished for at least 10 million years, mostly in Europe. At that time, the Arabian

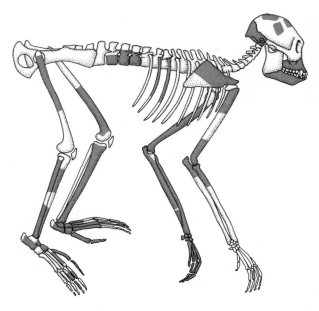

Figure 17.3 **Skeletal reconstruction of** *Proconsul,* an early Miocene ape. Dark color indicates fossil remains.

continued to recede and yield to the grasslands, the dryopithecines became extinct. Some species of this ape may have evolved into the still-surviving orangutan, gorilla, and chimpanzee. Other Miocene apes include *Pliopithecus, Sivapithecus,* and *Gigantopithecus.*

Gigantopithecus was probably the largest primate that ever lived. Although there is considerable disagreement on how large this creature was, males may have been nearly 3 meters tall (over 9 feet) and may have weighed as much as 250 to 450 kg (roughly 550 to 1000 lb). It is believed that *Gigantopithecus* was highly sexually dimorphic, with females being much smaller than males. Size is difficult to project, as this fossil ape is known only from several jaw fragments and over a thousand teeth, which were discovered in China in the mid-twentieth century. Like the modern panda, it is believed that *Gigantopithecus's* primary food source was bamboo. The first remains of this mammoth primate were found in a Hong Kong apothecary shop, where the fossilized teeth and bone, known as "dragon teeth," were ground into medicinal powder. *Gigantopithecus* survived until the Pleistocene, making them contemporaries of members of the genus *Homo.* For this reason and its immense size, myths and fables of Yeti (the Abominable Snowman), Sasquatch, and Big Foot, all of which have no basis in fact, typically reference *Gigantopithecus.*

The modern apes are intimately tied to the steadily shrinking forest habitats and are likely to face imminent extinction in the wild. The population numbers of the great apes have dwindled, in part, because the apes have failed to adapt to niches beyond the humid tropical forests. Contrarily, many monkey species have thrived as they have successfully adapted to diverse habitats.

peninsula joined Africa and Eurasia together, and the dryopithecines ranged widely throughout the two great land masses. Then, after their remarkable spread and divergence, nearly all of these early apes became extinct.

Dryopithecus may not be far removed from the common stock from which apes and humans arose. Although primarily a tree dweller, *Dryopithecus* apparently wandered on the ground and profited from the vegetable foods in the open grasslands (fig. 17.4). The transition from tree dwelling to ground living might well have first appeared at this time. The Miocene epoch was characterized by expanding populations of varied species of monkeys and apes in the European and African forests. Population pressure probably contributed to the departure of oak-apes from the forest fringes to the savanna grasslands. These pioneer ground-dwelling apes apparently were able to establish themselves in the open bushy plains by subsisting mainly on plant foods and largely avoiding confrontations with the savanna carnivores. But as the forests

EARLY HOMINOIDS

The earliest common ancestor of the great apes and humans dates to about 8 million years ago.

Figure 17.4 **Reconstruction** of *Dryopithecus aficanus*, an apelike type that prowled East Africa 20 to 25 million years ago. This pongid apparently led an agile life both on and off the ground.

(Painting by Maurice Wilson; by permission of the Trustees of the British Museum—Natural History.)

Four recently discovered fossils shed light on the early ancestry and descent of the hominoids (fig. 17.5).

Toumai (meaning "hope of life" in the local Tebou language) is more properly known by the scientific name *Sahelanthropus tchadensis* and was discovered in 2001 in the Central African country of Chad. *Toumai* is known from a skull that dates to approximately 6 to 7 million years ago and may be the oldest human ancestor. Because only the skull is available for study, there is an ongoing debate as to whether *Toumai* was bipedal. Whether *Toumai* is a direct descendant of humans or an ape ancestor is also a

matter of conjecture that awaits the discovery of other fossils from the same time period. Because *Toumai* dates to or before the time when the great apes and the line leading to humans diverged, it is not surprising that *Toumai* has a mixture of ape and human characteristics.

Another recent fossil find is *Orrorin tugenengis,* sometimes referred to as Millennium Man, which was found in the Tugen Hills of Kenya in 2001. *Orrorin* is currently the second oldest relative of humans, and its fossils date to around 6 million years ago. Known from roughly a dozen fossils from several individuals, the chimp-sized *Orrorin* was likely to have walked bipedally when it was not climbing trees. Whether *Orrorin* was an ape or a hominid remains uncertain, but regardless, it is an important fossil find that is helping extend our knowledge of human and ape evolution.

Ardipithecus (meaning "ground ape") is the genus name of two species (*ramidus* and *kadabba*) of basal *hominoids* (living and extinct humans and apes) that lived in what is now Ethiopia between 4.4 and 5.8 million years ago. *A. kadabba* was found in the Awash region of Ethiopia, and *A. ramidus* comes from a nearby site named Aramis. Both are named as though they are at the base of the human lineage; *kadabba* is Afar for "basal family ancestor," and *ramidus* is also Afar (and means "root"). Members of this genus are ancestral (fig. 17.5) to the australopithecines (members of the genus *Australopithecus*) described in the next section on the "Australopithecines". *Ardopithecus ramidus* was discovered in 1994 by a team lead by University of California anthropologist Tim D. White. *Ramidus* dates from about 4.4 million years ago, and when first discovered it was thought to be a member of the australopithecines. Although its name ("ground ape root") suggests that this species is at the base of the divergence between humans and the African apes, we now believe that this fossil is the oldest human ancestor. The older *Ardopithecus* species, *A. kadabba*, was first discovered in 1997 and lived around 5.8 to 5.3 million years ago. *Kadabba* is notable for its

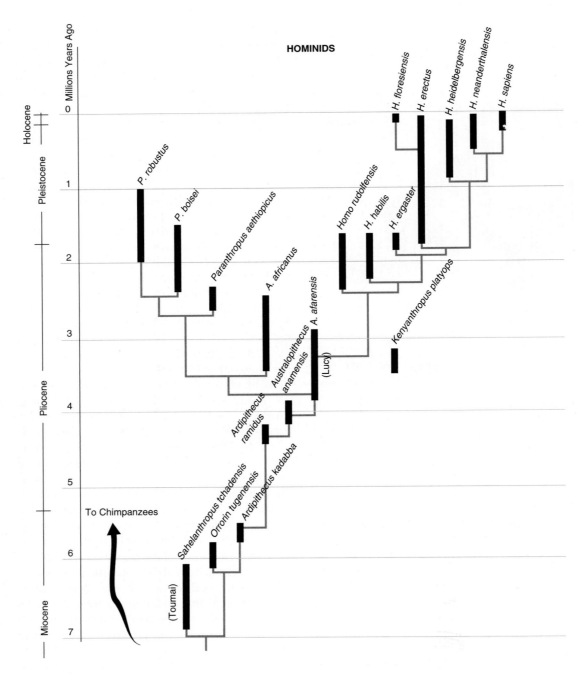

Figure 17.5 Human Ancestors: Evolutionary and historical relationship between fossil human ancestors. Throughout much of human fossil history, more than one hominid lived contemporaneously.

prominent canine teeth, a feature it shares with the older *Sahelantropus* and *Orrorin* fossils. Notwithstanding its prominent shearing canine teeth, *kadabba's* overall dental pattern is intermediate between humans and the African apes, although most of its dental features are more apelike than human-like.

In the fossil record that follows *Ardipithecus, Sahelanthropus* and *Orrorin* are members of the genus *Austalopithecus.*

AUSTRALOPITHECINES: THE FIRST HOMINIDS

The australopithecine stage is a relatively long phase of adaptive evolution, in which several species of early hominids apparently coexisted (figs. 17.5 and 17.9). The australopithecine era commenced 4 million years ago and ended about 1 million years ago. At least six species are currently recognized falling into two main groups, the robust australopithecines *(Paranthropus robustus, P. boisei,* and *P. aethiopicus)* and the gracile australopithecines *(Australopithecus anamensis, A. afarensis,* and *A. africanus).*

In 1924 the Australian anatomist Raymond A. Dart announced the discovery of an unusual small skull from a Pleistocene limestone quarry near the village of Taung in the Transvaal region of South Africa. The fossilized skull was that of a child of about six years. The little Taung skull bears some resemblance to the skull of a young chimpanzee, but many of its components, notably the teeth, show pronounced affinities to humans. Certain striking features of the Taung skull suggest that the child had walked upright. The remarkable skull was designated by the formidable name of *Australopithecus africanus* (*austral,* for "south," *pithekos,* for "ape," and *africanus,* for "from Africa"). Dart was confident that *Australopithecus* was related to the ancestral stock of humans rather than to the great apes.

Dart's declaration, which several scientists initially derided, was fortified by findings in the

1930s by the late Robert Broom, a Scottish paleontologist. Adult skulls of *Australopithecus* were dug out from caves in Sterkfontein, Kromdraai, and Swartkrans in South Africa. The adult skulls confirmed the hominid anatomical pattern seen in the juvenile Taung cranium. The several new fossil forms were originally given different names. However, in recent years it has become customary to refer to the South African fossils collectively as the australopithecines. They were short, four to five feet in height, with a small ape-size brain (cranial capacity range of 450 to 600 cubic centimeters). Nonetheless, the australopithecines stood upright, walked bipedally, and dwelt in open country (fig. 17.6). These circumstances

Figure 17.6 Reconstruction of *Australopithecus,* a hominid of about 4 million years ago who stood upright, walked bipedally, and dwelt in open country. (Painting by Maurice Wilson; by permission of the Trustees of the British Museum—Natural History.)

nullify the popular view that humans were intelligent animals when they first came down out of the trees. It seems clear that erect bipedal locomotion on the ground preceded the development of a large complex brain. Moreover, the upright posture conflicts with popular iconography that typically illustrates a hunched-over apelike ancestor that becomes progressively upright over a stretch of time.

The australopithecines are decidedly early representatives of the hominid lineage. However, it is not clear which of the australopithecines occupies a prominent place in the direct ancestry of humans. The initial fossil findings fostered the notion that there were at least two distinct species of australopithecines—the light-jawed, slender ("gracile") *Australopithecus africanus* and the heavy-jawed, robust *Australopithecus robustus* (now usually referred to as *Paranthropus robustus; paranthropos,* meaning "parallel to man", a name given to this fossil by Dart) with extremely large grinding molars. *Africanus* was not over four feet tall and weighed no more than 60 pounds, whereas the *robustus* species was a foot taller and at least 30 pounds heavier. Ecologists have observed that when two contemporaneous species occupy the same habitat, the potential competitors become differentially specialized to exploit different components of the local environment. Thus, direct competition for food resources is minimized and the two species are able to coexist. It is thought that *Australopithecus africanus* increasingly supplemented its diet with animal food. The dietary difference is supported by the finding that *A. africanus* had smaller molars than *A. robustus.*

If, as appears likely, the two (or more) *Australopithecus* species did coexist at the same time in the same region, then only one of the two could have been the progenitor of a more modern species of hominid. Paleontologists had earlier suggested that the vegetarian *A. robustus* perished without leaving any descendants, and that *A. africanus* was the forebear of a more advanced hominid. This was based on the assumption that a

dietary change to carnivorism was one of the more important steps in transforming a bipedal ape into a tool-making and tool-using human. Current thinking does not place *A. africanus* in the direct line of ancestry of humans (figs. 17.5 and 17.9).

The South African findings have been supplemented and extended by fossil discoveries by the late Louis Leakey and his wife Mary at the 25-mile-long Olduvai Gorge in Tanzania. Olduvai Gorge is situated in a volcanically active region in eastern Africa and has invaluable sediments of lava from prehistoric volcanic activity. Because lava contains argon 40, the daughter isotope of potassium 40, the rate of decomposition from potassium to argon permits an accurate dating of the Olduvai layers. Like a miniature Grand Canyon, the sides of Olduvai Gorge display different strata laid bare by the cutting of an ancient river.

In an exposed stratum of the gorge, the Leakeys in 1959 uncovered bony fragments of a robust australopithecine, characterized by extremely massive jaws. This heavy-jawed fossil form with enormous teeth was called *Zinjanthropus,* or the Nutcracker Man. Fossil remains of *Zinjanthropus* were found in strata judged, by the potassium-argon dating method (instead of the conventional uranium-lead technique), to be about 1.8 million years old. Most scientists today agree that *Zinjanthropus* is essentially an eastern African variety of *Australopithecus robustus. Zinjanthropus* has more exaggerated, or coarser, features than *A. robustus* and warrants recognition as a separate species, *Australopithecus boisei,* and thus represents yet another species in the australopithecine complex.

One might suppose that an older, as yet undiscovered australopithecine (from Pliocene sediments) was at the base of the hominid lineage. The candidate for such an ancestor was thought to be the fossil hominid recovered in 1974 by a team led by the anthropologist Donald C. Johanson, now at the Institute of Human Origins at Arizonia State University. The remains of this hominid were found at a formation in the Hadar region of Ethiopia, a formation that has been dated at 3.2 to 3.5 million years ago.

The fossil beds at Hadar have yielded a remarkably rich paleontological collection. The constellation of fossils includes a social group of 13 individuals, a group that has come to be known as "the First Family." The family members apparently died together at one spot. The most complete adult skeleton uncovered is the now-famous female hominid called "Lucy." Her name was borrowed from the Beatles' song "Lucy in the Sky with Diamonds." She was only 3.5 to 4.0 feet tall,

weighed less than 66 pounds (30 Kg), and died when she was in her early twenties (figs. 17.7, and 17.8). Other Hadar individuals are relatively large; the variation perhaps reflects differences due to sex. Parenthetically, it may be noted that an analysis of Lucy's pelvis casts doubt on whether she was a female!

In 1978 Mary Leakey discovered a series of footprints made by human ancestors 3.6 million years ago at Laetoli in northern Tanzania. Apparently the nearby volcano Sadiman experienced a mild eruption followed by a rainstorm just before

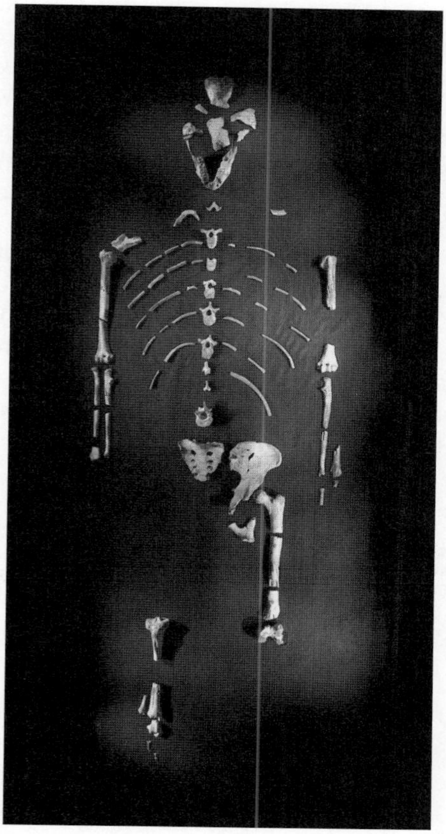

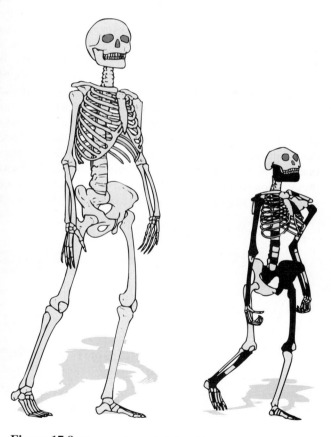

Figure 17.7 "**Lucy,**" a small adult female (less than 4 feet tall) who lived close to 3 million years ago. This famous skeleton was discovered by the paleoanthropologist Donald Johanson, and placed in a new species, *Australopithecus afarensis.*
© Dave Einsel/Stringer/Getty

Figure 17.8 Comparison of skeletal features of a modern human to a reconstruction of the famous *Australopithecus afarensis,* "Lucy" (fig 17.7). The fossil components are shown in black. The reconstruction is based on mirror images of known parts of the skeleton and on other fossils.

at least two hominids walked across this wet cementlike surface. When the volcano later covered the footprints with additional volcanic ash, the footprints were preserved. After the sediments eroded, the footprints were reexposed. The Laetoli footprints are remarkably similar to modern human feet. Evidently upright bipedalism was established as early as 3.6 million years ago.

Johanson contends that the hominid remains from Hadar and Mary Leakey's Laetoli fossils are so similar morphologically as to belong to a single human lineage. In 1978 Johanson and Tim D. White (an anthropologist from the University of California at Berkeley) assigned the Hadar and Laetoli specimens to the same species and delineated them as a new species, *Australopithecus afarenis*. The new species derives its name from the Afar locality in Hadar, Ethiopia, which yielded the most numerous specimens. *Australopithecus afarensis* was then pronounced as the root stock from which all later hominids sprang. At present, it is safe to say that the *afarensis* remains, dated between 3 and 4 million years ago, constitute one of the earliest definitive members of the family Hominidae (fig. 17.5). There are critics who contend that the bones and teeth of the Hadar

individuals provide evidence for two distinct species. If true, this would push the search for a common ancestor back even farther in time.

As paleontologists continue to search for fossils, new finds continue to emerge and the reexamination of older specimens calls for further revisions. Enter *Australopithecus anamensis,* the oldest of the australopithecines, dated to about 4 million years ago. Although the first fossil was discovered in 1965, it was not appreciated for what it was until 1989 and was not assigned its current name until 1995. *Anamensis* was named for the Turkana name *anam,* which means "lake," because many of the fossils were discovered near Lake Turkana in Kenya. Like *A. afarensis, A. anamensis* was a biped, a diagnostic character in separating the line of hominids from the African apes. As shown in figure 17.5, *A. anamensis* may well be a direct ancestor of *A. afarensis.*

In essence, an enormous burst of diversity characterized the early stages of hominid evolution (figs. 17.5 and 17.9). There were evidently multiple species of hominids, some of which appear to have coexisted. The next challenge is to trace the line of descent of our own genus, *Homo,* from the broad australopithecine family tree.

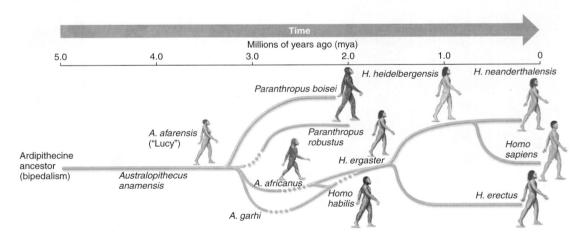

Figure 17.9 Diversity in the human family. Several hominids lived contemporaneously with one another throughout much of human history, but only one species, modern *Homo sapiens,* survived to modern times.

HUMANS EMERGE: *HOMO HABILIS*

Members of our genus, *Homo,* evolved in Africa from australopithecine ancestors approximately 1.8 million years ago. We are continually gaining new insight into how the genus *Homo* evolved and how its members came to live with, and eventually replace, the australopithecines. The increasingly diverse fossil record permits us to portray human evolution as a flourishing bush containing various species. Although only a solitary human species *(Homo sapiens)* exists today, several species coexisted for much of our history.

Early in the 1960s, Louis and Mary Leakey found remains of a light-jawed hominid that they claimed was more advanced and more human-like than any known australopithecine. Estimates of the brain capacities of the skulls averaged 637 cubic centimeters. This light-jawed type of Olduvai Gorge hominid was said to be the first civilized or humanized being, deserving of the rank of *Homo,* namely *Homo habilis* (fig. 17.5). The specific name *habilis* means "handy," from the inferred ability of this hominid to make stone tools. The recognition of *Homo habilis* indicates that this primitive human being was evolving alongside the less hominized australopithecines and lived side by side with them. The coexistence of *Homo habilis* and the australopithecines is generally accepted.

Spirited discourse surrounds son Richard Leakey's 1972 discovery of an unusual skull from the desiccated fossil beds of Lake Turkana in Kenya. This nearly complete skull was initially placed at 2.8 million years old but is currently acknowledged as being only 1.8 million years old. The specimen was cautiously designated only by its museum identification number—"1470." The cranial capacity of the "1470" skull measures 780 cubic centimeters, which is significantly larger than any australopithecine specimen. Skull "1470" clearly belongs to the genus *Homo.* Some anthropologists believe that "1470" is sufficiently different from *Homo habilis* to have earned its own species name, *Homo rudolfensis.* In the most current configuration,

Homo rudolfensis is viewed as a distinct species that evolved earlier than *Homo habilis* (fig. 17.5).

GATHERING-HUNTING WAY OF LIFE

The lower Pleistocene hominids *(Homo habilis)* fabricated crude chopping tools by striking a few flakes from a cobble or large pebble. The stone tools enabled *habilis* to cut scavenged meat into chewable size pieces, as well as smash large bones to get at the fat-rich bone marrow. The open African savanna offered the early plains-dwelling hominids a wider selection of foods than were available in the tropical forests. Tools were used to gather a variety of plant materials (nuts, seeds, fruits, tubers, and roots) and small animals (lizards and rodents). It is unlikely, as popularly imagined, that tools were first contrived as efficient spears to hunt large game or that our early ancestors were voracious meat-eating, predatory primates. The human species became accomplished hunters of the large animals only later in evolutionary history.

The early hominids on the savanna were most likely primarily gatherers of plants and small animal life, and both males and females were opportunistic in devising tools for digging, processing plant foods, and scavenging meat. The gathering of food was particularly critical to females with dependent young. There may have been some stalking and killing of game, but the uncertainty of success made hunting secondary or supplementary to the collection of plant food with its assurance of success. The exploitation of food sources by both males and females on the savanna fostered the collaborative interactions of individuals. Cooperative behavior promoted the sharing of food for the first time. The sharing of food regularly is a social achievement unique to humans; only rarely do apes share food. The exchange of food is considered to be the earliest expression of human social reciprocity.

Homo habilis apparently had little propensity to migrate beyond Africa. This early human thrived for nearly 500,000 years before it became extinct. The hominid that followed had an even greater brain size and a proclivity to migrate.

HOMO ERECTUS: THE EXPLORER

The earliest and best-known representative of *Homo erectus* is the famous Java fossil, first described as *Pithecanthropus erectus* ("upright ape-man"). This primitive human being was discovered at Trinil, Java, in 1894 by Eugene Dubois, a young Dutch army surgeon. Dubois had been profoundly influenced by the writings of Charles Darwin and was taken with the idea that he could find the origins of humans. He surprised the world with the discovery of this early human. Curiously, Dubois in his later years inexplicably doubted his own findings and contended that *Pithecanthropus erectus* was merely a giant manlike ape. In the 1930s, additional fossil finds of *Pithecanthropus* were unveiled in central Java by the Dutch geologist G. H. R. von Koenigswald. The fossil specimens date from between 1 million years and 700,000 years ago. The newer findings confirmed the human status of the *pithecanthropines.*

The pithecanthropines lived during middle Pleistocene times, between 1.3 million and 300,000 years ago *or less!* They arose during a period of shifting climates that turned much of forested Africa into cooler and dryer open grasslands. These low-browed hominids were toolmakers and hunters who had learned to control fire. They probably had some powers of speech. Their ability to exploit the environment is reflected in the expanded size of the brain. The cranial capacity of the pithecanthropines was in the range of 775 to 1,000 cubic centimeters. Their advanced tool kit (termed *Acheulean*) included finely worked, teardrop-shaped hand axes, various sharp cleavers, finger-sized scrapers, and cutting flake tools. The pithecanthropines were travelers and explorers that migrated successfully through the continents, from the tropical regions of Africa to Asia and Europe.

Current evidence suggests that at least two species deserve recognition. Many writers now reserve the name of *Homo erectus* for the Eurasian pithecanthropine specimens and accord the name of *Homo ergaster* for African finds. In this scheme, *Homo erectus* evolved from the African *Homo ergaster* and migrated out of Africa about 1 million years ago, if not earlier. In 1996, an interdisciplinary team of scientists re-examined two major fossil sites along the Solo River in Java and found surprisingly that *Homo erectus* apparently persisted in southeast Asia until about 53,000 years ago. If proven correct, this would mean that *Homo erectus* survived some 250,000 years after it had been surmised to have become extinct. This remnant population of *Homo erectus* would have existed at the same time that two other human species, *Homo neanderthalensis* and *Homo sapiens* roamed the Earth (fig. 17.10). Confirmation of the dates for the Java specimens would provide support for the new paradigm that more than one hominid species could have existed at any one time-level.

In the 1920s, Canadian anatomist Davidson Black's elaborate excavation of caves in the limestone hills near Peking, China, led to the discovery of another primitive human, *Sinanthropus pekinensis,* or "Peking man" (fig. 17.11). The cranial capacity in the sinanthropines varied from 900 to 1,200 cubic centimeters. They fashioned tools and weapons of stone and bones, and they kindled fire. A strong suspicion was once held that *Sinanthropus* was cannibalistic and savored human brains, for many of the fossil braincases show signs of having been cracked open from below. This view is now considered untenable; the damaged braincases are regarded as the work of scavengers or other natural events. Most of the original fossil specimens of Peking man were lost during World War II. Fortunately, photographs and measurements were made of the original fossils. Plaster casts made of the original fossils still exist.

The Java and Peking hominids were originally each christened with a distinctive Latin name, *Pithecanthropus erectus* and *Sinanthropus pekinensis,* respectively. There is, however, no justification for

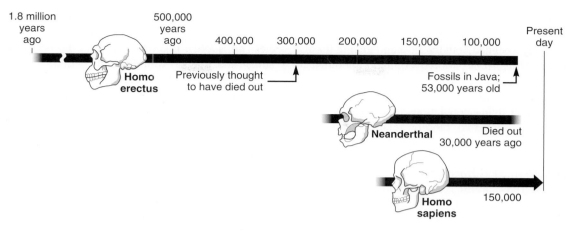

Figure 17.10 **Fossil skulls** found in Java suggest that *Homo erectus,* thought to have perished hundreds of thousands of years ago, coexisted on Earth with the Neanderthals and modern *Homo sapiens* as recently as 30,000 years ago.

(Source: New York Times, 12/13/96, front page, *3 Human Species Coexisted on Earth, New Data Suggest* by John Noble Wilford.)

Figure 17.11 **Peking humans** left remains about 400,000 years ago in limestone caves in northern China, kindled fire, and fashioned tools of stone and bone.

(Painting by Maurice Wilson; by permission of the Trustees of the British Museum—Natural History.)

recognizing more than the single genus of humans, *Homo.* Accordingly, modern taxonomists have properly assigned both the Java and Peking fossils to the genus *Homo.* Moreover, the morphological differences between these two fossil humans are readily within the range of variation that we observe in living populations today. These forms thus represent two closely related geographic races (subspecies) of the same species. Both Java and Peking hominids are usually placed together and classified as *Homo erectus.*

In 1984, an impressive complete skeleton of an adolescent boy dating from 1.6 million years ago was discovered from Kenya's Turkana basin. Dubbed the "Turkana boy," this young male fossil was 5 feet 3 inches tall and between 9 and 11 years of age. It is estimated that his adult height would have been 6 feet, much taller than previous hominids. He possessed a surprisingly modern human body structure, suggestive of modern gait. He was tall, slender, and had long limbs, traits that would appear to be ideally suited for traveling long distances in harsh climactic conditions. This Kenyan specimen has several distinctive features that distinguish it from the Asian *Homo erectus.* Accordingly, the Turkana boy and his kind in Africa have been set apart as a separate species, the aforementioned *Homo ergaster.* This species has been proposed as a credible ancestor for all successive humans.

FLORES MAN

In 2003 the bones and artifacts of a miniature "hobbit"-size human-like being were found in caves on the island of Flores in Indonesia, roughly 370 miles east of Bali. Discovered by paleoanthropologists Peter Brown, Michael Moorwood, and colleagues from the University of New England in New South Wales, Australia, the Flores find has changed our perception of recent human evolution. Named *Homo floresiensis* for its island place of origin, the discovery of the Flores people was both unexpected and controversial. Nowhere

in the history of humans, or their direct ancestors, is there any evidence of a diminutive species (fig. 17.12). Likewise, there are no previous human-like remains from this time period.

Estimated to be no bigger than an australopithecine, Flores people stood only a meter in height, weighed around 25 kilograms, and had a 380-cc brain (modern humans average 1400 cc, the australopithecines average ~ 500 cc). Remains date from roughly 13,000 to 94,000 years ago. However, the lore of the contemporary inhabitants of Flores, the Manggarai, speaks of diminutive people who survived in caves until the sixteenth-century arrival of European traders. It appears that about 12,000 years ago, a volcanic eruption resulted in the extinction of the Flores people.

Notwithstanding their relatively young ancestry, the Flores people's bones resemble *Homo erectus* more than modern *Homo sapiens.* However, the stone tools of the ancient Floresians are unlike any associated with *Homo erectus* and include such items as blades and other cutting and chopping tools as well as points that may have been used to hunt. There is also evidence that the Floresians hunted cooperatively and used fire.

On islands like Flores, evolutionary forces often select for either dwarfism or gigantism (or both). In the case of Flores, giant lizards and small elephants are thought to have once lived on this island. Although some critics have argued that the Flores people are small-headed *microcephalics,* a medical condition characterized by abnormally small heads, the evidence to date is compelling that very small human-like creatures once inhabited this island.

EVOLUTION OF HUMAN SOCIETY

As *Homo erectus* wandered from place to place, the hunting of game most likely became intensified. Vegetable foods continued to be important in subsistence, but the hunting habit was the way of life for nomadic *Homo erectus.* There is no evidence that the female participated in the hunting

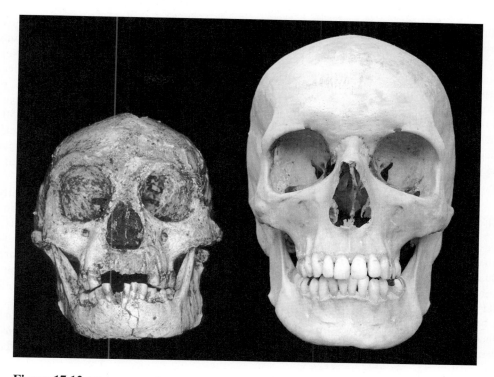

Figure 17.12 Flores man. Comparison of the skulls of miniature hominid, *Homo floresiensis* (left) who inhabited the Indonesian Island of Flores 13,000 to 94,000 years ago with modern human, *Homo sapiens.*
© AFP/Getty Images

of large game; the adult female was increasingly encumbered with a fetus or by the care of the young, or by both. The long period of dependency of the young strengthened mother-child bonds but also restricted the mobility and activity of the woman. It appears likely the female remained at the home base as a food gatherer while the male engaged in hunting. The immobility of the female and the prolonged immaturity of the young, coupled with the limitations imposed on the male in the number of females he could possibly support, apparently transformed a basically polygamous society into a monogamous structure.

A primary human innovation was relatively permanent pair-bonding, or monogamy. Sustained pair-bonding proved to be advantageous in several respects. It served to reduce sexual competition among the males. The prolonged male-female

pairing increased the probability of leaving descendants. The heterosexual pair-bonding relationship became fortified as the estrus cycle of the female, became modified into a condition of continuous sexual receptivity. The sustained sex interests of the partners made possible by the obliteration of estrus in the female increased the stability of the family unit and facilitated the development of permanent family-sized shelters for rest, protection, and play. In essence, strong interpersonal bonds between a male, a female, and their children became the basis of the uniquely human family organization. Critics of the pair bonding scenario assert the provocative notion that *Homo erectus* females were advantaged (though lower levels of male infanticide and greater food sharing) by having multiple male sexual partners.

Human language was fostered as males and females recorded experiences with each other and transmitted information to their mates and children. Speech favored cooperation between local groups and the fusion of small groups into larger communities. Speech also fostered the successful occupation of one geographical area after another. The exchange of ideas over wide areas permitted human cultures of great complexity to develop.

STATUS OF THE NEANDERTHALS

The classic Neanderthal was first unearthed in 1856 in a limestone cave in the Neander ravine near Düsseldorf, Germany (fig. 17.13). This group of fossils derives its name from the picturesque Neander Valley or Neanderthal ("thal" means valley, but the silent "h" has been dropped in the modern German to "tal"). The Neanderthals or Neandertals (both spellings continue to be used) are one of the best known of fossil hominids, having been found at numerous widely separate sites in Europe, particularly in France. The Neanderthals were cave dwellers, short (about 5 feet) but powerfully built, with prominent brow ridges. They had large brains with an average capacity of 1,450 cubic centimeters, as opposed to 1,350 cubic centimeters in modern humans. The Neanderthals occupied Europe, the Middle East, and western Asia from at least 230,000 years ago until about 30,000 years ago.

The Neanderthals were once popularly portrayed as brutish and dull-witted human creatures. On the contrary, the Neanderthals made complex stone tools, were accomplished hunters of large game, compassionately cared for their sick and infirmed, and withstood the rigors of the bitter cold climate of the last glaciation. There is evidence that the Neanderthals buried their dead with various ritual objects. The tools of the Neanderthals are known as *Mousterian* and are characterized by flint scrapers and points.

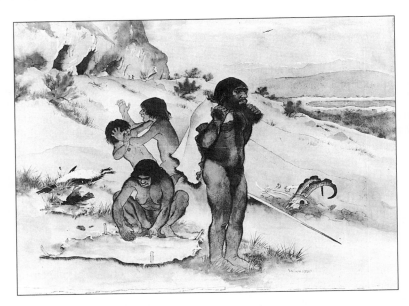

Figure 17.13 Neanderthal, a rugged cave-dweller who roamed Europe and the Middle East about 75,000 years ago.
(Painting by Maurice Wilson; by permission of the Trustees of the British Museum—Natural History.)

The Neanderthals roamed over Europe during the Upper Pleistocene until about 30,000 years ago and then dramatically disappeared. They were replaced by humans of a modern type, much like ourselves, which have been grouped under the common name of Cro-Magnon. The Cro-Magnons are decidedly representatives of our own species, *Homo sapiens.*

There is evidence that modern humans displaced the Neanderthals in the Middle East by 40,000 years ago. Modern humans spread into Europe about 40,000 years ago. From 40,000 to 30,000 years ago, a span of 10,000 years, the Neanderthals and Cro-Magnons appear to have lived side by side in Europe. Some authors have suggested that the Cro-Magnons conquered and destroyed the Neanderthals. Others have endorsed the view that the anatomically modern human migrants brought new diseases to Europe for which the Neanderthals had no resistance. There

is also the opinion that modern humans and the Neanderthals interbred and assimilated their respective gene pools. This opinion apparently has been dispelled by mitochondrial DNA studies, which reveal that the Neanderthals became extinct without contributing mitochondrial genes to modern humans.

MODERN HUMANS: THE CRO-MAGNONS

The Cro-Magnons, a representative of our own species, *Homo sapiens,* can be traced back about 40,000 to 10,000 years ago in Europe (figs. 17.14, 17.15 and 17.16). The Cro-Magnons were tall and slender: males reached 6 ft. (1.8 m) and weighed 155 lbs. (70 kg) while females reached 5 ft. 6 in. (1.7 m) and weighed 120 lbs. (55 kg). The Cro-Magnons were culturally superior to the

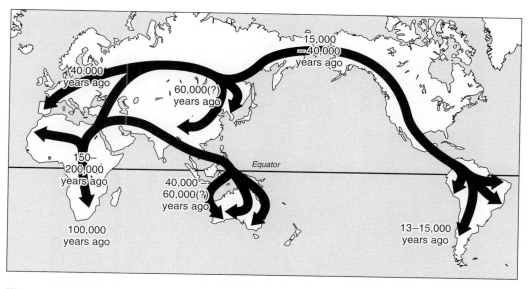

Figure 17.14 Modern humans (*Homo sapiens*) can be traced to sub-Saharan Africa from about 150,000–200,00 years ago. They migrated to southern Africa by about 100,000 years ago and into Europe by about 40,000 years ago. Westward, humans reached India by about 60,000 years ago, the far East and Australia by 40,000–60,000 years ago, and across the Bering Strait to North America 15,000–40,000 years ago. South America was inhabited around 13,000–15,000 years ago.

Neanderthals. Their refined *Aurignacian* tool kit included improved hunting weapons (spears, nets, harpoons, and hooks), elaborate clothing from animal skins and furs, and sophisticated art and culture (fig. 17.15). Magnificently engraved bones, paintings (fig. 17.16), and sculptures have been recovered from caves in Spain and France. Modern-day artists view with awe the brilliant Cro-Magnon paintings and carved statues, often located deep within caves.

ORIGIN OF MODERN HUMANS

The weight of evidence favors the thesis that modern humans were cradled in Africa. Most investigators would claim that modern humans with their distinctive features evolved in Africa between 100,000 and 150,000 years ago and displaced their predecessors as they migrated to various regions of the globe. Subsequently, regional ("racial") characteristics emerged in different geographical populations. This thesis is most often referred to as the *Out-of-Africa theory,* although a variety of notations have proliferated—the "single-origin model," "monogenesis theory," "Noah's ark model," and the "replacement theory."

Africa was apparently the center from which at least two waves of migrations occurred: one took place about 1 million years ago by *Homo erectus* and the other about 150,000 years ago by *Homo sapiens.* In the first wave, the African *Homo erectus* spread over Europe and Asia. In the second wave, modern *Homo sapiens* expanded from its continent of birth (Africa) and replaced *Homo erectus* and its descendants across the world.

There is a school of thought, chiefly identified with the paleoanthropologists Alan G. Thorne and Milford H. Wolpoff, that conceives of the modern human geographical populations as descending from different ancient hominid lineages evolving

Figure 17.15 Cro-Magnons, representatives of our own species, *Homo sapiens,* can be traced back about 40,000 years ago in Europe.

(Painting by Maurice Wilson; by permission of the Trustees of the British Museum—Natural History.)

Figure 17.16 Lascaux Cave Paintings. Horses and giant extinct bulls called aurochs (in the background) were common themes of Cro Magnon cave paintings. In recent years, the 17,000 year old bestiary at the Lascaux cave in southwestern France have been plagued by fungus and mold that threatens destroy these Paleolithic relics.

independently of one another. Thus, as seen in figure 17.17, the Indonesian *Homo erectus* ("Java man") was the early progenitor of the present native inhabitants of Australia, the Chinese *Homo erectus* ("Peking man") gave rise to modern Asians, the European *Homo erectus* gave rise to present-day Europeans, and the African *Homo erectus (Homo ergaster)* gave rise to present-day African populations. According to this view, gene flow between the different geographical groups was sufficient to permit human populations to be maintained as one biological species. Simultaneously, the populations in different regions were sufficiently isolated to permit the development of distinct anatomic identities. This state of affairs is known as the *Multiregional hypothesis* (also called "regional continuity hypothesis" or even the "candelabra hypothesis"). The Thorne-Wolpoff School would contend that geographic separation of the different early hominid branches did *not* lead to reproductive isolation, as might be expected of long-standing populations that are spatially separated and that differentiate along independent lines. It is,

however, exceedingly difficult to imagine how several hominid geographical assemblages, diverging in different parts of the world, could evolve independently and yet repeatedly in the same direction leading only to one species, *Homo sapiens*. The parallel pattern of evolution associated with the absence of speciation is not hopelessly out of the question, but if it occurred, it must have been the very rare exception to the normal process.

There is scarcely any disagreement that the world of humans today is a single large neighborhood. The once distinguishing features of geographical groups have become increasingly blurred by the interminglings and intermixings of peoples. Our present-day species lives in one great reproductive community.

GENOMES OF THE CHIMPANZEES AND HUMANS

Chimpanzees *(Pan troglodytes)* and humans *(Homo sapiens)* are two of the most thoroughly studied

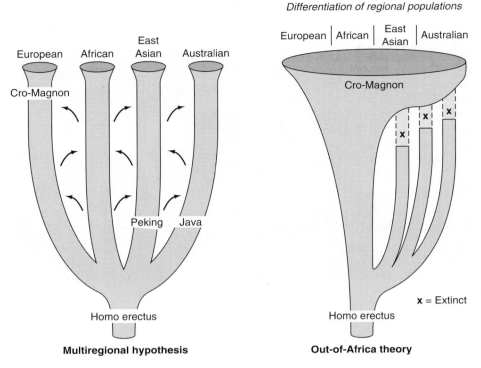

Figure 17.17 Origin of modern humans. The *multiregional hypothesis* envisions a distant separation of the principal regional populations. Each regional population that stemmed from *Homo erectus* evolved independently and in parallel fashion over hundreds of thousands of years. Although each regional group evolved along its own distinctive direction, gene flow *(arrows represent gene exchange)* between neighboring geographical groups was sufficient to permit human populations to be maintained as one biological species. According to the *Out-of-Africa theory,* regional differentiation occurred only after modern humans (Cro-Magnon) arose about 150,000 years ago in one place (Africa), and supplanted their predecessors as they migrated to various regions of the world.
Source: Modified from *Biology and Human Concerns,* Volpe, 4th ed., 1993:440.

species, at both the organismal and molecular levels. In recent years, a variety of protein molecules of the two species have been analyzed by amino acid sequencing, immunology, and electrophoresis. All the molecular data agree in showing that the chimpanzees and humans are remarkably similar.

The two species have completely identical fibrinopeptides, cytochrome *c,* and alpha, beta, and gamma chains of hemoglobin. With respect to myoglobin and the delta chain of hemoglobin, only a single acid replacement separates the

human polypeptide chain from that of the chimpanzee. Based on protein analyses, the sequences of human and chimpanzee polypeptides are, on average, more than 99 percent identical. Stated another way, the average human polypeptide is less than 1 percent different from its chimpanzee counterpart. The information on proteins has been reinforced by nucleic acid studies, which have revealed that our nuclear DNA is 98 to 99 percent identical to that of the chimpanzee. It would seem that the exceedingly meager molecular differences, at both the protein and DNA levels, are much too small to account for the substantial ana-tomical and behavioral differences between the two species.

It has been suggested that the major organis-mal differences between the two species are largely based on changes in the regulatory sequences that control the expression of structural (protein-coding) genes rather than on changes in the structural genes themselves. In particular, the complex biological differences between humans and chimpanzees may stem primarily from mutational changes in regulatory units, such as *promoters* (a gene sequence that controls gene expression), *enhanc-ers* (control elements that increase the rate gene transcription), and *transcription factors* (pro-teins that control the transfer of DNA to RNA in protein synthesis). A subtle change in a regula-tory sequence that codes for a particular tran-scription factor might result, for example, in the

prolongation of the growth period of the brain of the human fetus, thereby permitting additional time for the development of greater complexity of the brain.

Gross chromosomal rearrangements (inver-sions and translocations) may play a prominent role in shifting a given regulatory gene from its normal position to a new location in another chro-mosome. Despite appreciable homology of the two chromosomal sets, there is a profound differ-ence in the number of sets of chromosomes found in chimps (and gorilla) and humans. Humans have 23 ($2n = 46$) pairs of chromosomes, and chimps (and gorilla) have 24 pair ($2n = 48$).

On the tips of normal chromosomes are **telomeres,** a region of repetitive DNA that protects the chromosomes from damage. In the middle of chromosomes is a region central constriction called the **centromere.** (fig. 17.18A). During cell division,

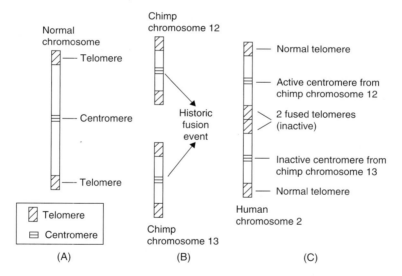

Figure 17.18 Origin of human chromosome 2. This diagram illustrates the historic fusion event between chimp chromosomes 12 and 13 to form modern human chromosome 2. (A) Normal chromosome with two terminal telomeres and one focal centromere. (B) Chimp chromosomes 12 and 13. (C) Human chromosome 2, composed of remnant molecular and cytological signatures of the historic fusion of chimp chromosomes 12 and 13, containing two normal telomeres, two fused inactive telomeres, one active centromere derived from chimp chromosome 12, and one inactive centromere derived from chimp chromosome 13.

Modified from Kenneth R. Miller, "Only a Theory" Viking Press, 2008. ISBN 978-0-670-0188-6, pp. 106–107.

the spindle fibers that orchestrate the movement of chromosomes attach to each chromosome via the centromere. Like the telomere, the centromere contains unique repetitive sequences of DNA. These repetitive sequences, along with other characteristics, provide these chromosomal elements with identifiable *molecular signatures*. Telomeres and centromeres are essential parts of eukaryotic chromosomes. Because of their unique molecular signatures, it has been possible to solve the riddle of why humans have 23 ($2n = 46$) pairs of chromosomes and chimps (and gorilla) have 24 pair ($2n = 48$).

For decades, people have compared chimp and human chromosomes by every means available. During that time, chromosome 2 has aroused attention, but only recently have the hard data been available to prove the origin of chromosome 2. By using the molecular signatures of the chromosomal structural elements of the telomere and the centromere, and sophisticated DNA sequencing, scientists have demonstrated that human chromosome number 2 is a fused chromosome of two chimp chromosomes, and thereby solved the puzzle of why these otherwise very closely related hominids have a different number of sets of chromosomes (23 vs. 24).

As shown in figure 17.18C, human chromosome 2 has the genetic signature of a fused chromosome. On close inspection, it is possible to identify two normal telomeres and two fused telomeres within the chromosome (fig. 17.18B,C). Likewise, it is also evident that chromosome 2 has both an active and an inactive centromere (fig. 17.18).

Somewhere during the course of evolution, in the line diverging from human and the African apes, the ape chromosomes 12 and 13 fused to form human chromosome 2 (fig. 17.18B,C). Because of the unique structure of the centromere and the telomere, it is now possible to show that this event took place. Based on the ability to sequence DNA and detect genes, it is possible to demonstrate that the genes on human chromosome 2 match the genes found on chimp chromosomes 12 and 13. The match is so unambiguous that chimp specialists have renamed chimp chromosome 12 and 13 as chromosomes 2A and 2B. The discovery of the chromosomal relationship between humans and the African apes is yet another line of evidence supporting evolution. The implications are obvious: humans and the African apes, especially the chimpanzee, are extremely closely related by common descent.

NATURAL SELECTION, SOCIAL BEHAVIOR, AND CULTURAL EVOLUTION

In his provocative book entitled *Sociobiology,* Edward O. Wilson of Harvard University delves imaginatively into the social interactions of all animals, including humans. The cardinal theme is that the social interplay of animals, no matter how complex, has evolved by natural selection. Accordingly, sociobiologists attempt to place social behavior on sound Darwinian principles. The doctrine of natural selection is thought of as central to the biological understanding of sociality.

If the behavioral attributes of animals are the products of the same evolutionary forces that shape morphological and physiological traits, then social behavioral patterns should be adaptive. That is to say, behavioral dispositions should be optimally designed to confer reproductive success upon the individual. There are, however, behavioral acts that are detrimental to the individual performing the act but promote the reproductive advantage of other members of the population. A female worker bee, for example, completely forsakes procreation and labors ceaselessly to enhance the reproductive fitness of the queen. Is it possible to select for behaviors that are individually disadvantageous but

beneficial to the species as a whole? This is one of the searching questions in sociobiology.

GROUP SELECTION

In 1932, the English geneticist J. B. S. Haldane speculated on the possibility that a trait may be selected that confers an advantage for the group but is costly to the individual. He used the term *altruism* for such a trait and defined an altruistic act as one that decreases the personal fitness of the organism performing the act but is beneficial to the population as a whole. Thus, a prairie dog that emits a loud warning call (or alarm) when it spots a coyote improves the chances that his fellow *conspecifics* will survive. In protecting the group, the prairie dog (alarmer) attracts attention to itself and places itself in immediate danger of being captured by the predator. If the prairie dog were to remain silent, its presence to the coyote would not be betrayed. The warning act by the alarmer is essentially self-sacrificial; the term *altruism* may be equated with self-sacrificial or selfless. Haldane acknowledged the difficulty in explaining, in a mathematical model, the establishment of an altruistic trait that apparently diminishes one's own chances of survival and reproduction.

The Scottish ecologist V. C. Wynne-Edwards evoked the concept of *group selection* to account for altruism as it relates to territorial behavior. Wynne-Edwards views territoriality as a method of population control. When a given population becomes excessive, many individuals cannot find territories and therefore cannot breed. Wynne-Edwards suggests that the territorial system has evolved by natural selection as part of a mechanism to stabilize the population density at a level that can be supported by the available food resources of the area. Wynne-Edwards' thesis assumes that natural selection operates for the benefit of the group as a whole. The implication is that many individuals are genetically or internally programmed to refrain from reproducing so as to not endanger the welfare or survival of the stock. In other words, natural selection has fostered both territorial "winners" and "losers" for the good of the species as a whole. Stated another way, a "loser" promotes the reproductive advantage of the "winner" at its own expense.

The curtailment of reproductive activity in "losers" is inconsistent with the notion of individual selection. An abridgment of reproductive behavior certainly does not benefit the individual. Natural selection characteristically operates on individuals (*not* groups) to augment (*not* decrease) individual reproductive capability. Since selection presumably operates solely to maximize the reproductive success of each individual, it is difficult to imagine how individuals can be selected to save the group at their own individual expense.

From a theoretical standpoint, can reproductive curtailment or restraint on the part of an individual evolve in a population so as to confer a reproductive advantage upon the group as a whole? Let us suppose that gene a_1 promotes reproductive capability and its allele, a_2, tends to curtail the reproductive capacity of an individual. Can the a_2 allele persist, or even spread, in a population when its effect is to impair the reproductive fitness of its possessor in the present and

succeeding generations? The a_2 allele is selected against since the possessors of the alternative allele, a_1, obviously leave more offspring than the possessors of the a_2 gene. Ultimately, the self-sacrificial a_2 allele will be replaced by the reproductively advantageous a_1 gene.

KIN SELECTION

Altruism apparently cannot evolve without violating the principle of individual selection, since the individuals displaying self-sacrificial behavior are less biologically fit than their selfish colleagues. Yet altruistic behaviors are clearly evident, particularly among birds and mammals. A female bird behaves altruistically when she protects her brood against predation; a male baboon will emerge from the heart of a troop to attack a leopard that threatens the group; and a human will place his or her life in jeopardy to rescue a drowning person. W. D. Hamilton proposed an alternative route for the evolution of altruism that is not founded on the indefensible premise of group selection.

Hamilton explained altruism as the outcome of a selective process called *kin selection*. The concept is based on the fact that close relatives share a high proportion of the same genes. Altruistic behavior can be favorably selected if the probability is high that the beneficiaries of the altruistic act also have the same genes as the self-sacrificial altruist. Under this view, kin selection is a special manifestation of *gene* selection. It is the gene coding for a particular behavior that is optimized or favorably selected. A given allele a_1 will spread if the behavior associated with this gene adds a greater number of a_1 alleles to the next generation than in the preceding generation. This can happen only if the reproductively successful individuals are close relatives of the altruist, thereby increasing the probability that they carry the same a_1 allele.

An instructive phenomenon is the special chirp, or warning call, in bird populations, by which one member alerts the flock to the danger of a predator. We may assume that the warning note is governed by an a_1 allele. A warning call originating as a signal from a parent to its offspring during the breeding season is selectively advantageous because the degree of relatedness between the caller and recipient is high. Gene a_1 will persist in the population, since the alarm call by the parent (bearing gene a_1) increases the fitness of sufficient numbers of offspring also possessing a_1. By protecting the offspring, the parent invests in its own genetic representation in subsequent generations. In an evolutionary sense, the offspring that are protected by the parent are, in part, genetically the parent itself.

The lifetime of an organism is to be viewed as a strategy for perpetuating the organism's genes. An individual can transmit its genes directly through its own reproduction as well as by proxy through its close relatives who share genes by common descent. An individual is said to maximize an *inclusive fitness*—its own reproductive fitness plus the reproductive fitness of close relatives. The degree of genetic relatedness has a crucial bearing on the likelihood that two individuals will behave altruistically toward one another. The extent to which two individuals share genes by descent from a common ancestor is referred to as the degree of relatedness, or coefficient of relationship (designated r). In diploid species, the degree of relatedness of an individual to his or her full siblings is 1/2; to half-siblings, 1/4; to children, 1/2; and to first cousins, 1/8. Stated another way, a given individual shares 50 percent of his genes with a full brother or sister but only 12.5 percent of his genes with a first cousin. By this logic, grandparents and grandchildren are related by 1/4.

Kin selection molds a form of altruism that is channeled to genetic relatives. An altruistic gene will be perpetuated if the reproductive benefit gained by the recipient of the altruistic act exceeds the cost in reproductive fitness suffered by the altruist. Figure 18.1 depicts kin selection with respect to two brothers. The altruist leaves no offspring, but his sacrificial act enables his brother to leave more

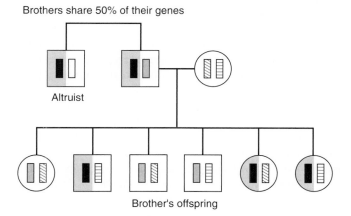

Brothers share 50% of their genes

Altruist

Brother's offspring

Figure 18.1 Sibling altruism. The coefficient of relationships between two brothers is 1/2. Thus, the probability is 0.50 that a gene (or chromosome) present in one individual also occurs in his brother. For simplicity, only a pair of chromosomes is depicted. The individual's altruism to his brother is selectively advantageous because the altruistic act enables his brother to transmit greater numbers of replicas of the shared chromosome to the next generation than the individual himself might have transmitted.

offspring than he would have otherwise. The altruist's genes have been effectively removed from the population by his failure to reproduce, but the genes that he shares with his brother are more than restored by three of the offspring sired by his brother. The altruist has actually gained genetic representation in the next generation. Indeed, he has gained in *inclusive fitness,* even though he has lost fitness in the classical sense.

Kin selection provides an explanation for the baffling social behavior in ants, bees, and wasps. Curiously, whole castes of sterile females devote their entire existence to the welfare of the queen. Females are diploid individuals that develop from fertilized eggs with maternal and paternal sets of chromosomes. Males are the haploid products of unfertilized eggs and possess only the maternal set of chromosomes (fig. 18.2). The startling outcome is that the sibling daughters of a queen are more closely related to each other than they would be to any of their own daughters!

Specifically, female workers share three-fourths of their genes with their sisters ($r = 3/4$). This is the case because sisters share the same set of paternal genes and share additionally, on the average, one-half of the maternal genes (fig. 18.2). Thus, sisters are related by the average of 1 (for paternal genes) and 1/2 (for maternal genes), or 3/4. If a daughter were to produce her own offspring, that daughter would share only one-half of its genes with any of her offspring. Accordingly, a female worker actually contributes more to her Darwinian fitness by assisting her mother in raising offspring (3/4 relationship) than by rearing her own offspring (1/2 relationship). Natural selection has fostered an unusual form of cooperative behavior among sisters.

The male bees, or drones, show a pronounced lack of concern over the welfare of their sisters. This is not surprising, since drones share only 1/4 of their genes with their sisters. Since a drone possesses only maternal genes, a sister cannot share any paternal genes with her brother (fig. 18.2). The total genetic relationship of a sister to her brother is the average of zero (for paternal genes) and 1/2

(for maternal genes), or 1/4. The drone is notoriously lazy ("lazy as a drone"), since his inclination is to produce daughters (who share all of his genes) rather than help his sisters. In turn, sisters invest more energy in raising sisters than brothers and characteristically drive their brothers from the hive in the early summer. The selfishness of the drones and the diligence of the female workers are predictable aspects of kin selection.

KIN SELECTION IN HUMANS

A variety of altruistic behavioral dispositions in the human species—the sharing of food, the sharing of implements, and the caring for the sick—probably evolved by kin selection in the early hominid hunting bands. These bands consisted largely of tightly knit groups of close relatives. In fact, in primitive cohesive societies, social behavior appears to have been dominated by kin selection.

A striking phenomenon in some human societies is the strong parental attitude exhibited by a brother to his sister's children, far out of proportion to his degree of relatedness to his nephews and nieces. However, as predicted by kin selection, a brother's sense of responsibility to his sister's children increases as confidence in paternity of his own children diminishes. A brother is always genetically related to his sister's children, whereas he may be totally unrelated to his wife's children if the probability of paternity is low. Apparently, the certainty of a genetic relationship between brother and nephew (or niece) outweighs the dubious or tenuous relationship between father and alleged son (or daughter). In essence, a high probability of paternity is a necessary antecedent for extended parental investment by the male.

It has been suggested that homosexual behavior is a product of kin selection. Although the genetic factors for homosexuality lower reproductive fitness, the production of offspring occurs by proxy through the homosexual's immediate kin. In this vein, homosexuals may be compared to sterile worker bees who help raise their close

Mother-offspring relatedness

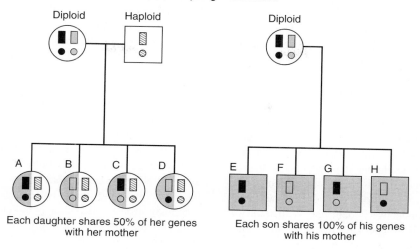

Sister-sister relatedness

Sister-brother relatedness

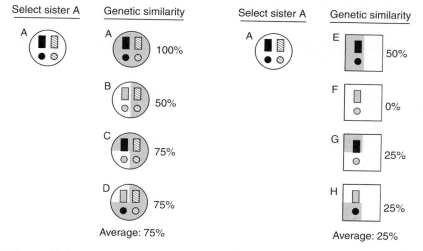

Figure 18.2 Degree of genetic relationship of a female bee to her ofspring *(top)* and the genetic relatedness of sisters to sisters *(lower left)* and sisters to brothers *(lower right)*. Females develop from fertilized eggs and have two sets of chromosomes (diploid state), whereas males develop from unfertilized eggs and have only one set of chromosomes (haploid state). For simplicity, the haploid set is represented by only two chromosomes. Sisters are more closely related to one another than they would be to their own offspring. The coefficient of relationship between mother and daughter is 1/2, but the coefficient of relationship between two sisters averages 3/4. Sisters have relatively few genes in common with their brothers, the coefficient of relationship averaging 1/4.

genetic relatives. The behavior of the homosexual increases the fitness of genetic relatives more than it decreases his or her own fitness. This view is consistent with kin selection, but there are as yet no supporting or confirmatory data.

Kin selection may have shaped the phenomenon of menopause in the female. The middle-aged woman enhances her fitness by caring for her children's children. As the menopausal woman loses her capacity to have her own children, she regains inclusive fitness by devoting her efforts to the rearing of her grandchildren. It is interesting that the duration of fecundity is greater in women of economically advanced countries. Menarche occurs earlier and menopause is deferred. With improved nutrition and better personal hygiene, the female increases the duration in which she can invest in her own children.

THE SELFISH GENE

The alarm calls of birds are often cited as an excellent example of kin selection. It is generally the case that a close kin is sufficiently near the caller to benefit from the warning call. However, some authors contend that the habit of warning cries evolved by conventional individual selection rather than by kin selection. A warning call may *not* serve to draw the attention of the predator to the alarmer. The warning call may *not* expose the caller to greater danger. Rather, the call note is intended to mislead the predator as to the position of the prey.

Under this view, the alarmer utters a call that the predator cannot easily locate. Since the predator cannot readily detect the location of the call, it would appear that the alarmer issues the warning to protect itself and not its colleagues (conspecifics). Suppose a flock of birds is feeding in a meadow and a hawk flies past in the distance. The hawk has not seen the flock, but there is the danger that he will soon spot the flock. The noisy rummaging activity of members of the flock could attract the hawk's attention. The alarmer who first detects the hawk issues a warning call to its companions but does so only to curtail the noise of the flock. Accordingly, the caller reduces the chance that the flock will inadvertently summon the hawk into his own vicinity. In large measure, the caller is not acting *altruistically* but rather *selfishly*.

The same situation may be viewed in another selfish vein. Suppose the caller fails to signal, quietly taking flight by itself without warning its unsuspecting conspecifics. In this event, the hawk might be easily attracted to a single bird flying off. The issuance of an alarm would be better than no alarm at all. An alarmer that issues a call curtails his risks by flying up into the tree—and simultaneously arouses the other members to fly with him into the tree. In corralling the conspecifics, the caller ensures protective cover for himself in flight. The conspecifics benefit from the caller's act, but in increasing his own safety, the caller benefits even more!

The portrait of a biological individual that emerges from these examples is one of self-serving opportunism. It seems that an individual is programmed to care about itself. It may be, as several investigators contend, that acts of apparent altruism are actually selfishness in disguise.

RECIPROCAL ALTRUISM

When a self-sacrificial parent saves his own child from some potentially tragic event, the parent acts to protect his share of genes invested in the child. If, however, an individual saves the life of an unrelated party or nonrelative, kin selection is ruled out. In this instance, the individual may have behaved altruistically with the expectation that the beneficent behavior will be reciprocated by the stranger at some future occasion. A person who saves another person from drowning, for example, hopes in turn for help when he or she is in danger. This is the concept of *reciprocal altruism,* first proposed in 1971 by Robert L. Trivers of Rutgers University. An altruist who places himself in danger for a biologically unrelated person incurs a reduction in fitness because of the energy consumed and the risks involved. However, a future reciprocal act by the recipient is likely to bring returns that are equal to, or greater than, the altruist's original expenditure.

There is definite survival value to both donor and recipient when altruistic acts are mutually exchanged. Each participant benefits by increasing his or her fitness. If, however, a recipient fails to reciprocate when the situation arises, that recipient would no longer have the benefits of future altruistic gestures to him. The nonreciprocator would then be at a selective disadvantage since subsequent adverse effects on his life would not be overcome by altruistic acts that might be life-saving. The ultimate effect of nonreciprocation would be the restriction of altruism to faithful fellow altruists, with the consequence that genes for reciprocal altruism would be perpetuated through the generations. In fact, the chances of selecting for altruistic genes are improved as more altruistic acts occur in the lifetime of the altruist. By the continual exchange of beneficent acts, altruists accrue more fitness in the long run than the nonreciprocators. Overall, altruists gain fitness, rather than lose fitness, from selfless acts.

"Stepmothering" may be considered a form of reciprocal altruism. When the biological mother dies or leaves her children, the male may remarry, and the stepmother assumes the care of the offspring. The basis of the reciprocity is best seen in the behavior of western gulls off the coast of California. An unmated gull, by raising the offspring of a previous mate, establishes a pair-bond relationship with the male and guarantees herself both a mate and a breeding site for subsequent years. Thus, in a population of gulls where there is a surplus of breeding-age females, the acquisition of a breeding territory complete with a mate is a compelling incentive for a female gull to behave altruistically by rearing another female's offspring on one occasion. By so doing, she increases her personal reproductive fitness for future years with only a relatively small initial investment. Such behavior is fostered by natural selection, since the final effect for the stepmother is an increase in her own genetic material in succeeding generations.

PARENTAL REPRODUCTIVE STRATEGIES

In the placental mammals, the young develop slowly and the prolonged period of caring for the young is almost exclusively restricted to the female. The males of most mammalian species do not establish even a semblance of a permanent relationship with the pregnant and nursing female. In its extreme form, such as in cats and bears, the female actually ejects the male from any contact with the young. It is as if the mammalian male were biologically superfluous after insemination. Apart from insemination, most mammalian females have little use for the males—except possibly to protect them from other males. It is only exceptionally, as in wolves and humans, that the male plays an important role in the protection and care of the mother and child.

Each sex, without conscious intent, strives to attain maximal reproductive success. The reproductive strategies of males and females are clearly different. Each sex may be viewed as an investor in which the capital invested is in terms of reproductive effort. The female undoubtedly invests more per offspring than the male. The overproduced, lightweight sperm are inexpensive compared with the thriftily produced, energy-rich eggs. The prolonged internal gestation and the protracted period of maternal care of the newborn place rigorous demands of energy and time on the female. Accordingly, the female has a much greater stake in any one reproductive act. Since a reproductive mistake is much more severe for the female, she is more discriminating than the male in the choice of a mate.

A male's basic reproductive strategy is to achieve as many fertilizations as possible. Males are less selective in the choice of acceptable sexual partners and more aggressive in excluding other males from the opportunity to mate. Females are choosier and carefully evaluate their options. Males almost universally act competitively as "sexual advertisers," and females act cautiously as "comparison shoppers." The task of critical discrimination of a sexual partner and the avoidance of an unproductive pregnancy resides almost wholly with the female.

The female selects that male as a sexual partner whose appearance and behavior signify that he will transmit a superior set of genes to the offspring. Mammals are generally polygynous—that is, a single male mates with more than one female. As previously mentioned, most mammalian males do not establish any durable association with the female beyond the sexual act. If a reproductively fit mammalian male has little to offer except his superior genes, then it is to the female's advantage to mate with that suitable male, no matter how often he may have already mated. In polygynous mammalian species, the fittest males (at least those that are behaviorally the fittest) have the greatest reproductive success. For example, in breeding colonies of elephant seals, the males clash for dominance status, which is associated with mating rights. Less than one-third of the males in residence during a season copulate, and as few as five dominant males account for 50 percent of the mating in a given season. Clearly, females who share an oft-mated male of unusual competitive ability are attracted to a "winner" with "proven" genetic qualities.

The situation is different with birds, in which monogamy is the general rule. One male and one female form a breeding pair and remain together, at least for the rearing of one brood. In some species of birds, the same pair-bonding may persist for several successive broods (as in songbirds) or even for a lifetime (as in geese and swans). In these situations, the male's contributions to the female and her offspring are much more than a mere complement of genes. The males of many species of birds defend a territory, provide protection against predators, assist in the building of nests, incubate the eggs, and furnish nourishment for the female and her young. Thus, where the male provides numerous indispensable services, it is to the female's advantage to maintain pair-bonding. Stated another way, we can expect enduring pair-bonding only where the male plays a substantial paternal role.

The tendency toward polygyny in human males apparently has been dampened by strong selective pressures. The human male engages in an elaborate courtship ritual, which is in itself a substantial investment for sustained pair-bonding. Robert Trivers has suggested that the prolonged male courtship serves as insurance for the male that he is not establishing a bond with an already inseminated female. The greater the male investment, the greater the importance of paternity knowledge. To assure accurate identification of his own offspring, the male jealously guards against "cuckoldry"—that is, against the possibility that he might invest in offspring that are not his own. Human males universally treat their mates as possessions and sequester them from the adulterous interests of other males. It is only when the female remains faithful that the male can overcome his ever-present uncertainty about paternity. Nevertheless, the other side of the male's strategy is his inclination to cuckold other males. There have evolved, however, female counterstrategies to improve the female's ability to hold her mate's attention and care.

With the notable exception of humans, the female of mammalian species becomes sexually aroused only at certain seasons. At specific times the female comes into heat, or *estrus,* and only during these restricted periods is the female receptive to the male. The onset of heat is typically synchronized with ovulation, or the release of the ripe egg. In simultaneously advertising sexual receptivity and ovulation, the mammalian female provokes intense competition among the dominant males and increases her likelihood of securing a male of exceptional genetic constitution. The females are monopolized by the males only at ovulation.

In the human female, there are generally no outward or conspicuous signs of ovulation. Human females may be sexually receptive during any part of the ovulatory cycle. Several authors have pointed out that the loss of the phenomenon of estrus in human evolution was an event of great importance. The concealment of ovulation by the female serves to maximize her mate's confidence of paternity, in that potentially competing males are not flauntingly apprised of the ovulation event. Stated another way, if the human female were to glaringly advertise ovulation, it might cost her the protection and parental care of her mate by attracting competing males who threaten her mate's confidence of paternity. Moreover, her potentially philandering mate is restrained in seeking copulation with other fertile females who, like her, conceal the event of ovulation.

When the pair-bonding relationship fails or is dissolved, the female typically experiences difficulty in finding a substitute male to aid in the rearing of the offspring. We would expect strong selection pressure against "stepfathering," since the substitute father is called upon to invest in some other male's genes. The literature is replete with examples in varied mammals in which a male eliminates the offspring of a previous mating and replaces them with his own offspring. When the male lions of a new group depose the resident males of a pride, the new regime often kills the cubs of the displaced males. In another instance, a pregnant female mouse responds to the odor of a novel male by aborting her fetus and returning to estrus in preparation for propagating the genes of the new male. Among the Yanomama Indians of Venezuela, a man will order his newly acquired wife to kill her infants of a previous marriage. The biological urge to invest only in one's own offspring is evidently powerful and pervasive.

PARENT-OFFSPRING CONFLICT

Robert Trivers defines parental investment as the amount of care and assistance rendered by the parent that increases an offspring's chance of future reproductive success at the cost of limiting the parent's capacity to invest in other offspring. Each offspring is of equal value to a parent, since 50 percent of the parent's genes are distributed in each offspring. But each offspring is completely related to

itself ($r = 1$), and related by one-half to each of its sibling ($r = 1/2$). Each offspring, then, would expect to receive twice as much parental care for itself as for each of its siblings. Thus, it is inevitable that conflicts will arise between parent and offspring.

In particular, a mother will normally apportion her investment among her offspring to maximize the number that survive to reproductive maturity. She must necessarily limit the amount of care that she can give to any one child. On the other hand, if the child is to maximize its own chances, the child will attempt to extract more than its share of parental care. The parental investment in an offspring eventually reaches a point of diminishing returns. That point is reached when the mother's investment in a given child costs her more than she gains and where it costs the child more than the child gains. Once the point of diminishing returns is met, the parent-offspring bond should be severed.

MEMES

Derived from the Greek *mimema,* meaning something imitated or mimicked, the term *meme* was coined by Richard Dawkins in his 1976 book *The Selfish Gene.* Memes include such things as ideas, rituals, gestures, beliefs such as religion, political theories, fads, songs, and technological innovations. Memes are transmitted from person to person and may spread vertically across generations or horizontally within a single generation. According to Dawkins, memes are "a unit of cultural transmission, or a unit of *imitation*" important to humans' capacity for cultural evolution, and are the cultural equivalent of genes. Unlike genes, memes can be relayed across time and space, can be transmitted between nonrelatives and even backward through generations, as when a child conveys an idea to a grandparent.

Meme proponents maintain that memes and *memetics,* the study of the transmission of memes within an evolutionary context, provides a theoretical basis for discussing cultural evolution in a manner analogous to our previous discussion of biological evolution via genes. Like recessive genes, memes can stay unexpressed for long periods (generations) and then reappear. Using the genetics analogy, memes have the capacity to self-replicate and can be heritable and subject to mutation, variability, reproductive success, and extinction. Because memes spread like viruses and other contagious diseases, they are capable of quick diffusion across populations and cultures. Memes can infect cultures, religions, and even politics.

Not surprisingly, the concept of memes has its critics and its supporters. One major criticism of the theory of memes is that unlike genes, memes cannot be empirically defined or readily subjected to statistical analyses. Paul Ehrlich of Stanford University has argued that "memetics has not led to any real understanding of cultural evolution." Alternatively the concept of memes may serve as a bridge integrating the social and natural sciences toward the study of evolution.

CULTURAL EVOLUTION

Some 10,000 years ago, humans gave up the precarious hunting and gathering way of life for a more settled and secure existence based on agriculture and the breeding of animals. This initial trial at cultivating the land ushered in the so-called Neolithic revolution. When we domesticated plants and animals, we took a major step in controlling nature rather than being at its mercy. We placed nature at our service with our ideas, discoveries, and inventions. We began to control our own food supplies, to congregate into more stable communities, and to establish a distinctive civilization.

We have since modified our external surroundings at an ever increasing rate. We have undergone an industrial revolution and are now witnessing a technological revolution. The rapidity and efficiency with which we have dominated the environment reflects our capacity for learning and transmitting our accumulated knowledge. This capacity for cultural evolution has had a profound influence on our way of life and on our destiny.

There are important differences between biological evolution and cultural evolution, one of the primary differences being the tempo of change (table 18.1). Unlike biological evolution, which is driven by mutation, natural selection, and the other factors and forces we have discussed thus far, cultural evolution is uniquely non-Darwinian. Likewise, while biological evolution can be glacially slow, cultural evolution has the capacity to advance and progress at astonishing rates. Cultural transmission is rapid, limited only by the efficiency of communication methods and our inventiveness. The acquisition and transmission of learned ideas occurs through the generations from mind to mind rather than through the germ cells. A given cultural change can be passed on to large groups of unrelated individuals, whereas a given genetic modification can be transmitted only to direct descendants, who are generally few. Biological evolution is necessarily slow because it depends on accidental mutational changes in the DNA molecule, and it can take many generations before a

TABLE 18.1	**Comparisons of Biologic and Cultural Evolution**	
	Biologic Evolution	**Cultural Evolution**
Agents	Genes	Ideas
Rate of change	Slow	Rapid
Direction of change	Random mutations; subject to selection	Usually purposeful and directional
Nature of new variants	Often harmful	Often beneficial
Transmission	Parents to offspring	Wide dissemination by many means
Distribution in nature	All forms of life	Unique to humans

chance genetic change can become established in the population under the force of natural selection.

In chapter 1, we dismissed Lamarck's concept of inheritance of acquired characteristics as an early concept of evolution that was specious and fanciful. However, cultural evolution is implicitly Lamarckian. As acquired characteristics or knowledge is obtained, cultural evolution can result in a straight-line, purposeful change. In this way, cultural evolution sharply contrasts with biological evolution and can be, and often is, directional, purposeful, and progressive.

In a 10-year period or less, what was innovative becomes passé and often irrelevant. Medicine, computer technology, various branches of science, including genetics and molecular biology, are evolving at ever increasing rates. Through global interconnectivity via the Internet and other modalities, ideas spread like viruses from one end of the Earth to another. Hence, cultural evolution has the capacity to build upon its immediate past, without the aid of genes or DNA to transmit itself and build upon itself.

Exemplary of the directional, purposeful, and progressive capacities of cultural evolution were the dramatic and ambitious pronouncement made by President John F. Kennedy's on May 25, 1961, that the United States would put a man on the Moon before the end of the decade. In the years that followed, deliberate steps (Projects Mercury, Gemini, and Apollo), each designed to build on the previous, resulted in the realization of Kennedy's goal when on July 20, 1969, *Apollo 11*'s commander, Neil Armstrong, stepped off the lunar module's ladder onto the Moon's surface. Biological evolution is not capable of this type of purposive, deliberate, vector-like change. Today, with the Internet and other forms of communications and technology, cultural evolution has the capacity to adapt even faster than it did in the 1960s.

The human capacity for rapid cultural adaptation provides great hope for the future of our species and our planet. Our world is filled with seemingly impossible dilemmas that threaten our very existence. Nevertheless, a directed commitment by people could very quickly abate regional and global tribulations. Let us hope that when we are called to act, the arrow is still in the bow or at least that its trajectory can still be influenced.

EPILOGUE

THE CRUCIBLE OF EVOLUTION

Charles Darwin's (fig. E.1) explanation of evolution has been debated ever since the first appearance of his book, *The Origin of Species*. The entire first edition of 1,250 copies was sold out on the very day it first appeared, November 14, 1859. Some press reviews were favorable: others were scathing and satiric. The most favorable comments appeared in the London *Times,* authored by the distinguished scientist and Darwin advocate Thomas Henry Huxley. Huxley's bold support of Darwin's ideas earned him the reputation of being "Darwin's bulldog," and his candid reaction to *The Origin of Species* was to conjecture, "How extremely stupid of me not to have thought of that."

Darwin's theory appeared to many scientists to be refreshingly simple: some chance variations better adjust individuals to their environment, and such variant individuals tend to survive and transmit their favorable characteristics to their descendants. As we have discussed, this is the essence of natural selection, and it seemed harmless enough until Darwin hinted that humans may have evolved from "lower" forms of life. This notion was heretical because it contradicted the story of divine creation as told of in the Bible. The biblical rendition is that all organisms were created as separate species, all ruled over by humans, who were created themselves in the image of God. Religious opposition to Darwin's oblique suggestion that humans were not the unique crown of divine creation came from evangelists in Victorian England, who believed staunchly in the inerrancy of the scriptures.

In Darwin's day, the prevailing worldview was that all that was natural was the work of the Creator and the role of science was to explain the phenomenon produced by the Creator. Because of this perspective, Darwin first characterized his theory to friend and botanist Joseph Hooker as "like confessing a murder." Darwin knew that his work would challenge the established dogma of his day and revolutionize the way people thought about themselves and their position relative to the rest of life on Earth.

Today the roots of opposition to Darwinism spring largely from fundamentalist sects that insist on the literal interpretation of the Bible or other origin books. Mainstream ministers, rabbis, priests, and even Pope Paul II have accepted the validity of the theory of evolution, while also maintaining their religious beliefs. To preserve their absolutist concept of the Bible, some radical Protestant fundamentalists have woven Genesis into a spurious theory called *"scientific creationism"* and a newer form of the same ideology called *"intelligent design"* (ID). The intent of scientific creationism and ID is to give the story of Genesis a veneer of scientific respectability. The creationists have pressed for the teaching of their supernatural brand of evolution in the public schools. Broadly speaking, a devotee of creationism is one who rejects modern science in favor of a belief in a supernatural explanation for life. ID asserts that the universe was purposefully created by an intelligent being.

Figure E.1 Charles Darwin (1809–1882) in 1880, two years before his death, at his house at Down House, Kent, England.

Scientific creationism and ID are a mockery of the objectivity of science and debase conventional religion. There are many devoted theologians who accept evolution. Likewise, there are many esteemed scientists who are deeply committed to their religious faith. Religious faith and an acceptance of the tenets of Darwinian evolution are in no way incompatible. They are different realms that are non-overlapping. Religion is a matter of faith, and science is an objective portrayal of the world we live in. The bitter conflicts that have arisen are unwarranted.

Scientific creationism and ID are not scientific alternatives to Darwinian evolution. They demand absolute acceptance of views not subject to testing or revision. ID and scientific creationism extend beyond the simple suggestion that a divine being could have created the universe. Fundamentalist creationists would assert, for example, that the Earth is less than 10,000 years old. This indefensible view is not found in mainstream religions of any denomination. It is, however, a view that certain fundamentalist sects choose to read into

the book of Genesis. The fact that the Earth is over 4 billion years old is not disputed by any serious scholar, be they religious scholar or scientist.

Furthermore, scientific creationism is an inflexible view not open to empirical testing. Science necessarily rejects certainty and predicates acceptance of a concept on objective testing and the possibility of continual revision. The claims of creationists are unverifiable (or unfalsifiable) and, hence, inherently unscientific. Creationism starts with a conclusion, accepts it as revealed truth not open to empirical testing, and then tries to "prove" the contention—not by abducting positive evidence but by attempting to undermine evolutionary evidence.

The public has reacted to issues of scientific creationism and ID with confusion. The average person on the street has been bewildered by the complex arguments of both sides. Creationists challenge the validity of data supporting evolutionary change. Evolutionists acknowledge the uncertain nature of science and that the body of evolutionary knowledge is continually growing and increasing in accuracy. Unfortunately, most people are not conversant with the concepts of evolution and lack the confidence to weigh the issues. It is therefore not surprising that many citizens perceive the controversy as a dispute between two groups of passionately committed individuals and supposes that one outlook is as good as the other. When creationists ask for equal time in the science classroom, the person on the street views the situation as a reasonable request for fair play. But the request is unreasonable because creationism only masquerades as science, and the integrity of science cannot be violated in the science classroom. Science does not resort to miraculous explanations. Likewise, there are many good reasons to discuss theistic beliefs about creation in a school curriculum, but religious beliefs should be discussed in appropriate courses dealing with theology or contemporary religion. The biblical account of creation, however disguised, should not be taught as if it were science—which it decidedly is not.

Changing Worldviews

The reasons for the ongoing debate about evolution are complex, but in many respects they revolve around the ways in which Darwin's ideas changed our worldview. This is why the evolution debate festers and refuses to go away. For many people, Darwinian thinking is an assault against their own worldview and is therefore stridently opposed.

By any measure, it is not an exaggeration to call Darwin's legacy revolutionary. Our post-Darwinian framework for knowing and perceiving the natural world is fundamentally different from the pre-Darwinian framework. In many respects, the extent to which our worldview has changed since Darwin's Victorian times is greatly underappreciated. Below are some of the ways in which Darwin's ideas resulted in a profound paradigm shift.

Darwin rejected the supernatural as the cause of our existence and instead established natural selection as the mechanism by which evolution occurs. This did not deny the existence of God or negate the belief in God, but proffered that natural selection explained the existence of the natural world.

Darwin argued that variation was fortuitous, due to chance and not ordained by any divine plan; that life was not stable and invariable, but instead was in constant flux. Darwin described a world where species are forever changing and modifying themselves, some evolving into new species and some going extinct. This was an entirely new way of thinking about biology and contradicted the previous conception of an inflexible and stable world. Under this new perspective, organisms and populations are viewed as unique and constantly changing, having both a history and a future.

Darwin dispelled the notion that evolution was linear and replaced this with the concept of branching, the tree metaphor (fig. 14.7), whereby branching implied common descent and a unified concept of all life being interrelated from a single common ancestor. More than any other idea,

descent with modification from a common ancestor defines Darwinian thinking.

Prior to the time when Darwin lived, people viewed the Earth as young, as only thousands of years old. We now understand that the Earth and life on our planet are billions of years old. And while we know that evolution can occur at a rapid pace, the true age of the Earth provides sufficient time for many of the geologic changes modern science ascribes to the natural world.

Before Darwin, adaptation and variation were viewed as the work of a divine Creator who decided on structure and function. A post-Darwinian view sees adaptation and the random interplay of environmental and genetic variation as the causes of variation and adaptation.

Though his own scientific work, Darwin established the notion of a scientific method of hypothesis testing. This way of thinking was at variance with the notion that observations simply substantiate the prevailing worldview. Modern science is based on the use of the scientific method (chapter 1).

Finally, Darwin argued that humanity evolved by the same forces as every other form of life. In the metaphor of the branching evolutionary tree, the human species is just another twig on the tree of life, no different, no better or worse, than any other branch. This proposition was most difficult for the general public to accept and forms the basis of much of the current debate regarding evolution.

Scientific Literacy

As scientifically literate citizens, we are often called upon to articulate what we know and engage in thoughtful discussions with others. Yet when asked about evolution, many U.S. citizens do not have the knowledge and background to fully engage. As a society, we are unacquainted with the facts that form the basis of Darwinian evolution. The following statistics epitomize the societal level of Darwinian literacy in the United States.

Surveys reveal that one-third of U.S. adults believe that evolution is "absolutely false." When asked if "the earliest humans lived at the same time as the dinosaurs," only 50 percent of U.S. adult citizens answered "false." Merely 45 percent answered "true" to the statement that "human beings, as we know them today, developed from earlier species of animals." In a 20-year (1985–2005) national survey of U.S. adults, 55 percent were tentative in their acceptance of evolution. One recent poll found that an even greater percentage (70 percent) support the teaching of creationism in schools. Perhaps the most troubling statistic comes from the comparison of adults in 32 European countries, the U.S., and Japan regarding their acceptance of evolution. In that study (fig. E.2), U.S. adults ranked 33 out of 34. Only Turkish adults were less accepting of evolution. Over the 20-year study period, the percentage of U.S. adults who accept evolution declined from 45 percent to 39 percent. The percent that were "not sure" about evolution rose from 7 percent to 21 percent. In Japan, 78 percent of adults accept evolution and in Denmark, Iceland, France, and Sweden, 80 percent or more accept evolution.

One of the reasons for the discrepancy between the United States and Europe and Japan is that the subject of evolution has been politicized in the United States and defined as a partisan issue. Unlike the United States, no major European or Japanese political party has embraced either creationism or evolution. Furthermore, surveys in both the United States and Europe find that there is a positive correlation between overall literacy in genetics and acceptance of evolution. Our society is no better versed in modern genetics than it is in evolution.

These statistics should disturb political leaders, educators, and citizens alike. Science teachers from middle school through college recognize that present-day scientific instruction of evolution is substandard and ineffective. Attitudes and beliefs regarding the instruction of evolution in the United States must change if twenty-first-century citizens

Figure E.2 Public acceptance of evolution in 34 countries. The United States ranked 33rd out of 34 in a recent (2005) survey of industrialized countries regarding their acceptance of the concept of evolution. Numbers in parenthesis are the number of individuals surveyed in each country. Respondents were asked to answer true, false, or not sure to the statement "Human beings, as we know them, developed from earlier species of animals."

are to have the tools needed to address the challenges ahead.

DARWIN IN THE COURTS

In the United States, there is a long history of the issue of teaching evolution in public schools ending up in the courts. The reasons are both historical and contemporary. In the past, most legal confrontations have been about the teaching of evolution. As that issue was resolved by the courts, the more recent clamor to the bench has come from religious fanatics who wish to disguise religious doctrine and pass it off as science. Compounding this are the dual issues discussed above: a citizenry that is poorly educated regarding evolution and the fact that evolution is at variance with a comforting anthropocentric worldview. These factors, in combination with an effective appeal by a vocal minority that both sides of the controversy should be taught in science classes, has kept the evolution wars on the front burners of our legal system.

The most famous court case regarding the teaching evolution in the public schools is the 1925 case of John Scopes in Dayton, Tennessee. The Scopes case tested the Butler Act, which made it unlawful "to teach any theory that denies the story of Divine Creation of man as taught in the Bible, and to teach instead that man has descended from a lower order of animals." Many historians consider this case a turning point in the U.S. creation-evolution debate. Scopes was a high school teacher of biology who intentionally violated the Butler Act by teaching about evolution. The lawyer for the prosecution was Williams Jennings Bryant, a three-time presidential candidate, former secretary of state, and congressman. The defense was represented by Clarence Darrow, a nationally prominent defense attorney and trial lawyer. The goal of the defense was not to have John Scopes acquitted, but to have a higher court, such as the U.S. Supreme Court, declare that laws forbidding the teaching of evolution were unconstitutional.

After an eight-day trial, the jury deliberated for only nine minutes before finding Scopes guilty. Scopes was ordered to pay a $100 fine. The case was appealed to the Tennessee Supreme Court, where the plaintiffs challenged the ruling on several grounds, including that the statute was vague, violated Scopes right to free speech, was in violation of the Tennessee State Constitution, and violated the Establishment Clause of the U.S. Constitution (see below). The Tennessee court rejected all of these arguments and found the statute to be constitutional. Ironically, while the Butler Act was found to be constitutional, the conviction of John Scopes was set aside because of the legal technicality that the jury should have decided the fine, not the judge. The Butler Act was not repealed until 1967. In the aftermath of these Tennessee court cases over a dozen states considered some form of anti-evolution statute. Most of these efforts were defeated, but Mississippi and Alabama did enact anti-evolution legislation.

Over 40 years after the Scopes trial, in 1968 (table E.1), the U.S. Supreme Court finally declared laws forbidding the teaching of evolution unconstitutional. In Epperson v. Arkansas, the Court invalidated an Arkansas statue that prohibited the teaching of evolution. In its decision, the Court held that the Establishment Clause of First Amendment of the U.S. Constitution (see below) does not allow a state to require that learning be customized to any particular religious sect or doctrine.

In the 1982 case of McLean v. Arkansas Board of Education, a federal court declared that "creation science" is not a science. Furthermore, it held that a "balanced treatment" law violated the Establishment Clause of the U.S. Constitution (see below) and was therefore unconstitutional. More specifically, Judge William Overturn made clear that "creation science" is religion. This ruling had a profound impact on the creationist movement in the United States, which did not challenge this decision until 1987.

In 1987 the U.S. Supreme Court held, in Edwards v. Aguillard, that the Louisiana "Creationism Act" was unconstitutional. That law prohibited the teaching of evolution in public schools, except when it was taught in conjunction with "creation science." In its ruling the Court found that the Louisiana law was deliberately designed to advance a particular religion. To assist the defense, over 70 Nobel laureates filed "friends of the court" briefs describing creation science as a religious doctrine.

TABLE E.1	**Significant Court Decisions Regarding the Teaching of Evolution, Creationism and Intelligent Design**		

Case	Year	Judicial Venue	Major Finding(s)
Kitzmiller *v.* Dover, PA	2004	U.S. District Court Middle District Pennsylvania	Ruled that intelligent design (ID) is another form of creationism or "creation science" and therefore is a religious doctrine that violates the Establishment Clause of the U.S. Constitution.
Rodney LeVake *v.* Independent School District et al.	2000	Minnesota Supreme Court	Ruled that a high school biology teacher's First Amendment right to free speech did not entitle him to teach "evidence for and against the theory" of evolution. Supreme Court refused to hear appeal.
Frieler *v.* Tangipahoa Parish Board of Education	1997	U.S. District Court Eastern District of Louisiana	Rejected policy requiring teachers to read a disclaimer when teaching about evolution. Seen by court as an endorsement of religion. Recognized "intelligent design" (ID) as "creation science."
	1999	Fifth Circuit Court of Appeals	Affirmed decision
	2000	U.S. Supreme Court	Declined to hear appeal, thus allowed lower court's ruling to stand.
Peloza *v.* Capistrano School District	1994	Ninth Circuit Court of Appeals	Upheld lower court ruling that teacher's First Amendment rights to free exercise of religion is not violated by school district's requirement that teacher be required to teach evolution in biology classes.
Webster *v.* New Lenox School District	1990	Seventh Circuit Court of Appeals	Affirmed district court ruling that " teaching creation science for any reason was a form of religious advocacy" and thus violated the Establishment Clause of the First Amendment.
Edwards *v.* Aguillard	1987	U.S. Supreme Court	Held that Louisiana's "Creationism Act," prohibiting the teaching of evolution except when "creation science" was also taught, was unconstitutional (violated the Establishment Clause).
McLean *v.* Arkansas Board of Education	1982	Federal Court	Declared that "creation science" is not science and that the notion of a "balanced treatment" requiring that both "evolution-science and "creation-science" be taught was a violation of the Establishment Clause.
Segraves *v.* State of California	1981	Sacramento Superior Court	Court rejected parent's contention that teaching about evolution violated their First Amendment rights to freely exercise their religion.
Epperson *v.* Arkansas	1968	U.S. Supreme Court	Invalidated an Arkansas law prohibiting teaching of evolution. Establishment Clause violation. State cannot require teaching anything that fosters a specific religion.

Compiled from: Brian Alters, "Teaching Biological Evolution in Higher Education" (2005),
Barbara Forrest and Paul R. Gross, "Creationism's Trojan Horse" (2004) and court proceedings.

In 2005 creationist proponents tried to change the words from "creation science" to "intelligent design." However, Judge John E. Jones III, the presiding judge in the trial of Kitzmiller et al. *v.* Dover Area School District et al., ruled that it is unconstitutional to require that "intelligent design" be presented as an alternative to evolution. Tammy Kitzmiller was one of several parents who sued their school board when the teachers were asked to alter their teaching of evolution and to make students aware of the "gaps in the theory" and to teach intelligent design as a reasonable "explanation for the origin of life." Creation science had already been established to be a form of religion by the courts in McLean *v.* Arkansas Board of Education (table E.1). Judge Jones ruled that "intelligent design" is simply another form of "creationism" or "creation science" and is therefore a religious doctrine that violates the Establishment Clause of the U.S. Constitution.

The Jones ruling, contained in a 139-page finding of fact, is viewed by many as one of the most powerful and important decisions in the history of evolution court cases. Others believe that Jones's decision is flawed and that the matter will soon resurface in another venue. Regardless of the merits of the Dover ruling, the leaders of the "wedge" movement are fervently determined to force their position on American education.

The "wedge strategy" is a political manifesto aimed at reversing the teaching of evolution with religious doctrine. Authored by the conservative evangelical think tank called the "Discovery Institute," the wedge strategy is a metaphor for the way in which a metal wedge is used to split logs. By advancing a small opening into the public debate about evolution, the wedge strategy hopes to create an opportunity for promoting their political and social agenda.

The *Establishment Clause* refers to the first statements of the First Amendment to the U.S. Constitution, which say that "*Congress shall make no law respecting an establishment of religion or prohibiting the free exercise thereof.…*". It is aimed at preventing the establishment of a state religion and thwarting the passage of laws that give preference to any one religion. As discussed below, the Establishment Clause is paired to a second clause, called the Free Exercise Clause, which prohibits restricting the free exercise of religion. As summarized in table E.1, this portion of the U.S. Constitution has formed the basis of many court rulings. What this makes clear is that the issue of teaching evolution in schools is not a question of fairness, but a matter of the highest possible legal principles.

Some court cases have been brought because a party argued that there was a violation of the *Free Exercise Clause,* which is the second part of the Establishment Clause ("or prohibiting the free exercise thereof"). In the 1981 case Seagraves *v.* the State of California (table E.1), the parents of three schoolchildren sued the state, accusing them of violating the Free Exercise Clause of the First Amendment of the U.S. Constitution by teaching about evolution. In this case, the California court judge ruled that California's State Board of Education's policy provided adequate accommodation to the views of the Seagraves in contradiction to Mr. Seagraves's contention that the class discussion of evolution prohibited his and his children's free exercise of religion.

Other important court cases are summarized in table E.1. In each instance, the courts were used as the final arbiter in the ongoing controversy between religious zealots and science education. Efforts to wedge religion into science classes or remove evolution from the curriculum have formed the basis of most cases. Every court ruling has supported the teaching of evolution in schools and against the teaching of religious doctrine in science classes. By whatever name (creationism, creation science, intelligent design), the courts have been unanimous in affirming that religion has no place in science class.

CLOSING COMMENTARY

Ultimately, the effective and widespread teaching of the principles of evolution is in the best interests of the greater society. Darwinian evolution is, as we have repeatedly discussed, the unifying principle in biology and the other natural sciences. Many scholars have articulated the view that if our global society is to effectively address our present and future environmental problems, coincidental with our exponential population expansion of the last millennium, the disciplines of ecology and evolution must be understood by citizens, scientists, and politicians alike. The Darwinian worldview, while disconcerting to some, defines a path to a sustainable future. In our increasingly rapidly changing world, what could be more important than an appreciation for how and why change occurs? Our future existence may well depend on how well we come to appreciate the history of our species and our planet.

GLOSSARY

The following is a non-exhaustive list of definitions for terms found in this text. For those terms not found in this glossary, the reader should refer to a good scientific dictionary.

Aberrant: Not normal or usual.

Acidophile: Archaeabacteria in the domain Archaea, which are capable of living under conditions of extreme acidity.

Adaptation: A characteristic that promotes survival and reproduction in a particular environment.

Adaptive radiation: The diversification of species into new ecological niches.

Adenine (A): One of the two purine nitrogenous bases. Pairs with thymine in DNA and uracil in RNA.

Admixture: In population genetics, a measure of the mixing of genes from different populations.

Aerobic: Activity or function that requires air. Compare to anaerobic.

Albino: An organism lacking the pigment melanin. In humans, the condition of lacking any pigment coloration.

Algae: Diverse group of photosynthetic plants.

Allantois: The extra-embryonic membrane of vertebrate eggs that functions primarily in respiration and the storage of wastes.

Allele: One of two or more alternative forms of a gene.

Allelic frequency: A measure of the occurrence of an allele in a population expressed as a proportion of the entire population.

Allopatric: Being geographically isolated. Compare to sympatric.

Allopatric speciation: The formation of new species when two groups physically separate. Also called geographic speciation.

Allopolyploid: An organism or species containing the sets of chromosomes from two different organisms.

Altruism: Behavior of self-sacrifice for the benefit of others.

Amerind: The indigenous peoples of the Americas.

Amino acid: The twenty building blocks of proteins coded for by the genetic code.

Amnion: The innermost fluid-filled extra-embryonic membrane of a vertebrate egg.

Amniotes: Living and extinct vertebrates (e.g., reptiles, birds, and mammals) possessing an amniotic sac.

Amniotic egg: The type of egg of a vertebrate that possesses an amnion, a chorion, yolk sac and an allantois as extraembryonic membranes.

Amniotic sac: Fluid-filled membrane surrounding a vertebrate embryo that provides protection and aids in respiration.

Amphibia: Member of the class Amphibia, which includes the frogs, toads, and salamanders.

Analogous: Serving the same function but not evolving from the same common ancestor. *See* convergence.

Anagenesis: The evolutionary process whereby one species evolves into another. Evolution within a single lineage.

Anaerobic: Conditions without the presence of oxygen. Compare to aerobic.

Anamniotes: Vertebrates without an amnion; includes the cyclostomes (hagfish and lamprey), fish, and amphibians.

Angiogenesis: The growth of new blood vessels in the body.

Angiosperm: A flowering plant.

Anthropoid: Among the primates, the monkeys, apes, and humans. Apelike.

Antibody: A protein that is produced in response to a foreign substance (antigen).

Anticodon: The triplet sequence of nucleotides on transfer RNA that is complementary to a codon on messenger RNA.

Antigen: A foreign substance that stimulates the immune system. (See antibody).

Anthropoid: Primates that include living and extinct monkeys, apes, and humans.

Aneuploid: Having one or more extra or missing chromosomes. *See* polyploidy.

Arboreal: Living in the tress.

Archaeobacteria: Members of the domain Archaea.

Archosaur: Diaspid amniotes that were the ancestors of the dinosaurs and include the modern crocodilians and birds.

Artificial selection: Selection through the intentional actions of humans encouraging the production, via selective breeding, of individuals with desired characteristics.

Asexual reproduction: Reproduction in the absence of sex.

Aurignacian: An early Upper Paleolithic cultural phase (named after a cave in Aurignac, France) beginning in Europe about 40,000 years ago and lasting until about 22,000 years ago.

Australopithecines: Members of the genus *Australopithecus,* some of the earliest hominids.

Autopolyploid: A species or organism that contains duplicated chromosomes (polyploid) derived from the same species.

Autosome: The nonsex chromosomes. In humans, chromosomes 1–22.

Autotroph: An organism that has the ability to harness energy from the sun or from inorganic sources. Usually photosynthetic green plants.

Azoospermia: The absence of sperm.

Background extinction: The normal extinction rate found throughout the geological history of Earth. Compare to mass extinction.

Balanced polymorphism: Phenomenon by which two or more alleles are maintained in a population over the course of many generations through selection (heterozygote advantage) rather than mutation.

Base (nucleotide): The nitrogenous component of a nucleotide consisting of either a purine or a pyrimidine.

Base pair: Two complementary pairs of nucleotides in DNA (A–T; G–C) or RNA (A–U), connected by hydrogen bonds.

Binomial: Literally, "two names." The Linnaean two-name system of naming organisms where the genus name is followed by the species S name. E.g., *Homo sapiens.*

Biogeography: The study of the geographic distribution of living and extinct species.

Biological species concept: The concept of a species as a reproductively isolated, freely interbreeding population(s) that produces fertile offspring.

Biosphere: The regions of the Earth's surface and atmosphere where living organisms exist.

Bipedal: Walking on two hind limbs (legs) for locomotion.

Brachiation: An arboreal locomotion pattern used by gibbons and siamangs in which they swing from tree limb to tree limb using only their arms.

Cambrian explosion: Massive expansion and diversification of life that occurred roughly 540 million years ago.

Carcinogen: A substance that causes cancer.

Carrier: An individual with one copy of a recessive allele and one copy of a dominant allele who is unaffected by the recessive allele.

Carrying capacity: Based on the available resources, the maximum sustainable population of a particular environment.

Central dogma: A representation of the flow of genetic information as DNA→RNA→protein.

Centromere: The central constriction of a chromosome; binds to spindle fibers during cell division.

Chemoautotroph: Organism capable of producing energy from chemicals (vs. the sun).

Chloroplast: Oval-shaped cellular organelle found in the cytoplasm of greens plants and green algae.

Chorion: The outermost of the extra-embryonic membranes in terrestrial vertebrates; contributes to the formation of the placenta in mammals.

Chromatid: One of a pair of a duplicated chromosome held together by a centromere.

Chromosomal aberration: Changes to the arrangement of parts of a chromosome such as an inversion, duplication, or translocation.

Chromosome: Structures in living organisms containing genes that are passed from one generation to the next. In humans and most eukaryotes, chromosomes are linear mixtures of DNA and proteins. In bacteria, chromosomes are circular.

Cladogenesis: Evolution that results in the splitting of a lineage; the splitting of one phyletic lineage into two.

Class: A taxonomic category between a phyla and an order. A phylum contains classes that contain orders.

Cline (clinal): A gradient of change (usually physical or physiological) in a phenotype or genotype in populations of related organisms, usually along a line of environmental or geographic transition.

Clonal: Cells that are descended from and genetically identical to an original cell.

Co-dominant: Condition in which the heterozygote produces both gene products. In sickle-cell anemia, for example, a heterozygote "carrier" with one sickle gene and one normal gene produces roughly half sickle-shape red blood cells and about half normal-shape red blood cells.

Coding strand: The strand of DNA that is used to make messenger RNA transcript during transcription.

Common descent: Having a common ancestor. A major principle of Darwinian evolution, which states that all organisms are related through common ancestry.

Competitive exclusion (Gause's principle): Principle that under a particular set of environmental conditions, competition between species seeking the same ecological niche results in one species surviving while the other perishes, resulting in each species occupying a distinct niche.

Congenital: Present at birth.

Conjugation: The temporary union of unicellular organisms, such as bacteria, during which genetic material is transferred from one cell to another.

Consanguinity: Close genetic relationships. For example, brother-sister, mother-son, father-daughter, cousins. Usually used to indicate mating between close relatives.

Conspecific: Members of the same species.

Convergent evolution: The independent development (evolution) of similar characters (structures or genes) in genetically unrelated organisms as a response to similar environments.

Creationism: The belief that organisms were independently created by a supernatural force.

Creole: Used variously to describe the people and languages of the descendants of various ethnic backgrounds who interbred with French settlers to Louisiana.

Crossing over: The exchange of chromosome segments by nonsister chromatids during synapsis of meiosis I.

Crossopterygian: The lobbed-finned fish of the Devonian period that may have developed into the first tetrapods. Include the modern day coelacanth.

Cuckold: The male mate of an unfaithful female mate; from the cuckoo's habit of laying eggs in the nest of other birds.

Cultural selection: Application of Darwinian concept of selection to culture.

Cussorial: Adapted for running.

Cyclostome: Primitive jawless fish such as hagfish and lamprey.

Cytochrome *c:* A mitochondrial protein involved in energy transport.

Cytoplasm: Part of the cell enclosed by the plasma membrane that contains organelles.

Cytosine (C): One of the pyrimidines that pairs with guanine (G) in DNA and RNA.

Darwinism: Charles Darwin's theory that biological evolution led to many different highly adapted species via the mechanism of natural selection.

Darwin's finches: The species of finches native to the Galápagos Islands.

Decaploid: Having ten complete sets of chromosomes.

Descent with modification: Darwin's theory that existing life forms are the product of modifications from pre-existing life forms.

Deduction: Deriving a conclusion inferred by logical reasoning.

Degenerate (redundant): A property of the genetic code in which more than one codon specifies a particular amino acid.

Deleterious gene: A gene associated with a debilitating condition or disease.

Deleterious mutation: Harmful mutation. Mutation that is not adaptive.

Deletion: The loss of a portion of a chromosome.

De novo: Arising anew or from the beginning.

Deoxyribonucleic acid (DNA): The set of nucleotides that serve as genetic information.

Deoxyribose: Five-carbon sugar of DNA.

Devonian: Geological period in the Paleozoic which lasted from 417-354 million years ago that is known as the "Age of Fishes".

Directional selection: Selection acting to reduce or eliminate one extreme form from a distribution of phenotypes.

Diapsid: Group of living and extinct amniotes with two temporal openings in their skull.

Differential reproduction: Preferential reproduction of those individuals who are best adapted to a particular environment over those who are less well adapted to that environment.

Differential survival: Preferential survival of those individuals who are best adapted to a particular environment over those who are less well adapted to that environment.

Dihybrid cross: A cross between individuals that differ with respect to two different traits.

Dimorphism (as in sexual dimorphism): The presence or two morphologically distinct types.

Diploid: Having two sets of chromosomes.

Directed panspermia: The notion that life was purposefully seeded on Earth.

Directional selection: When selection favors one extreme of the distribution of traits, causing the distribution to shift in one direction.

Disruptive selection: When the extreme of a distribution of traits has the highest fitness, and the intermediate values are most disadvantageous.

Divergent evolution: Change leading to differences between lineages.

DNA polymerase: Enzyme responsible for linking nucleotides to form DNA strands. Primary enzyme of DNA replication.

Dollo's law: Hypothesis that evolution is not reversible.

Double helix: Spiral pattern of DNA in which two complementary nucleotide strands coil around a common helical axis.

Domain: In scientific classification schemes, the highest order of grouping.

Domain Archaea: In the three-domain system of taxonomy, the group that contains the diverse unicellular prokaryotes called Archaea.

Domain Bacteria: In the three-domain system of taxonomy, the vast group of unicellular organisms called Bacteria.

Domain Eukarya: In the three-domain system of taxonomy, the group that contains the eukaryotic organisms, including the fungi, protists, plants, and animals.

Dominant Allele: Allele that exerts a phenotypic effect in the heterozygote.

Drosophila: Genus name for fruit flies.

Duplication: A mutation that results in the duplication of a portion of a chromosome.

Dyad: Two chromatids found in late meiosis I and early meiosis II.

Ectothermic: An organism that has a body temperature that varies with the environment. Sometimes erroneously called "cold-blooded."

Electrophoresis: The movement of charged particles under the influence of an electric field. Usually conducted in some medium, such as a gel.

Embryo: A stage in early development of a multicellular organism. In humans, *embryo* usually refers to the first nine weeks of gestation, after which the embryo is called a fetus.

Endemic: A species or population confined to a specific geographic locality.

Endogamous: Mating between a restricted group of individuals.

Endoplasmic reticulum: Cellular organelles made up of tubules, vesicles, and sacs that perform specialized functions in cells.

Endosymbiosis: Theory that eukaryotic cells evolved from a symbiosis between different prokaryotic species.

Endothermic: Organism that generates heat to maintain its body temperature, typically above the temperature of its surroundings. Sometimes erroneously called "warm-blooded."

Enhancers: Genetic control elements that increase the rate of transcription.

Environmental modification: Changing the environment.

Enzyme: Protein that catalyzes or speeds up an organic reaction.

Eon: A geologic unit that includes several eras, which is in turn composed of periods, which are composed of epochs.

Epoch: A unit of geological time that is a subdivision of a period and is itself divided into ages.

Equilibrium: A balance between two or more forces.

Era: A category of geologic time between and eon and a period.

Erythroblastosis fetalis: Also known as Rh disease. A type of anemia in which the red blood cells (erythrocytes) of the fetus are destroyed by the maternal immune system.

Escherichia coli: A common form of bacteria, usually abbreviated as *E. coli.*

Establishment Clause: Part of the First Amendment to the U.S. Constitution that says that *"Congress shall make no law respecting an establishment of religion or prohibiting the free exercise thereof"*.

Estrus: Sometimes called "heat"; when the female is sexually receptive. Also describes the female ovulatory cycle.

Eugenic: The hereditary improvement of the human race.

Eukaryote: A cell or organism that possesses a clearly defined nucleus.

Evolution: Biological change over time.

Evolutionary developmental biology (evo-devo): The field of biology in which developmental processes are studied from an evolutionary perspective.

Evolutionary divergence: Accumulation of differences in populations or species, leading to the formation two or more new species.

Extinction: The end of the existence of a species or group of species.

Extremophile: An Archaean organism capable of living in very extreme environments.

Family: A taxonomic unit of similar species above the level of a genus and below the level of an order.

Fecundity: The ability to reproduce.

Fertilization: Union of two gametes, such as an egg and sperm, to form a zygote.

Fitness: Darwinian fitness is reproductive success; the number of offspring produced who survive to reproduce.

Flagella: Tail-like structure on certain single-celled organisms.

Fossil: Geological remains from the past.

Fossorial: Adapted for living underground.

Founder effect: Related to genetic drift. When a small part of a larger population migrates and brings a nonrandom sample of founder population's genes.

Frameshift mutation: Mutation involving the addition or deletion of a nitrogenous base.

Free Exercise Clause: Second part of the Establishment Clause of the First Amendment that says *"or prohibiting the free exercise thereof"*.

Galápagos: Islands off the coast of Ecuador where Charles Darwin landed in his voyage of the *Beagle*. Home of the Galápagos finches and tortoises.

Gamete: Specialized haploid (*n*) reproductive cell such as an egg or a sperm with one complete set of chromosomes. Male and female gametes fuse during fertilization to form a diploid (2*n*) zygote, also known as a fertilized egg.

Gemmules: Imaginary particles of inheritance that Darwin proposed for the theory of pangenesis that involves miniature body parts discharged into the blood that concentrates in the gametes.

Gene: A unit of heredity composed of a sequence of DNA that codes for a protein.

Gene clusters: A set of two or more genes that serve to encode for the same or similar products.

Gene flow: Changes in the genetic composition of populations as a result of migration and immigration.

Gene pool: All of the genes in a population.

Genetic code: The sequence of messenger RNA triplets (codons) that specifies the sequence of amino acids in a polypeptide or protein. In general, the genetic code is universal (identical in all organisms) and redundant (degenerate).

Genetic drift: Random fluctuation in gene frequencies due to chance.

Genetic load: A measure of the impact of deleterious genes on a population compared to if it had the most favored genotype.

Genotype: Genetic constitution of an individual.

Genotypic frequency: A measure of the occurrence of a genotype in a population expressed as a proportion of the entire population.

Genome: The complete genetic constitution of a cell or organism.

Genomics: The comparative study of genomes.

Genus: Taxonomic unit above a species and below a family.

Germ cell: Cell involved in reproduction, such as an egg or a sperm.

Germinal: Relating to germ (sex) cells or sexual tissue (ovaries and testis)

Germinal mutation: Heritable mutations to germinal tissue such as ovaries or testis.

Germ plasm: Cells that give rise to gametes (e.g., eggs or sperm).

Genus (pl. genera): A taxonomic grouping above a species and below a family (e.g., *Homo*).

Group selection: The concept that, under certain circumstances, genes will be selected for because they benefit the overall success of the group rather then the individual.

Guanine: One of the 5 nitrogenous bases found in nucleic acids (DNA and RNA). Binds with cytosine in DNA and RNA.

Gymnosperm: Meaning "naked seed", this large group of seed bearing plants that includes the conifers and cycads.

Gynogenesis: A form of parthenogenesis found in some salamanders in which the embryo develops from maternal DNA only which was activated by contact with sperm from a male.

Halophile: An extremophile (archaebacteria) organism that thrives in environments with very high concentrations of salt.

Handicap principle: Principle proposed to explain how evolution may lead to honest or reliable signaling between animals that have the motivation to bluff or deceive one another.

Haploid: Having one set of chromosomes.

Haplotype: A group of alleles of different genes on a single chromosome that are closely linked and inherited as a unit.

Hardy-Weinberg equilibrium: An equation and a concept that specify conditions under which it is possible to estimate gene frequencies from one generation to the next and describe how, when the specific conditions are met, that gene frequencies do not change from one generation to the next.

Hemoglobin: Component of red blood cells that carries oxygen.

Heterotroph: Organisms that cannot produce their own food and must obtain it from other organisms.

Heterozygous: Possessing unlike alleles. E.g., "Aa."

Heterozygote: Individual possessing unlike alleles.

Heterozygote advantage: Phenomenon in which the heterozygote has a higher Darwinian fitness than either homozygote.

Hexaploid: Having six complete sets of chromosomes.

Histone: A type of protein that complexes with DNA in eukaryotes.

Holocene: The geological epoch that began about 10,000 to 11,000 years ago, characterized by the development of human civilization.

Homeobox: A 180-base-pair DNA sequence that involves the regulation of a pattern of development.

Homeotic genes: Genes that determine which parts of the body form what body parts.

Homeodomain: A domain in a protein that is encoded by a homeobox.

Hominoid: The living and extinct members of the superfamily Hominoidea. Modern members include the apes (gibbons, chimpanzees, and gorillas) and the humans.

Hominid: Members of the family Hominidae. Modern members are gorillas, orangutans, chimps, and humans.

Hominin: Members of the subfamily Homininae. Modern members are gorillas, chimps, and humans.

Hominini: Members of the tribe Hominini. Modern members are chimps and humans.

(NOTE: The taxonomy of the terms hominoid, *hominid, hominin,* and *hominini* is not agreed upon universally. The above terminology follows figure 17.1).

Homology: Similarity in biological features (genes, structures, behaviors) due to descent from a common ancestor.

Homologous structure: Structures that have similarities due to common ancestry and common descent.

Homozygote: Individual with two of the same alleles.

Homologous genes: Genes derived from the same ancestral gene.

Homologue or homologous chromosome: In a diploid (2*n*) organism, a pair of chromosomes, such as both chromosomes 1.

Hox gene: A class of genes involved with pattern formation in early embryos.

Hydrocarbon: A simple organic compounds containing hydrogen and carbon such as methane.

Hydrothermal vents: Openings in the ocean floor at or near mid-ocean ridges that release hot, mineral-rich water that often supports rare and unusual forms of life that rely on chemoautotrophs for energy.

Hyperendemic: Term applied to malaria when transmission occurs, usually throughout the year at high intensity, and the disease burden is high in young children.

Hypothesis: A proposed explanation for a natural phenomenon based on previous observations or experiments. An unproven scientific conclusion.

Hypothetico-deductive method: Scientific method whereby science sets up testable hypotheses and then tries to falsify them.

Ichthyosaur: The aquatic Mesozoic dinosaurs.

Inbreeding: Mating among genetically related individuals.

Inclusive fitness: The fitness of an allele or genotype measured by its effects on related individuals that also possess it (see kin selection).

Independent assortment: Mendel's principle that during gamete formation (meiosis) homologous chromosomes (or genes located on them) assort independently of each other.

Industrial melanism: The evolutionary process in which initially light-colored organisms become darker as a result of natural selection.

Inheritance of acquired characteristics: Lamarck's hypothesis that the environment modifies traits that are then passed on to the next generation.

Insertion sequence: A short DNA sequence that acts as a transposable element.

Intelligent design: Assertion and belief that life and the universe were purposefully designed by an intelligent being. Also sometime referred to simply as ID.

Intergenic sequences: DNA sequences located between clusters of genes that contain few or no genes.

Intersexual selection: Sexual selection between members of the opposite sex.

Interspecific: Between species.

Intrasexual selection: Sexual selection between members of the same sex.

Intraspecific: Within species.

In utereo: Within the uterus.

Inversion: A 180-degree change in the direction of the genetic material in a single chromosome.

Isolating mechanisms: Biological mechanisms that serve as barriers to gene flow or interbreeding between populations.

Jumping genes or transposon: A mobile genetic element.

Karyotype: Picture of a complete set of chromosomes arranged in a standard format.

Kin selection: Evolutionary mechanism that selects for genes (or behaviors) that lowers an individual's own fitness but enhances the reproductive success of a relative. Selection that benefits the individual's relatives rather than the individual (e.g., an alarm call).

Kingdom: A taxonomic unit that contains phyla.

Labyrinthodont: The main group of fossil amphibians that contains forms closely related to certain fish.

Lamarck: Jean Baptiste de Lamarck developed the theory of the inheritance of acquitted characteristics. Although incorrect, it was one of the first cohesive concepts regarding natural causes for changes in organisms.

Lamarckian: Having to do with Lamarck, specifically having to do with the notion of the inheritance of acquired characteristics.

Law: In science, something known with absolute certainty, such as the laws of thermodynamics. The term *law* is not applicable to biology.

Lethal gene: A gene that is incompatible with life. When dominant, it is lethal by itself. When recessive, it requires the presence of two alleles to be lethal.

Lethal selection: One hundred percent (100%) selection against a gene.

Linked genes: Genes on the same chromosome.

Linkage group: Genes on the same chromosome that tend to be transmitted together.

Locus: The physical location of a gene on a chromosome.

Macroevolution: The pattern of evolutionary change at or above the level of species.

Macromutation: A large-scale mutation that produces a characteristic.

Mammal: Vertebrate animals with hair and mammary glands.

Malignant: A medical term meaning "progressing toward death." Applied to cancers (neoplasm) to mean a rapidly growing aggressive growth.

Marsupial: Mammal with a marsupial pouch, such as the opossum.

Mass extinction: A relatively sudden sharp decline a large number of species over a short period of time, affecting major taxonomic groups.

Master genes: Genes that control the expression of many other genes.

Meiosis: Cell division resulting in the formation of sex cell or gametes (e.g., sperm and eggs), each with half the genetic material of the parental cell.

Meiosis I : The first meiotic division, in which homologous chromosomes separate.

Meiosis II: The second meiotic division, in which duplicated chromosomes separate.

Meme: Cultural unit of evolution analogous to a gene.

Memetics: The study of the transmission of memes.

Mendelian: Having to do with the concepts described by Gregor Mendel.

Messenger RNA (mRNA): The RNA that is transcribed from complementary DNA and later translated into a protein at the ribosome.

Metacarpal: The tubular bones between the wrist and the base of the fingers.

Metaphase: Stage of mitosis (or meiosis) following prophase.

Metastasis: Process by which cancer cells move from their point of origin to another place in the body.

Methanogen: Anaerobic archaebacteria that produces methane.

Microcephalic: An abnormally small head often associated the genetic condition called microcephaly.

Micromutation: A small mutation.

Microevolution: Evolution below the level of a species, such as the change of allele frequencies (i.e., due to selection, mutation, genetic drift, or migration) within a population.

Missense mutation: A base substitution mutation that results in the alteration of a single amino acid.

Mitochondria: Tubular-shaped organelle found in the cell cytoplasm of eukaryotic cells that is responsible for energy production in the cell.

Mitochondrial DNA (mtDNA): DNA that is located in mitochondria. Mitochondrial DNA is inherited maternally, from mothers to their off-spring, and from their female offspring to subsequent generations.

Mitochondrial Eve: Concept of a population of women who are the matrilineal most-common ancestor of all currently living humans.

Mitosis: Cell division that forms two identical cells.

Molecular clock: The concept of using the number of differences in a DNA sequence or protein variation to measure relative time between species evolution. Application of this concept allows for the estimation of the dates or relative divergence between taxa.

Molecular signature: A feature of a cell or subcellular structure, such as a centromere, that makes it identifiable.

Monohybrid cross: A cross between individuals that differ with respect to a single trait.

Monotreme: Egg-laying mammals, such as the duckbill platypus.

Monotypic: A single type. Compare to polytypic.

Morphology: The anatomical form and structure of an organism.

Mousterian: Flint tool style primarily associated with Neanderthals.

Mutagen: A substance or force that causes a mutation.

Mutation: A heritable change in DNA.

Natural sciences: The sciences that have to do with nature such as biology, zoology, botany, geology, biochemistry, physics, and so forth.

Natural selection: Process described by Charles Darwin by which individuals reproduce to a greater or lesser degree based on the presence of traits that are favorable or not to survival and/or reproduction.

Na-Dene: The Native Americans who are the descendants of the second major migratory wave into North America.

Neutral mutation: A mutation has no effect on Darwinian fitness.

Niche: The habitat supplying the factors necessary for the existence of an organism or species or the ecological role of an organism has in a community regarding food and energy consumption.

Niche diversification: Ecological principle that no two species occupy exactly the same niche.

Nitrogenous base: One of the building blocks of DNA and RNA: cytosine (C), guanine (G), adenine (A), thymine (T), and uracil (U).

Nonsense mutation: A single base substitution mutation that changes a coding sequence into a stop sequence, causing a protein to be truncated.

Nuclear genes: Genes on the chromosomes in the nucleus (as compared to mitochondrial DNA).

Nuclear envelope: Membrane that encloses the cell's nucleus.

Nucleic acid: Organic molecule composed of nucleotides. DNA or RNA.

Nucleolus: Eukaryotic site of ribosomal RNA (rRNA) synthesis.

Nucleotide: Unit of nucleic acid composed of a phosphate, a five-carbon sugar, and a nitrogenous base (purine or pyrimidine).

Nucleus: The membranous organelle that contains chromosomal DNA.

Octaploid: Having eight complete sets of chromosomes.

Order: A taxonomic level of classification above a family and below a class.

Ornithischian dinosaurs: Herbivorous dinosaurs with a hip structure similar to birds.

Organelle: A small subcellular structure that has a unique function.

Paleopolyploid: An ancient polyploid.

Paleontology: The study of extinct fossil organisms.

Pangenesis: Theory of inheritance proposed by Darwin that suggested that minute, body parts or gemmules were the agents of inheritance.

Panspermia: Hypothesis that life was initiated by material from outer space brought to Earth by a meteor or comet or cosmic dust.

Parthenogenesis: Egg development in the absence of fertilization.

Pathogenic: Capable of causing disease.

***Pax-6* genes:** Master control gene for the development of eyes and sensory organs, and certain neural and epidermal tissues.

Period: The geologic unit of time that is a division of an era.

Phalanges: Segmental bones of the fingers and toes.

Phocomelia: Series of deformities caused by prenatal exposure to thalidomide.

Phyletic gradualism: Hypothesis that species change slowly over millions of years.

Phylogeny: The evolutionary history of an organism.

Phylum (pl phyla): A taxonomic rank below kingdom and above class.

Pithecanthropine: Resembling members of the extinct genus *Pithecantropus,* now *Homo.*

Placental mammal: The large group of mammals with a placenta. Compare to monotremes and marsupials.

Plasmid: A small circular extra chromosomal ring of DNA that is not in a chromosome but is capable of autonomous replication.

Plasmodium falciparum: Protozoan parasite that is transmitted by female Anopheles mosquito to cause malaria in humans.

Pleistocene overkill: Hypothesis that humans were responsible for the extinction of the megafauna and flightless birds within a short time after they colonized different continents.

Pelycosaur: Primitive synapid reptiles of the late Paleozoic such as *Dimetrodon.*

Plesiosaur: Carnivorous aquatic reptile of the Jurassic through the Cretaceous.

Point (or gene) mutation: A mutation involving a single nucleotide.

Polygyny: A mating system in which a male mates with multiple females.

Polymer: Two or more compounds formed by polymerization.

Polymerization: The process of changing a compound so as to form a new compound.

Polymorphism: The presence of two or more genetic or phenotypic variants in a population.

Polynucleotide: Sequence of nucleotides.

Polypeptide: Sequence of amino acids.

Polyploid: An organism with three or more sets of chromosomes.

Polytypic: Many types. Compare to monotypic.

Population: A group of organisms occupying a specific area with a common gene pool.

Postzygotic isolating mechanism: Reproductive isolation in which a zygote is produced but is unable to develop into an adult.

Preadaptation: A characteristic that was adaptive under a previous set of conditions and later provides the initial stage for the evolution of a new adaptation under new conditions.

Prezygotic isolating mechanism: Reproductive isolation in which the formation of a zygote is prevented.

Principle of independent assortment: Mendelian concept that during meiosis, homologous chromosomes segregate independently of each other and of other homologous pairs of chromosomes.

Principle of segregation: Mendelian concept that during meiosis, homologous chromosomes segregate.

Proofreading: The correction of an error in DNA replication.

Prophase: The first phase of mitosis.

Prophase I: The first phase of the first meiotic division, when chromatids are tetrads and crossing-over occurs.

Prophase II: The first phase of the second meiotic division.

Prokaryote: Any organism that lacks a distinct nucleus and other organelles due to the absence of internal membranes. Bacteria are among the best-known prokaryotic organisms.

Promoters: Gene sequences that control gene expression.

Prosimian: Suborder of primates that includes lemurs, lorises, and bush babies.

Protein synthesis: Process by which proteins are assembled at the ribosome according to the sequence specified by the genetic code. Transcription and translation. DNA→RNA→protein.

Proteomics: The study and comparison of protein's structure and function.

Pseudogenes: Defunct relatives of known genes that have lost their protein coding ability or are otherwise no longer expressed.

Psilophyte: Various extinct vascular rootless and often leafless plants of the Devonian and Silurian periods.

Punnett square: Graphical representation of a mating or cross, named after British geneticist R. C. Punnett.

Punctuated equilibrium: The view that evolutionary change goes through long periods of stasis and short intervals punctuated by rapid speciation.

Purine: One of two families of nucleotides with two chemical rings; eg., adenine and guanine.

Pyrimidine: One of two families of nucleotides with a single chemical ring; e.g., thymine, cytosine, and uracil.

Quadrupedal: Walking on four limbs.

Race: A biological race is a group of organisms that differ in ecological, behavioral, geographic or physiological characteristics from other groups in that species. Same as subspecies.

Random drift: Fluctuations in the frequency of a trait due to chance.

Random mating (panmixia): Mating without respect to phenotype or genotype of the sexual partner.

Recessive allele: Allele that is only expressed when found in a double dose and whose expression is otherwise masked by the presence of a dominant gene.

Recessive trait: Traits that require two recessive alleles in order to be expressed.

Reciprocal altruism: Mutually beneficial exchanges of altruistic behaviors between individuals.

Recombinant DNA: DNA that has been altered by joining genetic material from two different sources, usually from two different species.

Recombination: Process by which chromosomes are broken and rearranged to form novel genetic combinations during meiosis. A result of meiotic cross-over.

Redundant (degenerate): Property of the genetic code whereby more than one codon specifies a specific amino acid.

Repetitive DNA: DNA sequences that are repeated many times in the genome but that do not code for proteins.

Replica plating: A method of transferring bacterial colonies from one plate to another.

Reproductive isolating mechanism: The various mechanisms by which mating and/or hybridization between two species is prevented.

Restriction endonucleases: An enzyme that cleaves DNA at a particular base sequence.

Retinoblastoma: An inherited cancer of the retina or light-sensitive cells of the eye, most common in children.

Reverse transcription: Process by which single-stranded RNA makes double-stranded DNA with the enzyme reverse transcriptase.

RhoGAM : Drug used to suppress the immune response of Rh-negative mothers who have an Rh-positive child soon after the delivery to prevent Rh disease from occurring in the woman's next Rh-positive child.

Ribonucleic acid (RNA): One of two forms of nucleic acids. There are three forms of RNA, messenger RNA (mRNA), transfer RNA (tRNA), and ribosomal RNA (rRNA), and each functions in protein synthesis.

Ribose: Five-carbon sugar of RNA.

Ribosomal RNA (rRNA): The RNA that makes ribosomes, which are the sites of protein synthesis.

Ribosome: Site of protein synthesis.

Ribozyme: RNA molecule that catalyzes a chemical reaction.

RNA polymerase: Enzyme that synthesizes RNA during transcription. Primary enzyme of transcription.

RNA World: A theoretical period of primitive Earth when enzymatic activity and information needed for life was accomplished by RNA exclusively.

Same-sense mutation: An alteration in a single base that causes no change in the amino acid sequence due to the redundancy in the genetic code.

Saurischian dinosaurs: Carnivorous dinosaurs with lizard-like hips.

Secondary sex characteristics: Traits found in males and females that are associated with sexual maturation and sexual selection.

Segregation: Mendelian principle that during gamete formation (meiosis), homologous chromosomes segregate from each other.

Selection coefficient: A measure of the contribution of one genotype relative to the contributions of other genotypes.

Selectively neutral: A trait or gene that is not selected for or selected against.

Semi-lethal: Mutant allele that causes semi-sterility.

Semi-sterile: When the recessive homozygote has half the reproductive capacity of the normal individual.

Sewall Wright effect: Same as genetic drift.

Sex chromosomes: Chromosomes controlling sex determination. In humans, the X and the Y chromosomes. 46XX = female; 46XY = male.

Sexual dimorphism: Gender differences in physical characteristics that mediate sexual responsiveness.

Sexual selection: A type of natural selection that acts on an organism's ability and success in reproduction.

Sickle-cell anemia: Disease due to a mutation in the hemoglobin gene in which sickle-shaped red blood cells are produced, which causes a series of ailments.

Sickle-cell trait: Having one copy of the sickle-cell gene. Being a carrier of the sickle-cell anemia allele.

Signaling macromolecule: A large molecule produced by a signaling cell that transmits information from one cell to another.

Sister chromatids: Genetically identical duplicated chromosomes held together by a centromere.

Species: A reproductively isolated group of interbreeding organisms capable of producing fertile offspring.

Sphenodontia: Order of lizard-like reptiles that includes only one living genus, the tuatara (*Sphenodon*).

Speciation: The formation of a new species.

Spontaneous generation: Concept that life can spontaneously come about from nonlife.

Somatic: Relating to the body and not the gamete (germinal).

Somatic mutations: Mutations to somatic or body tissue.

Spontaneous generation: Theory that living organisms can be produced from nonliving matter.

Spontaneous mutation: Mutation occurring in the absence of a mutagen.

Squamata: The largest recent order of reptiles, including lizards and snakes.

Stabilizing selection: When selection favors the mean of the distribution of phenotypes.

Start codon: The three-base sequence (usually AUG) that initiates translation.

Stop codon: One of three triplet base sequences (codons) (UAA, UAG, and UGA) of the genetic code that do not code for an amino acid, but signal the end of translation.

Subvital: A recessive gene that when homozygous, impairs reproductive fitness to less than 100 percent (but not less than 50 percent) of normal.

Subspecies: A taxonomic unit below the species often distinguished by specialized phenotypic characteristics and/or by geographic region. Same as race.

Symbiosis: A close ecological relationship between the individuals of two (or more) different species.

Symbiotic: Relationship in which two or more organisms are in direct contact with each other.

Sympatric: Organisms living in the same location. See allopatric.

Sympatric speciation: Speciation that occurs when species diverge while inhabiting the same place.

Synapsis: During meiosis, the aligning of four chromatids to form a tetrad.

Synapsids: Early vertebrates with a pair of temporal openings behind the eye socket; included the ancestors of mammals.

Taxa: Evolutionarily related groups of organisms, such as a species or genus.

Taxonomy: Field of biology concerned with classifying living and extinct organisms.

Telomere: Terminal section of a chromosome.

Template strand: The DNA strand that is used as a template for producing the messenger RNA used in transcription.

Teratogen: A substance that can cause a birth defect.

Terminator: Three sequences of the genetic code for the command to stop.

Tetrad: Four homologous chromatids bundled together early in meiosis I.

Tetrapod: A four-legged animal, including the amphibians, reptiles, birds, and mammals.

Tetraploid: Having four complete sets of chromosomes (4n).

Thalassemia: Inherited disorder of the hemoglobin molecule.

Thalidomide: A sedative that, when used during pregnancy, causes a number of mostly limb-related abnormalities known as phocomelia.

Theory: A well-tested, coherent explanation that holds true for a large number of facts and explains a wide range of observations about the natural world. Examples are the theory of gravity and the cell theory.

Therapsid: Extinct reptile of the order Therapsida; thought to be direct ancestor of the mammals.

Threshold: Minimum limit before which an effect is observed.

Thermophile: A type of extremophile (archaebacteria) that thrives at relatively high temperatures, between 45 and 80 °C.

Thymine: Pyrimidine base that binds with adenine (A) in DNA and RNA and is replaced by uracil (U) in RNA.

Transcription: Assembly of messenger RNA from DNA. Together with translation, the process called protein synthesis.

Transcription factor: A protein that influences the ability of RNA polymerase to transcribe genes.

Transfer RNA (tRNA): An RNA that carries amino acids used to translate mRNA into polypeptides (proteins).

Translation: Protein assembly at the ribosome. Together with transcription, the process called protein synthesis.

Translocation: A chromosomal mutation in which a portion of one chromosome breaks and becomes attached to another chromosome.

Transmutation: The term used Jean Baptiste Lamarck and Charles Darwin to mean the changing of one species into another. In Darwin's day, the word that meant evolution.

Transposable element: Mobile pieces of genetic material (also called transposon, jumping gene).

Transitional forms: Fossils or organisms are intermediate between an ancestral form and that of its descendants. Sometimes referred to as "missing links."

Transposon: A mobile genetic element (also called Jumping genes).

Trihybrid cross: A cross between individuals that differ with respect to three traits.

Trimester: One third of the time of a pregnancy. In humans, approximately three months.

Triplet: A group of three bases that function as a codon.

Triploid: Three sets of chromosomes ($3n$).

Trisomy: Having one extra chromosome ($2n+1$), such as trisomy 21 or Down syndrome.

Ungulate: The hoofed mammals.

Uracil: Pyrimidine base of RNA that binds with adenine.

Urea: Organic chemical compound that is the waste produced when the body metabolizes protein.

Vector: An organism that causes disease.

Vertebrate: An organism with a backbone.

Vertebrates: Animals with a vertebral column. Form the taxonomic unit called Vertebrata.

Vestigial organ (structure): Structure or organ that has lost its original function but resembles a structure in a presumed common ancestor. An evolutionary relic.

Virus: A small infectious particle that consists of nucleic acid and a protein coat.

Wild type: In genetics, the phenotype or genotype that is most common in the natural population.

Y chromosome: Chromosome with the genes that determine maleness.

Y-chromosome Adam: Male population that is the patrilineal most recent common ancestor to existing human beings.

Yolk sac: The extra-embryonic membrane that is responsible for nutrition and circulation in vertebrates. Is largely vestigial in mammals, but does provide some early blood before internal circulation begins.

Zeitgeist: Intellectual and moral characteristics of an age. The spirit of the time or world view.

Zygote: A fertilized egg; capable of growth and development to form a complete organism; a diploid ($2n$) cell resulting from the fusion of male and female gametes.

INDEX

Note: Page numbers in italics refer to figures; page numbers followed by *t* refer to tables.

A

A blood type, 113, 114
AB blood type, 114
Aber, Werner, 89
Acanthostega, 155, *157*
Achillea lanulosa, 122–123, *123*
achondroplasia, 56–57
acidophiles, 180
adaptations
  as evolutionary outcome, 21, 22
  for land life, 152, 154–155, *154, 155, 157*
  pre-adaptations, 155, *157*
adaptive radiation
  absence of competition and, 145
  of amphibians, 152–156, *153, 154, 155, 156*
  competitive exclusion and, 148
  defined, 142
  of humans, 217, 231, *231,* 232–233
  of mammals, 163–164
  of marsupials, 205, *206*
  mass extinction and, 163–164, 208
  of primates, 214–216, *216*
  of reptiles, 159–163, *160, 161, 162, 163*
adenine, in DNA and RNA, 47, *47, 48*
admixture of genes, 119–121, 120*t*
Aegilops species, crossed with wheat, 135, *137,* 139, 141

aerobic bacteria, 172, 174, 175–176, *175*
Africa
  admixture of European genes in, 119–121, 120*t*
  impact of extinctions, *211,* 212
  migration from, 231, *231,* 232–233
  most recent common ancestors from, 194–195, *195,* 196
  sickle-cell anemia in, 98, 100–101, *101*
African-Americans
  admixture of European genes in, 119–121, 120*t*
  sickle-cell trait in, 102
age of parents, congenital defects and, 56
agriculture, sickle-cell trait and, 100–101
Albemarle Island, 146, 147
albinism
  in American Indians, 113–114
  carrier frequency, 65
  segregation principle and, 24–27, *25, 26,* 28*t*
albumin, 185
alkaptonuria, 70
allantois, 157, 158, *159*
alleles
  defined, 7
  dominant, 7–8
  lethal, 69, *70,* 71
  meiosis and, 35, *35*
  population-specific, 120–121, 120*t*
  recessive, 7–8
  segregation principle and, 24
  selectively neutral, 109
Allison, Anthony, 100

allopatric populations, 128, 131
allopolyploidy, 135
Altman, Stuart, 171
altruism
  group selection and, 237–238
  in humans, 240, 242
  kin selection and, 238–242, *239, 241*
  reciprocal, 242–243
Alvin (submersible), 179
Ambystoma species, 138
American Indians, 113–114, 195
Amerinds, 195
amino acids
  artificial synthesis of, 168
  coding of, 54, *54,* 55*t*
  molecular evolution and, 182, 183–185
  overview, 50–51
Amish people
  Ellis-van Creveld syndrome in, 8–9, *9,* 116–117
  family names frequency, 116*t*
  genetic drift and, 116–117
amnion, 157, *159*
amniotes, 158, *159*
amniotic cavities, 157
amniotic eggs, 157–159, *159*
amniotic fluid, 157, *159*
amphibians
  emergence of, 152–156, *153, 154, 155, 156*
  polyploidy in, 138
  vulnerability of, 3
  See also specific types
anaerobic bacteria, 172
anagenesis, 133

analogous appendages, 165
anamniotes, 158, 159
anapsids, 159, *160*
anemia, from Rh disease, 107–108
   See also sickle-cell anemia
angiogenesis, 85
angiosperms, 200–201, 203
animals
   cytochrome *c* composition, 184*t*
   fossil record of, 201–203, *203*
   kingdom of, 176, 177
   polyploidy in, 136*t*, 138
   See also specific types
Anopheles gambiae, 100
Anthropoidea, 215
antibiotic resistance, 85–89, *86, 87, 88,*
   91–93
apes. *See* great apes
Apollo 11, 246
appendages, similarities among species,
   164–165, *164*
archaea domain, 178, 179–180, *179*
Archaeopteryx, 162, *162*
archosaurs, 162
Ardipithecus species, 219, 221
Argentine pampas specimens, 14–15, *14*
Armstrong, Neil, 246
artificial selection, 16–17, *16, 17*
Ashkenazi Jews, 102–103
Asphidoscelis species, 138
atmosphere, of primordial Earth,
   167–168, 172
australopithecines, 221–224, *221, 223, 224*
autopolyploidy, 135
autosomes, 31
autotrophs, 172
azoospermia, 104

B

B blood type
   in American Indians, 113
   in Dunker sect, 114
   European distribution of, 118–119, *118*
babies. *See* infants
background extinction, 208, 209
bacteria
   antibiotic resistance of, 85–89, *86, 87,*
   88, 91–93
   domain of, 178, *179*
   emergence of, 172–173
   fossils of, 198
   organelle evolution and, 174–176, *175*

balanced polymorphism, 94–109
   cystic fibrosis and, 103–104, *105*
   heterozygote superiority and, 100–101
   overview, 97, 98–100
   relaxed selection and, 101–102, 102*t*
   Rh disease and, 104–109, *106, 107*
   selection against heterozygotes and,
     104–108
   selectively neutral alleles and, 109
   sickle-cell anemia and, 94–98, *95, 96,*
     97, 98, 102*t*
   Tay-Sachs disease and, 102–103
   types of, 99*t*
base pairs, 47, *48*
base substitution mutations, 50, *51*
Basque people, 119
bats, 206
Beagle, H.M.S., voyage of, 12–15, *13*
beak adaptations, 143, 144–145
The Beak of the Finch: A Story of
   Evolution in Our Time (Weiner), 144
bees, 240, *241*
begonias, 16
behavior. *See* social behavior
Big Bang, 166
bilirubin, 107
binomial nomenclature, 128–129
biological diversity. *See* diversity
biological evolution. *See* evolution
biological species concept, 126
birds
   emergence of, *153,* 162
   fossil record of, 203
   on Galápagos Islands, *143*
   reproductive strategies in, 244
   warning calls of, 238–239, 242
   See also specific types
Biston betularia, 82–84, *83*
Bitter Springs formation, 198
Black, Davidson, 226
black oak, 129, *130*
Blackburnian warblers, 149
blood transfusions, for Rh disease, 107
blood types, 113, 114, 118–119, *118*
blue-green algae, 172, 175–176, *175*
body fluids, similarity to seawater, 157
bonobos, 215
Boreostracon, 207, *208*
brachiopods, 202, 203
Brassica oleracea, 137–138
bread wheat, 135, *137,* 139, 141
breeding experiments, importance of, 5
Brenner, Sidney, 189
Broom, Robert, 221

Brown, Peter, 228
Bufo species, 130
bullfrogs, multilegged, *2*
   environmental hypothesis, 5, *6, 7*
   genetic drift and, 110
   genetic hypothesis, 7–8, *8,* 9, 11
   history of, 1
   hybrid inviability in, 130
   Mendelian inheritance of, 61–62, *62*
   Punnett square analysis of, 27–28, *29*
   selection and, 67–68, 68*t*
Burgess Shale, 198

C

Cajun people, 103
Calamites, 202
Camarhynchus species, *145,* 150, *151*
Cambrian period
   animal life, 201, *203, 204*
   plant life, 200, *201*
   rise of diversity in, 198, 209
camouflage, 82–84, *83*
cancer cell heterogeneity, 84–85, *84*
Carboniferous period, 153, 200,
   201, 202
carpenter bees, 123, *124*
carriers
   for albinism, 27, 65
   bullfrog case hypothesis, *8*
   defined, 8
   See also heterozygotes
carrying capacity, 150
catastrophism theory, 15–16
Cech, Thomas, 171
cells
   eukaryotic, 173–176, 174*t*
   origin of, 169–172, *170*
   prokaryotic, 172–173, 174*t*, *175,* 176
   selection at level of, 84–85
Cenozoic era, 153, 200, 203
centromeres, 235–236, *235*
Certhidea, 145, *151*
chance genetic fluctuations. *See* genetic
   drift
character gradients. *See* clinal variation
checkerboard analysis. *See* Punnett
   square analysis
chemicals, as mutagen, 4
chemoautotrophs, 179
chicory plant, 130
children. *See* infants; parent-child
   behavior

chimpanzees
  vs. humans, 233–236, *235*
  molecular evolution and, 185, *186,* 215
China, Rh disease in, 119
chloride ion transport, 104
chloroplasts, 174–176, *175*
cholera resistance, 104
chorions, 157, 158, *159*
chromatids, *30,* 32–33, *34*
chromosomal aberrations, 58*t,* 58–60,
  59, 60
chromosomes
  as basis for heredity, 28–31, *30,*
    35–36
  chimpanzees *vs.* humans, 235–236, *235*
  doubling of, 137, 139–140, *140,* 141
  homologous, 32, *34,* 35, *35*
  independent assortment principle
    and, *41*
  odd number of, 135
  second, in fruit flies, 75
cladogenesis, 134
classification
  binomial nomenclature for, 128–129
  of Darwin's finches, 150
  of organisms, 176–179
  of primates, 215–216, *216*
  of reptiles, 159, *160*
Clausen, Jens, 122
clinal variation, 122, *123,* 132
clonal tumor cells, 84
Cocos finches, *151*
codons, 54, 55*t*
coexistence, 148–150
colchicine, 141
competition
  adaptive radiation and, 145
  natural selection and, 17
  niche diversification and, 149–150
competitive exclusion, 148
complete selection, 67–70, 68*t,* 69*t,*
  69, 70
concealed variability, 75
congenital defects, 56–57
conjugation, 88
consanguinity
  Amish people and, 8–9, 116
  genetic drift and, 117–118
convergent evolution
  defined, 165
  of eyes, 193
  in fast-swimming predators, *165*
  in marsupials, 205, *206,* 207
  overview, 203, 205

Cook, L. M., 83
Cope's gray tree frogs, 138
Cordaites, 200, 201
cormorants, 143
Correns, Carl, 24
cotton species, 130
cotylosaurs, 153
creationism, 12
Creole people, 121
Crepis species, 130
Cretaceous period, 209, *209,* 210*t*
Crick, Francis H. C., 46, 47, 171, 180
Cro-Magnons, 231–232, *232, 233*
Crocodilia, 163
crossing over, 41–42, *41*
crossopterygians, 152–155, *153, 154,*
  155, 202
cryptic female choice, 79
cultural evolution, 245–247, *246*
cultural selection, 113–114
Cuna people, albinism in, 113
cursorial mammals, 164
Cuvier, Georges, 16
cyclostomes, 188
cystic fibrosis, 70, 103–104, *105*
cystic fibrosis transmembrane conductance regulator (CFTR) gene, 104
cytochrome *c,* 183–184, 184*t, 185*
cytosine, 47, *47, 48*

D

Daeschler, Ted, 153
Daphnia, 57
Dart, Raymond A., 221, 222
Darwin, Charles, *13*
  development of evolution theory,
    16–21
  on finches, 143
  on giant tortoises, 146
  on ginkgo tree, 213
  on inherited variability, 43
  on Lamarck, 7
  natural selection principle, 12, 17–21, *19*
  on origin of life, 166
  Punch's Almanack cartoon, *20*
  sexual selection concept, 78–79
  on species concept, 125–126
  tree diagram, 176–177, *177*
  voyage on H.M.S. *Beagle,* 12–15, *13*
Darwinian fitness. *See* reproductive
  effectiveness
Darwinius maisillae, 215, *217*

Darwin's finches
  adaptive radiation and, 143,
    144–146, *146*
  classification of, 150
  coexistence of, 148–150
  diversity of, *145*
  evolutionary tree of, *151*
  niche diversification of, 149, *150*
  overview, 15
daughter strands, *49*
Dawkins, Richard, 247
de Vries, Hugo, 24
deductions, scientific, 9
deep-sea vent hypothesis, 179–180
degeneracy of genetic code, 54
deleterious genes, 7–8
deletion mutations, 50, *51*
deletion of chromosomes, 58
deoxyribose, in DNA, *48*
*The Descent of Man and Selection in
  Relation to Sex* (Darwin), 78
descent with modification, 16, 165, 166
Devonian period
  as Age of Fishes, 202
  land adaptations during, *155*
  mass extinction during, 209, *209,* 210*t*
  overview, 152–155, *153*
diapsids, 159, *160,* 161
differential reproduction, 21
dihybrid inheritance, 36, *37, 38*
Dimetrodon, 163, *163*
dinosaurs, 161–162, *161,* 163–164
diploid cells
  defined, 32
  mitosis and, *31, 33*
  polyploidy and, 140
directed panspermia, 180
directional selection, 76, *76*
disorders
  balanced polymorphism and, 97, 99*t*
  dominant, 73–75, 74*t, 74,* 95
  mutation rates and, 55–57
  recessive, 7–8, *8,* 65, 110, 113, 117
  See also specific disorders
disruptive selection, *76,* 77
distal hyperextensibility, 114, *115*
diversity
  Cambrian explosion of, 198, 209
  of hominids, *224*
  mass extinction and, 208–209, *209*
DNA (deoxyribonucleic acid)
  chemical structure of, 46–49, *47, 48*
  of chimpanzees *vs.* humans, 234–235
  junk regions of, 189

DNA (deoxyribonucleic acid)—*Cont.*
  recombinant, 89–91, 90*t*, *90, 91*
  repetitive, 89
  replication of, *51,* 49–50, *49, 56, 57*
  RNA world hypothesis and, 171
  transcription of, 53–55, *54,* 55*t, 56, 57*
Dobzhansky, Theodosius, xi, 75, 131
doctrine of irreversibility, 162
Dollo, Louis, 162
Dollo's law, 162
domain system of classification, 178, *179*
domestication of species, 16–17, *16, 17*
dominant alleles
  carriers and, 27
  defined, 7–8
  selection against, 69, *70*
dominant disorders
  selection against, 73–75, 74*t, 74*
  sickle-cell anemia as, 95
double helix DNA structure, 46–47, *47*
doubling of chromosomes, 137,
    139–140, *140,* 141
Drosophila species. *See* fruit flies
drug-resistant organisms, 85–89, *86,
    87, 88*
Dryopithecus, 217–218, *219*
Dubois, Eugene, 226
Dunker sect, 114–116, *115*
dyads, 33

E

ear lobe differences, 114, *115*
Earth, formation of, 166–168, *173*
eastern gray tree frogs, 138
ecological destruction. *See* habitat
    destruction
ecological isolation, 129, *130*
ecological niches, 149
*Eco*RI and *Eco*RV, 89, 90*t*
eggs, 155, 158, *159*
Ehrlich, Paul, 247
Einkorn wheat, 135, *137,* 138–139, 141
Eldredge, Niles, 133, 134
electromagnetic spectrum, *45*
electrophoretic patterns of
    hemoglobins, *97*
Eleutherodactylus planirostris,
    110–112, *111*
Ellis-van Creveld syndrome, 8–9, *9,*
    116–117
embryonic development
  of amniote vertebrates, 157–159, *159*

environmental factors in, 3–7
  homeotic genes and, *191,* 192
Emmer wheat, 139, 141
endangered species
  apes, 218
  giant tortoises, 146, 148
  human overpopulation and, 150
  marsupials, 206
endogamous populations, 117
endoplasmic reticulum, 176, *176*
environmental modifications
  beneficial *vs.* harmful, 57–58
  bullfrog case study and, 3–4, *6*
  defined, 4–5
  Lamarckism and, 5–7
enzymes
  in DNA replication, 50, 171
  restriction endonucleases, 89–90,
      90*t, 90*
  ribozymes, 171
Eoraptor, 162
epochs, geologic, 198, 199*t*
equilibrium, genetic. *See* genetic
    equilibrium
Equus, 14
eras, geologic, 198, 199*t*
  See also specific eras
erythroblastosis fetalis, 104–109,
    106, 107
Escherichia coli, 88, 89, 90*t*
An Essay on the Principle of Population
    (Malthus), 17
estrus, 244–245
eukarya domain, 178, *179*
eukaryotic cells, 173–176, 174*t*
European genotypes, 118–121,
    118, 120*t*
eurypterids, 202, 203, 204
Eusthenopteron, 157
evolution
  clock model of, *200*
  convergent, 165, *165,* 193, 203, 205,
      206, 207
  cultural, 245–247, *246*
  defined, 21–22
  descent with modification, 16,
      165, 166
  doctrine of irreversibility, 162
  Hardy-Weinberg equilibrium
      and, 66
  macroevolution, 132–133
  microevolution, 132
  See also molecular evolution
expert agreement on species, 126

extinction(s)
  adaptive radiation and, 163–164,
      208
  background extinctions, 208, 209
  of amphibians, 3
  Holocene, 210*t,* 211–213, *211*
  human overpopulation and, 150
  of marsupials, 205
  mass, 163–164, 208–211, *209,* 210*t*
  of reptiles, 163–164
  of South American hoofed
      mammals, 208
  See also endangered species
extraembryonic membranes, 157, 158
extraterrestrial origin of life, 180–181
extremophiles, 178, 179–180, *179*
eye malformations, 192–193

F

family names, among Amish, 116*t*
female choice, 79, 244
female *vs.* male reproductive strategies,
    243–245
fermentation, 172
fertility, 78, *78,* 140
fertilization
  cryptic female choice and, 79
  as male reproductive strategy, 243–244
  overview, 5
fibrinopeptides, 184
Fig Tree sediments, 198
finches. *See* Darwin's finches
fish
  fossil record of, 202
  hemoglobin evolution in, 188
  lobe-finned, 152–155, *153, 154, 155*
  polyploidy in, 138
fitness. *See* reproductive effectiveness
Fitzroy, Robert, 13, 143
five-kingdom classification, 177, *178*
flamingos, 143
Fleming, Alexander, 92–93
Flores humans, 228, *229*
flowering plants, 200–201, 203
Ford, E. B., 82
fossils
  Argentine pampas specimens, 14–15, *14*
  of hominids, 221, 222–224
  of hominoids, 219, *220*
  of *Homo* species, 225, 226, *227*
  over geologic ages, 197, 198
  of primates, 215, 217, *217, 218*

punctuated equilibrium and, 134
as record of animals, 201–203, *203*
as record of plants, 200–201, *200*
See also missing links
fossorial mammals, 164
founder effect, 113–114
fowl, domesticated, 16, *16*
Fox, Sidney M., 180
frameshift mutations, *51,* 50
Franklin, Rosalind, 46
Freda, Vincent, 108
French Canadians, 103
frequency of recessive alleles, 68–70,
    69*t*, *69*
frogs
    gray tree, 138
    green, 130
    greenhouse, 110–112, *111*
    homeodomain amino acid
        sequences, 192*t*
    leopard, 124–125, *125*
    mink, 1, *3*, 11
    retinoic acid effect on, 193–194, *194*
    See also bullfrogs, multilegged
fruit flies
    concealed variability and, 75
    homeodomain amino acid
        sequences, 192*t*
    homeotic genes in, 189–192, *190*
    homeotic mutations in, *191*
    independent assortment principle, 36,
        37, 38, 40
    linkage principle experiments, 41
    morphological differentiation in, 130
    Pax genes and, 193
    reproductive isolating mechanisms
        in, 132
fungi, kingdom of, 177
fused chromosomes, 236

G

G6PD (glucose-6-phosphatedehydroge-
    nase) deficiency, 101
Galápagos Islands
    giant tortoises on, 15, *15*
    map of, *143*
    overview, 142–144
    See also Darwin's finches
Gallus gallus, 16
gametes
    crossing over and, 41–42, *41*
    monohybrid inheritance and, 35–36, *35*

single alleles in, 25
trihybrid inheritance and, *39*
gametic isolation, 130
Gamow, George, 51
gatherer-hunters, 225, 228–229
Gause, G. F., 148
Gause's principle, 148
Gehring, Walter J., 193
gemmules, 17
gene flow, 118–121, *118,* 120*t*
gene frequencies, 64, *64*
genes
    as carried by chromosomes, 35
    chemical nature of, 45–49, *47, 48*
    defined, 53
    deleterious, 7–8
    dominant, 7–8
    duplication of, 185–186, *187*
    homeotic, 189–192, *190, 191*
    Hox, 191, 192, 193–194
    jumping, 89
    lethal, 69, *70,* 71
    linkage of, 40–41, *41*
    overview, 5
    Pax, 192–193
    pseudogenes, 188–189, *189*
    segregation principle and, 24
    semi-lethal, 71, 72, 73*t, 73*
    subvital, 71–72, 73*t, 73*
    See also alleles; DNA; mutations
genetic code
    beginnings of, 170
    flow of information, 53–55, *54, 56, 57*
    overview, 50–51, 52, 55*t*
    universality of, 190
genetic disorders. *See* disorders
genetic drift
    Amish people and, 116–117, 116*t*
    defined, 110
    Dunker sect case study, 114–116, *115*
    founder effect and, 113–114
    greenhouse frogs case study,
        110–111, *111*
    molecular evolution and, 183
    theory of, 112–113, 112*t*
genetic equilibrium
    balanced polymorphism and,
        98–100, *99*
    of dominant disorders, 73–75, 74*t, 74*
    Hardy-Weinberg equilibrium,
        64–66, *65*
    overview, 61–64, *62,* 63*t, 64*
    of sickle-cell anemia, 98–100, *99*
genetic load, 75

genetic recombination, 38, 39*t,*
    40, *40*
genetic screening programs, 103, 104
genotypes
    Hardy-Weinberg equilibrium and,
        64–66, *65*
    independent assortment principle
        and, 39*t*
    vs. phenotypes, 27
genus, defined, 128
Geochelone nigra. See giant tortoises
geographic speciation, 126–128, *127*
geologic ages, 198–200, 199*t, 200*
Geospiza species. *See* ground finches
germ cells, 5, 45, *46*
germ-line *vs.* somatic mutations,
    45, *46*
giant tortoises, 15, *15,* 142,
    146–148, *147*
Gigantopithecus, 218
Gilbert, Walter, 171
ginkgo tree, 213
giraffes, Lamarck on, 7
Glass, Bentley, 114
Glass, William, 118
glutamic acid coding, 51, *52,* 53, 55, *57,*
    96–97, *98*
glyptodonts, 208, *208*
goat grass, crossed with wheat, 135, *137,*
    139, 141
Goin, Coleman, 110, 111, 112
Goldschmidt, Richard, 190
gorillas, 185, *186,* 215
Gorman, John, 108
Gossypium species, 130
Gould, Stephen Jay, 133, 134
Grant, Peter, 144
Grant, Rosemary, 144
grass, crossed with wheat, 135, *137,*
    139, 141
Gray, Asa, 79
gray tree frogs, 138
great apes, 185, *186,* 217–218, *217*
green frogs, 130
greenhouse frogs, 110–112, *111*
ground finches
    competitive exclusion and, 148
    diet of, 144–145
    diversity of, *145*
    evolutionary tree of, *151*
    niche diversification of, 149, *150*
ground sloths, 208, *208*
group selection, 237–238
guanine, in DNA and RNA, 47, *47, 48*

gulls, 243
gymnosperms, 200, 201, 202
gynogenesis, 138

H

habitat destruction, 150, 212–213, 218
habitat isolation, 129, *130*
Haemophilus influenzae, 89, 90*t*
Haldane, J. B. S., 237
halophiles, 180
halteres, 191
Hamilton, W. D., 238
handicap principle, 79–80
hands, six-digited, *9*
haploid cells, *31,* 32, *33, 34,* 140
haplotypes, 121
Hardy, Godfrey H., 61, 64
Hardy-Weinberg equilibrium, 61,
    64–66, *65*
Helix typicus, 126–128, *127,* 129
hemoglobin
    chimpanzees *vs.* humans, 234
    DNA coding for, 51, *52,* 53
    electrophoretic patterns of, *97*
    molecular evolution of, 186–189, *188*
    sickle-cell, 54–55, *57,* 95–97, *97*
    variants in, 56
Henslow, John, 12
heredity
    chromosomal basis for, 28–31, *30,*
        35–36, *35*
    dihybrid, 36, *37, 38*
    Lamarckism and, 5–7, 17, 246
    Mendelian, 27–28, *29,* 61–62, *62*
    mitochondrial DNA and, 194–195
    monohybrid, 35–36, *35*
    trihybrid, 36, 38, *39*
heritable variation
    concealed, 75
    genetic recombination significance in,
        38, 39*t,* 40, *40*
    independent assortment principle and,
        36–38, *37, 38, 39*
    linkage and crossing over effects,
        40–42, *41*
    mitosis and meiosis and, 31–35, *31,*
        32, 33, 34
    overview, 7–9
    Punnett square analysis, 27–28, *29*
    segregation principle and, 24–27, *25,*
        26, 28*t*
    with three or more gene pairs, 36, 38, *39*

Herrick, James B., 94
heterogeneity of cancer cells, 84–85, *84*
heterotrophs, 171
heterozygotes
    defined, 25
    frequency of, 65
    recessive alleles in, 70, 71*t*
    segregation principle and, 27
    selection against, 104–109, *106, 107*
    selection for, 97–101, *99*
Hiesey, William, 122
*Hind*II and *Hind*III, 89
histone IV, 184–185
history of life, 197–200, 199*t, 200*
hitchhiker's thumb, 114, *115*
Holocene extinction, 210*t,* 211–213, *211*
homeobox sequence, 192
homeodomains, 192, 192*t*
homeotic complex (HOM-C), 191–192
homeotic genes, 189–192, *190, 191*
Hominidae, forerunners of, 218–221,
    219, 220
hominids
    emergence of, 221–224, *221, 223, 224*
    molecular evolution and, 185, *186*
Hominoidae, 215–216, *216*
hominoids, 215–216, *216,* 219–221, *220*
Homo erectus, 226–230, *227,*
    232–233, *234*
Homo ergaster, 226, 228, 233
Homo floresiensis, 228, *229*
Homo habilis, 225–226
Homo neanderthalensis. See
    Neanderthals
Homo sapiens
    coexistence with other species,
        226, *227*
    emergence of, 231–233, *232, 233, 234*
    as single variable species, 132
    See also humans
homologous chromosomes, 32, *34,*
    35, 35
homologous limbs, 164–165, *164*
homology, 165
homosexual behavior, 240, 242
homozygotes
    defined, 25
    vs. heterozygous carriers, 70, 71*t*
    segregation principle and, 27
hoofed animals. *See* ungulates
Hooker, Joseph, 166
Hopi people, 113–114
Hoppe, David, 3
horses, 14, *14*

Hox genes, *191,* 192, 193–194
human growth hormone synthesis, 91
human species emergence, 214–236
    australopithecine stage, 221–224,
        221, 223
    coexisting species, 217
    great apes forerunners, 217–218, *217*
    hominid forerunners, 218–221, *219*
    Homo erectus and, 226–230, *227,*
        232–233, *234*
    Homo floresiensis and, 228, *229*
    Homo habilis and, 225–226
    molecular evolution and, 185, *186,*
        233–236, *235*
    most recent common ancestors,
        194–196, *195*
    Neanderthals and, 230–231, *230*
    primate adaptive radiation and,
        214–216, *216*
    Ramapithecus and, 185, *186*
    timelines of, *220, 224*
    See also Homo sapiens
humans
    adaptive radiation of, 217
    vs. chimpanzees, 233–236, *235*
    chromosomal aberrations in, 58*t,*
        58–60, *59, 60*
    chromosome complement of, 30–31, *30*
    cytochrome *c* composition, 184*t*
    embryonic development, 158–159
    extinction and, 150, 211–213, *211, 212*
    genetic load in, 75
    kin selection in, 240, 242
    mitosis and meiosis in, 32, *33*
    mutation rates in, 55–57
    ovulation in, 244–245
    pair-bonding in, 244–245
    polyploidy in, 138
    secondary sex characteristics in, *80,*
        81, 81*t*
    as single variable species, 132
    hunter-gatherers, 225, 228–229
    hybrid breakdown, 130
    hybrid inviability, 130, 131
    hybrid sterility, 130
    hybridization, polyploidy and, 140–141,
        140
    hydrogen, in early atmosphere, 168, 172
    hydrothermal vents, 179–180
    Hyla species, 138
    Hylobatidae, 215–216, *216*
    hypotheses, 9–11, *10*
    hypothetico-deductive reasoning,
        9–11, *10*

I

Ichthyostega, 155, *156, 157*
icthyosaurs, 162
Ida (fossil), 215, *217*
iguanas, 142, *144*
immunosuppressants, for
    Rh disease, 108
inclusive fitness, 239, 240
independent assortment principle, 36–38,
    37, 38, 39
independent evolution. *See* convergent
    evolution
individual *vs.* group selection, 238
induced mutations, 43, 44*t*
industrial melanism, 82–84, *83*
infants
    bonding with mothers, 229
    chromosomal aberrations in, 77–78, *78*
    optimal birthweight in, 77, *77*
    Rh disease in, 104–109, *106, 107,* 119
    thalidomide-related malformations, 4, *4*
inheritance. *See* heredity
insertion mutations, *51,* 50
insertion sequences, 88
instantaneous speciation, 135–141, 136*t*,
    137, 140
insulin synthesis, 90
interbreeding, 122, 128
intersexual selection, 79
interspecific selection, 79
intervals between pregnancies, 60
intrasexual selection, 79
intraspecific selection, 79
invariant amino acids, 183
inversion of chromosomes, 58
invertebrates, fossil record of, 201,
    203, *204*
See also specific types
isolated populations
    consanguinity and, 117–118
    giant tortoises as, 146
    religious isolates as, 8–9,
        114–116, *115*
    reproductive isolating mechanisms
        and, 129–132, 129*t, 130*

J

Jamaica, admixture of genes in, 121
Java man, 226, 228
Jewish people, 102–103
Johanson, Donald C., 222, 224

jumping genes, 89
junk DNA, 189
Jupiter, search for water on, 181

K

karyotypes, 30–31, *30*
Keck, David, 122
Kelsey, Frances O., 4
Kennedy, John F., 246
Kettlewell, H. B. D., 82, 83
Kihara, Hitoshi, 141
Kimura, Motto, 183
kin selection, 238–242,
    239, 241
King, Samuel, 117
kingdoms, 176, 177–179, *178*
Klebsiella pneumoniae, 88
Klinefelter syndrome, 78
Koopman, Karl, 131–132

L

labial genes, 190
labyrinthodonts, 153, 155–156, *159*
Lack, David, 144, 149
Laetoli footprints, 223–224
laissez-faire doctrine, 19–20
Lamarck, Jean-Baptiste de, 6, 7
Lamarckism, 5–7, 17, 246
land iguanas, 142, *144*
land life, transition to, 200
land snails, 126–128, *127,* 129
land vertebrates, 152–156, *153, 154,
    155, 156*
See also specific types
Landauer, Walter, 4
language development, 230
laws, in natural sciences, 11
Leakey, Louis, 222, 225
Leakey, Mary, 222, 223, 225
Leakey, Richard, 225
Lederberg, Joshua, 85
left-handedness, 114, *115*
leopard frogs, 124–125, *125,
    126,* 131
Lepidodendron, 200, 201, 202
lethal alleles, 69, *70,* 71
lethal selection, 67–70, 68*t,* 69*t,
    69, 70*
Levine, Philip, 105
Lewis, Edward B., 189, 190

life
    defined, 167
    history of, 197–200, 199*t, 200*
    See also origin of life
limbs, similarities among species,
    164–165, *164*
limpets, 77
linkage of genes, 40–41, *41*
Linnaeus, Carolus, 128, *128,* 176
living fossils, 163, 213
lizards, 138, 142, *144*
lobe-finned fish, 152–155, *153, 154,
    155,* 202
locus of alleles, 35
lollipop transposon configuration, 88, *88*
Lucy (fossil), 223, *223*
Lyell, Charles, 12–13, 18

M

M blood type, 114
Macrauchenia, 14, *14,* 208
macroevolution, 132
macromutations, 190
Maden, Malcolm, 193
Majerus, E. N., 84
malaria, 100–101, *101*
male–male competition, 79, 244
male *vs.* female reproductive strategies,
    243–245
Malthus, Thomas Robert, 17, 18
mammals
    adaptive radiation of, 163–164
    emergence of, *153*
    fossil record of, 203
    parental reproductive strategies,
        243–245
    placental *vs.* marsupial, 205, *206*
    plains-dwelling, 207
    See also specific types
Margulis, Lynn, 174
marine iguanas, 142, *144*
marine mammals, 164
Mars Exploration Rover (MER), 181
Mars Global Surveyor, 181
Mars, search for water on, 181
marsupials, 164, 205, *206,* 207
mass extinction, 163–164, 208, 209,
    209, 210*t*
maternal mitochondria, 194–195
mating
    Mendelian inheritance and, 61–62, *62*
    random, 62–64, 63*t*

mating calls, 125, 126
Mayr, Ernst, 126
McClintock, Barbara, 89
McCormick, Robert, 13
McFadden, E. S., 141
McKusick, Victor A., 8, 116
mechanical isolation, 130
medical advances, effect of, 71, *72,*
    104, *105*
Megatherium, 207, *207*
meiosis
    crossing over and, 41–42, *41*
    with elimination of extra
        chromosomes, 78
    in generalized life cycle, *32*
    in human sexual cycle, *33*
    overview, 32–35, *34, 35*
    polyploidy and, 140, *140*
    relationship with mitosis, *31*
melanin, 24, 25
melanism, industrial, 82–84, *83*
membranes
    cellular, 171
    extraembryonic, 157, 158
memes, 245
memetics, 245
Mendel, Gregor Johann, *24*
    independent assortment principle,
        36–38, *37, 38, 39,* 40–41
    pea plant experiments, 23–24, *24*
    segregation principle, 24–27, *25, 26,* 28*t*
Mendelian inheritance
    bullfrog case study, 61–62, *62*
    phenotypic ratios, 27–28, *29*
Mendelian laws, 7, 11
menopause, 242
MER (Mars Exploration Rover), 181
Mesozoic era, 153, 200, 201, 202, 203
messenger RNA, 53–55
metastasis, *84,* 85
Methanococcus jannaschii, 178–179
methanogens, 180
methicillin-resistant *Staphylococcus
    aureus* (MRSA), 92
microevolution, 132
mid-digital hair, 114, *115*
migration. *See* adaptive radiation
Miller, Stanley, 168, 169
mink frogs, 1, *3,* 11
miscarriage. *See* pregnancy loss
missense mutations, *51,* 50
missing links
    Archaeopteryx as, 162, *162*
    for primates, 215, *217*

punctuated equilibrium and, 134
    therapsids as, 163
    Tiktaalik as, 155
Mississipian period, 153
mitochondria
    DNA in, 194–195
    origin of, 174–176, *175*
mitochondrial Eve, 194–195
mitosis, 31–32, *31, 32, 33*
mole salamanders, 138
molecular clock, 184–185, *185*
molecular evolution
    chimpanzees *vs.* humans,
        233–236, *235*
    of cytochrome *c,* 183–184, 184*t, 185*
    gene duplication and, 185–186, *187*
    of hemoglobin, 186–189, *188, 189*
    homeotic genes and, 189–192,
        190, 191
    molecular clock and, 184–185, *185*
    neutral theory of, 182–183
    ocular malformations and, 192–193
molecular level of natural selection,
    170–171
molecular signatures, 236
Monera, 177
monkeys, 188, 215
monogamy. *See* pair-bonding
monohybrid inheritance, 35–36, *35*
monoploid cells. *See* haploid cells
monotremes, 164
monotypic species, 129
moon exploration, as cultural
    evolution, 246
Moore, John A., 124, 131
Morgan, T. H., 41
morphological differentiation, 130,
    133–134
Morwood, Michael, 228
mosquitoes, 100
most recent common ancestors,
    194–196, *195*
mother-child bonding, 229, 245
moths, melanic, 82–84, *83*
mouse species, *191,* 192–193, 192*t*
mRNA (messenger RNA), 53–55
MRSA (methicillin-resistant *Staphylo-
    coccus aureus*), 92
Muller, Hermann J., 44, 131
multiple antibiotic resistance, 87–89,
    87, 88
multiregional hypothesis, 232–233, *234*
Murchison meteorite, 180
mutagens, 44*t,* 45*t, 45,* 89

mutations
    base substitution, 50
    causes of, 43–45, 44*t, 45*
    defined, 43
    in drug-resistant bacteria, 85–87, *86*
    evolutionary consequences, 57–58
    homeotic, 189–190
    induced, 43, 44*t*
    interplay with selection, 70–71, *72*
    molecular evolution and, 182–183, 186
    point, *51,* 50
    rates of, 55–57
    spontaneous, 43, 44*t,* 50
myoglobin, 187–188, *188*
Myrianthopoulos, Ntinos, 103
myrtle warblers, 149

N

N blood type, 114
Na-Dene people, 195
NASA (National Aeronautics and Space
    Administration), 181
Nathans, Daniel, 89
Native Americans, 113–114, 195
natural selection, 67–78
    concealed variability and, 75
    Darwin and, 12, 17–21, *19*
    against dominant defects, 69, *70,*
        73–75, *74*
    gene duplication and, 185–186, *188*
    group selection and, 237–238
    industrial melanism and, 82–84, *83*
    kin selection and, 238–242, *239, 241*
    at molecular level, 170–171
    mutation interplay with, 70–71, *72*
    partial, 71–73, 73*t, 73*
    pregnancy loss and, 77–78, *78*
    against recessive defects, 67–70, 68*t,*
        69*t, 69, 70*
    relaxed, 101–102, 102*t*
    for resistance, 85–89, *86, 87, 88,*
        91–93
    selectively neutral alleles and, 109
    vs. sexual selection, 79–80
    types of, 76–77, *76, 77*
nautiloids, 202, 203, 204
Neanderthals, 226, *227,* 230–231, *230*
Neel, James V., 95
neutral theory of molecular evolution,
    182–183
neutralists, 183
New Orleans, admixture of genes in, 120*t*

New World monkeys, 215
niche diversification, 149–150
niches, ecological, 149
nitrogenous bases, 47, *47, 48,* 168
nomenclature, 128–129
nonsense mutations, *51,* 50
northern leopard frogs, 124–125,
    125, 126
nucleic acids, 169
    See also DNA; RNA
nucleotides, 88, 89, 169–170
nucleus, origin of, 176, *176*
Nutcracker Man, 222

O

O blood type, 113, 114
oak species, 129, *130*
ocular malformations, 192–193
Ohta, Tomoko, 183
Old World monkeys, 188, 215
Olduvai Gorge, 222, 225
*On the Tendencies of Varieties to Depart
    Indefinitely from the Original Type*
    (Wallace), 18
Oparin, Alexander I., 167
Opportunity (rover), 181
orangutans, 185, *186,* 215
Ordovician period
    animal life, 202, *204*
    mass extinction, 209, *209,* 210*t*
    plant life, 200, 201
organelles, 174–176, *176*
organic compounds, 168, *169*
origin of life
    deep-sea vent hypothesis, 179–180
    development of cells, 169–172, *170*
    extraterrestrial hypothesis, 180–181
    formation of universe,
        166–167
    heterotrophs *vs.* autotrophs,
        171–172
    organelles and, 174–176, *176*
    prokaryotic *vs.* eukaryotic cells,
        172–176, 174*t, 175*
    spontaneous generation and, 166
    timeline of events, *167, 173*
The Origin of Life (Oparin), 167
The Origin of Species (Darwin), 12, 18,
    43, 78, 176, 177
Ornithischia, 159, 163
Orrorin tugenengis, 219, 221
ostrich species, 15

Out-of-Africa theory, 195, *195,* 196,
    232–233, *234*
Owen, Robert, 15
oxygen, 172

P

pair-bonding, 229, 244–245
paleopolyploidy, 137–138
Paleozoic era, 198, 200, 201, 202, 203,
    204
Paley, William, 18
Pan species. *See* chimpanzees
Panderichthys, 155, *156, 157*
pangenesis, 17
panspermia, 180
Paramecium species, 148
Paranthropus species, 221, 222
parent-child behavior
    mother-child bonding, 229, 245
    probability of paternity and, 240
    stepfathering, 245
    stepmothering, 243
parent-offspring conflict, 245
parental investment, 245
parental reproductive strategies, 243–245
parental strands, *49*
parthenogenesis, 138
partial selection, 71–73, 73*t, 73*
Pasteur, Louis, 166
Pasturella pestus, 88
paternal ancestry, 195–196
Pauling, Linus, 95, 184, 188
Pax genes, 192–193
peacocks, 79–80
Peking humans, 226, *227,* 228
pelycosaurs, 163
penicillin, 92–93
Pennsylvanian period, 153
peppered moths, 82–84, *83*
Percival, John, 139, 141
periods, geologic, 198, 199*t*
    See also specific periods
Permian period, 209, *209,* 210*t*
phenotypes
    vs. genotypes, 27
    independent assortment principle and,
        38, 39*t,* 40
    sexual dimorphism and, 81
phenotypic ratios, 28, 36
Philosophie Zoologique (Lamarck), 7
phocomelia, 4
phosphate groups, *48*

phospholipid membranes, 171
photosynthesis, 172
phyletic gradualism, 133–134, *133*
Pinaroloxias, 151
pithecanthropines, 226
placenta, 164
placental *vs.* marsupial mammals,
    205, *206*
plains-dwelling mammals, 14, *14,*
    205, 207
plains leopard frogs, 125, *126*
plants
    fossil record of, 200–201, *201*
    kingdom of, 176, 177
    polyploidy in, 136*t,* 137–138
    See also specific types
plasma membranes, 171
plasmids
    as carriers of DNA, 89–91, *90, 91*
    resistant organisms and, 87–88, *87*
Plasmodium falciparum, 100–101
Pleistocene overkill, *211,* 212
plesiosaurs, 162
point mutations, *51,* 50
Pollack, William, 108
polygyny, 244
polynucleotides, 169
polypeptides
    chimpanzees *vs.* humans, 234
    genetic code and, 50–51
polyploidy
    in animals, 136*t,* 138
    defined, 135
    mechanism of, 135, 138–139, *140*
    in plants, 136*t,* 137–138
    species with, 136*t*
    in wheat, 136*t, 137,* 138–141, *140*
polytypic species, 129
population size
    consanguinity and, 117–118
    extinctions and, 150, 211, *212,* 213
    frequency of recessive alleles and, 110,
        113, 117
    religious isolates and, 8–9,
        114–116, *115*
population-specific alleles,
    120–121, 120*t*
populations
    allopatric, 128, 131
    endogamous, 117
    evolution as property of, 21–22
    mating systems and, 62–64, 63*t*
    sympatric, 129, 132
    See also genetic equilibrium

postmolecular evidence, 186
postzygotic isolating mechanisms, 129*t*
pre-adaptations, 155, 157
Precambrian period, 198
predictions, scientific method and, *10*
pregnancy loss
    from chromosomal aberrations, 58*t*,
        58–60, *59, 60*
    natural selection and, 77–78, *78*
    polyploidy and, 138
premolecular evidence, 186
prenatal screening, 103, 104
prezygotic isolating mechanisms, 125,
        126, 129*t*
primates
    adaptive radiation of, 214–216, *216*
    fossils of, 215, 217, *217, 218*
    hemoglobin evolution in, 188
    See also specific types
Principles of Geology (Lyell), 12–13
probability
    random mating and, 63
    trihybrid inheritance and, 36, 38
Proconsul, 217, *218*
prokaryotic cells, 172–173, 174*t, 175,* 176
Prosimii, 215
proteins
    artificial synthesis of, 168
    origin of life and, 169–171, *170*
    overview, 50–51
    translated from RNA, 54, *56, 57*
    See also amino acids
Protista, 177
pseudogenes, 188–189, *189*
psilophytes, 200
Pteridosperms, 200, 201, 202
pterosaurs, 162
pulmonary tuberculosis, 103
Punch's Almanack cartoon, *20*
punctuated equilibrium, 133–134, *134*
Punnett, R. C., 28, 61
Punnett square analysis
    bullfrog case study, 27–28, *29*
    dihybrid inheritance, 36, *37*
    Hardy-Weinberg equilibrium and,
        64, *65*
    trihybrid inheritance, 36
purines
    artificial synthesis of, 168
    in DNA and RNA, 47, *47, 48*
pyrimidines
    artificial synthesis of, 168
    in DNA and RNA, 47, *47, 48*
Pyrotherium, 14, *14,* 207

Q

Quercus species, 129, *130*

R

R plasmids, *87,* 88
race
    biological *vs.* cultural meaning, 132
    defined, 123–124
    self-identification with, 119
    speciation and, 126, 127, 128
radiation, as mutagen, 44–45, 45*t, 45*
Ramapithecus, 185
Lithobates berlandieri, 125, *126*
Lithobates blairi, 125, *126*
Lithobates catesbeiana. See bullfrogs,
        multilegged
Lithobates clamitans, 130
Lithobates pipiens, 124–125, *125, 126,* 131
Lithobates septentrionalis, 1, *3,* 11
Rana temporaria, 193–194, *194*
Lithobates utricularia, 125, *126*
random mating, 62–64, 63*t*
random union of eggs and sperm,
        64, *64*
reasoning, hypothetico-deductive,
        9–11, *10*
recessive alleles
    complete selection against, 67–70, 68*t*,
        69*t, 69, 70,* 72, 73*t, 73*
    defined, 7–8
    in heterozygotes, 27
    partial selection against, 71–73, 73*t, 73*
    replenishing of by mutations,
        70–71, *72*
recessive disorders
    bullfrog case hypothesis, 7–8, *8*
    heterozygote frequency and, 65
    population size and, 110, 113, 117
    See also specific disorders
reciprocal altruism, 242–243
recombinant DNA, 89–91, 90*t, 90, 91*
red jungle fowl, 16
red vizcacha rats, 138
relatedness, degree of, 239, 240, *241*
relaxed selection, 101–102, 102*t*
religious isolates, 8–9, 114–116, *115*
repetitive DNA, 89
replica plating, 85, *86*
reproduction
    differential, 21
    by gynogenesis, 138

parental strategies for, 243–245
    by parthenogenesis, 138
reproductive effectiveness
    vs. altruism, 237–238
    bullfrog case study, 9, 21
    defined, 21
    deleterious genes and, 71–72
    of heterozygotes *vs.* homozygotes,
        100, 104, 109
    inclusive fitness and, 239, 240
    sexual selection and, 78–81, *80,* 81*t*
reproductive isolating mechanisms,
        129–132, 129*t, 130*
reptiles
    adaptive radiation of, 159–163, *160,*
        161, 162, 163
    classification of, 159, *160*
    emergence of, *153,* 156–159, *158, 159*
    extinction of, 163–164
    fossil record of, 202
    See also specific types
resistance, natural selection and, 85–89,
        86, 87, 88, 91–93
restriction endonucleases, 89–90, 90*t, 90*
retinoblastoma, 75
retinoic acid, 193–194
reverse transcriptase, 171
reverse transcription, *57,* 171
Rh antigens, 114
Rh disease, 104–109, *106, 107,* 119
Rhea species, 15
RhoGAM treatment, 108
ribose, in RNA, *48,* 53
ribozymes, 171
Rio Grande leopard frogs, 125, *126*
RNA enzymes, 171
RNA (ribonucleic acid)
    chemical structure of, *48*
    codons of, 54, 55*t*
    origin of life and, 168, 169–171, *170*
    transcribed from DNA, 53–55, *54,* 55*t,*
        56, 57
    translation of, 54, *56, 57*
RNA world hypothesis, 171
Roberts, D. F., 118

S

saber-toothed tigers, 207, *208*
Sahelanthropus tchadensis, 219, 221
salamanders, 138
salmon, 138
Salmonella typhosa, 88

Sarich, Vincent M., 185
Saurischia, 161, 163
scarlet oak, 129, *130*
scientific method, 9–11, *10*
scientific nomenclature, 128–129
screening programs, genetic, 103, 104
seal limbs condition, 4
Search for Extraterrestrial Intelligence
    (SETI), 180–181
Sears, E., 141
seasonal isolation, 129
seawater, and body fluid similarity, 158
second chromosomes, 75
secondary sex characteristics, *80,* 81, 81*t*
segregation principle, 24–27, *25, 26,* 28*t*
selection
    artificial, 16–17, *16, 17*
    cultural, 113–114
    sexual, 78–81, *80,* 81*t*
    See also natural selection
selection coefficients, 72–73, 73*t*
selectionists, 182–183
selectively neutral alleles, 109
selfish gene, 242
The Selfish Gene (Dawkins), 247
semi-lethal genes, 71, 72, 73*t, 73*
semi-sterility, 71, 78, *78*
sensitization to Rh antigen, 106, *107*
SETI (Search for Extraterrestrial
    Intelligence), 180–181
Sewall Wright effect, 112
    See also genetic drift
sex chromosomes, *30,* 31
sexual cycle in humans, 32, *33*
sexual dimorphism, 81
sexual isolation, 130
sexual selection, 78–81, *80,* 81*t*
Seymouria, 156, *158*
shared ancestry, 193
shelled invertebrates, 201
Shigella, 87, 88
Shriver, Mark D., 119, 121
Shubin, Neil, 153
sibling altruism, 239–240, *239, 241*
sickle-cell anemia
    balanced polymorphism and, 94–98,
        95, 96, 97, 98
    as dominant *vs.* recessive, 95
    hemoglobin structure in, *52,* 53,
        54–55, *57,* 95–97, *97*
sickle-cell trait, 95, *97*
signaling macromolecules, 193
silent mutations, 50
Silurian period, 200, 201

Sinanthropus pekinensis, 226
sister chromatids, 32, *34*
Sivapithecus, 185
six-digited hands, *9*
skull 1470, 225
slavery, admixture of genes and,
    119–120, 121
Smith, Hamilton D., 89
Smith, Theobald, 108
snails, 126–128, *127,* 129
social behavior
    group selection and, 237–238
    of *Homo erectus,* 228–229
    kin selection and, 238–242, *239, 241*
    parental reproductive strategies,
        243–245
    reciprocal altruism and, 242–243
    sharing of food, 225
Sociobiology (Wilson), 237
solar system, formation of, 166–167
somatic *vs.* germ-line mutations, 45, *46*
southern leopard frogs, 125, *126*
speciation
    geographic, 126–128, *127*
    instantaneous, 135–141, 136*t, 137, 140*
    punctuated equilibrium *vs.* phyletic
        gradualism, 133–134
species
    concept of, 125–126
    defined, 128
    humans as single variable, 132
    polytypic *vs.* monotypic, 129
    reproductive isolating mechanisms
        and, 129–132, 129*t, 130*
Spencer, Herbert, 19, 20
Sphenodontia, 163
Spirit (rover), 181
spontaneous abortions. *See* pregnancy
    loss
spontaneous generation, 166
spontaneous mutations, 43, 44*t,* 50
Squamata, 163
stabilizing selection, 76–77, *76, 77,* 213
Staphylococcus aureus, methicillin-
    resistant, 92
stepfathering, 245
stepmothering, 243
sterility
    hybrid, 130
    partial, 71, 78, *78*
    of polyploid hybrids, 140
    sex chromosome aberrations and,
        78, *78*
Stern, Curt, 73

Stetson, B. E., 105
stop codons, 54, 55*t*
structural chromosomal aberrations, 58
sturgeons, 138
subclonal tumor cells, 84, 85
subspecies, defined, 129
substitution mutations, 50, *51*
subvital genes, 71–72, 73*t, 73*
survival of the fittest, 19, 21
    See also natural selection
survival until reproduction. *See*
    reproductive effectiveness
Sutton, Walter, 35
symbiosis, 174–176, *175*
sympatric populations, 129, 132
synapsids, 159, *160*
synapsis, 32
synthesis
    of organic compounds, 168, *169*
    using recombinant DNA, 90–91, *91*
Systema Naturae (Linnaeus), 128

T

tadpoles, retinoic acid effect on,
    193–194, *194*
taxonomy. *See* classification
Tay-Sachs disease, 102–103
telomeres, 235–236, *235*
teratogens, 4
terminating codons, 54, 55*t*
territorial behavior, 238
Testudines, 163
tetrads, 32–33, *34*
tetraploidy, 138, 141
tetrapod vertebrates, 153–156, *155,
    156, 157*
thalassemia, 101, 189
thalidomide-related malformations, 4, *4*
theories, hypothetico-deductive
    reasoning and, *10,* 11
therapsids, 163
thermophiles, 180
Thoatherium, 14, *14,* 207
Thorne, Alan G., 232
Thorne-Wolpoff hypothesis, 232–233
threatened species. *See* endangered
    species
three-domain classification system,
    178, *179*
thumb, hyperextensible, 114, *115*
thymine, in DNA, 47, *47, 48*
Tiktaalik, 153–155, *154, 156, 157*

Tinbergen, Niko, 83
toads, 130
tools, early use of, 225, 226, 228, 230, 232
tortoises, giant. *See* giant tortoises
Toumai, 219, 221
Toxodon, 14, *14*, 207
transcription factors, 191, 193
transitional fossils. *See* missing links
translocation of chromosomes, 58
transposons, 88–89, *88*
tree concept of ancestry, 176–177, *177*
tree-dwelling primates, 214, 215
tree finches, *145*, 150, *151*
tree shrew, 215
Triassic period, 209, *209*, 210*t*
trihybrid inheritance, 36, 38, *39*
trilobites, 202, 203, 204
triplets, in DNA coding, 50, 51, *51, 53*
triploidy, 58, 138
trisomy, 58
Tristan da Cunha (island), 117–118, *117*
Triticum spelta, 141
Trivers, Robert L., 242, 244, 245
Tschermak, Erich von, 24
tuataras, 163
tuberculosis, 103
tumor cell heterogeneity, 84–85, *84*
Tupaia, 215
Turkana boy, 228
Turner syndrome, 78
Typanoctomy barrerae, 138

U

uncertainty, in natural sciences, 11
ungulates, 14, *14*, 207
universe, formation of, 166–167

uracil, *48*, 53
urea synthesis, 168
Urey, Harold, 168
Ussher, James, 12

V

valine coding, *52*, 53, 55, *57*, 96–97, *98*
variant alleles. *See* recessive alleles
variation, heritable. *See* heritable variation
Vecht, J. van der, 123
vegetable kingdom, 176
vertebrates
    amniote *vs.* anamniote, 158, 159
    fossil record of, 203
    land, 152–156, *153, 154, 155, 156*
    limb similarities among, 164–165, *164*
    See also specific types
Victoria Island rocks, 201
von Koenigswald, G. H. R., 226
The Voyage of the Beagle (Darwin), 13, 143, 146

W

Wallace, Alfred Russel, 18
warbler finches, 145, *151*
warblers, 149
warning calls, 238–239, 242
water-dwellers, convergent evolution among, 165, *165*
water fleas, 57
water, possible extraterrestrial, 181
Watson, James D., 46, 47
Weinberg, Wilhelm, 61, 64
Weiner, Jonathan, 144
western gulls, 243

wheat, polyploidy of, 135, 136*t*, *137*, 138–139, 141
whiptail lizards, 138
White. Tim D., 219, 224
Whittaker, R. H., 177
wild grass, crossed with wheat, 135, *137*, 139, 141
Wilkins, Maurice H. F., 46
Wilson, Allan C., 185
Wilson, Edward O., 125, 237
Woese, Carl R., 178, 179, 180
Wöhler, Frederick, 168
Wolpoff, Milford H., 232
woodpecker finches, 145, *146*
Woolf, Charles M., 113, 114
woolly mammoths, 207, *208*
Wright, Sewall, 112
Wynne-Edwards, V. C., 238

X

X chromosome, *30*, 31, 78
x-rays, as mutagen, 44–45, 45*t*, *45*
Xylocopa nobilis, 123, *124*

Y

Y chromosome, *30*, 31, 78
Y chromosome Adam, 195–196
yarrow plant, 122–123, *123*
yolk sacs, 157, 158, *159*

Z

Zahavi, Amotz, 79
Zinjanthropus, 222
Zuckerkandl, Emile, 184
Zuni people, 113